ARTHROPOD NATURAL ENEMIES IN ARABLE LAND
III

Proceedings of the Third EU Workshop on Enhancement, Dispersal and Population Dynamics of Beneficial Insects in Integrated Agroecosystems: "Analysing and modelling of population dynamics of beneficial predators and parasitoids in agroecosystems: the individual, the population and the community", held at Rodney Lodge Conference Centre, University of Bristol, U.K., 23-25 November 1995

ARTHROPOD NATURAL ENEMIES IN ARABLE LAND

III

The Individual, The Population and The Community

Edited by Wilf Powell

ACTA JUTLANDICA LXXII:2
Natural Science Series 11

AARHUS UNIVERSITY PRESS

AARHUS UNIVERSITY PRESS
University of Aarhus
DK-8000 Aarhus C
Fax (+ 45) 8619 8433

73 Lime Walk
Headington, Oxford OX3 7AD
Fax (+ 44) 1865 750 079

Box 511
Oakville, Conn. 06779
Fax (+ 1) 860 945 9468

ANSI/NISO
Z39.48-1992

Foreword

This is the final book in a series of three volumes that form the proceedings of three workshops funded as a Concerted Action by the European Union. These workshops dealt with the biology of arthropod predators and parasitoids and their role as biological control agents for limiting pest insect populations in European agriculture. Particular attention was paid to the methodological problems associated with studying this subject. Each workshop involved a limited number of invited scientists from a range of EU countries and consisted of individual presentations of recent research progress interspersed with discussion sessions on selected themes. Following the workshops, participants prepared written research and review papers as a permanent record of the proceedings. All papers were critically refereed, predominantly by other participants, before publication. The common title of the three workshops is:

ENHANCEMENT, DISPERSAL AND POPULATION DYNAMICS OF BENEFICIAL PREDATORS AND PARASITOIDS IN INTEGRATED AGROECOSYSTEMS.

The first workshop was held at the University of Aarhus, Denmark on 21-23 October 1993 and discussed the topic:

Estimating population densities and dispersal rates of beneficial predators and parasitoids in agroecosystems

The second workshop was held in Wageningen, The Netherlands on 1-3 December 1994 and dealt with:

Estimating survival and reproduction of beneficial predators and parasitoids in relation to food availability and quality of the habitat

The final workshop, reported in this volume, was held in Bristol, U.K. on 23-25 November 1995 under the title:

Analysing and modelling of population dynamics of beneficial predators and parasitoids in agroecosystems

Over recent years there has been a growing appreciation of the important role played by naturally-occurring populations of arthropod predators and parasitoids in limiting agricultural pest populations. This has stimulated much research into their biology, especially their ecology and behaviour, with the aim of conserving or increasing their numbers in farmland and enhancing their impact as biological control agents. The stimulus for these workshops was the recognition that researchers throughout Europe were independently attempting to address the same methodological problems which were hindering progress in the study of natural enemies of pests within agricultural ecosystems.

The papers in this third volume are presented in three sections, reflecting the structure of the workshop which was conducted in three sessions. The first session dealt with the behaviour of individuals, especially foraging behaviour and local movement patterns, which could influence population dynamics and natural enemy efficiency in pest control. Secondly, methods of analysing and modelling predator and parasitoid populations were considered and the use of single-species spatially explicit models was discussed and subsequently reviewed. Finally, the community level was addressed and discussions highlighted the importance of considering interactions between the different natural enemies within predator/parasitoid communites associated with specific pests. This topic also led to the production of a review paper.

Special thanks should go to Kees Booij who acted as coordinator for the entire workshop series as well as organising the second meeting in Wageningen and editing its proceedings with the able assistance of Loes den Nijs. Thanks also go to Søren Toft and Werner Riedel for the organisation and proceedings editing of the first workshop as well as arranging publication of the series with Aarhus University Press. I am personally grateful to Harry Anderson of IACR-Long Ashton who helped me with the organisation of the third meeting in Bristol. Finally, the success of these workshops would not have been possible without the enthusiastic participation of all the scientists who took part in one or more of the meetings.

Harpenden, March 1997

Wilf Powell

Contents

THE COMMUNITY

DISCUSSION PAPER

List of participants

THE INDIVIDUAL

Biological control of aphids: the effectiveness of ladybird beetles is limited by their foraging behaviour

J.-L. Hemptinne & A.F.G. Dixon[1]

UER de Zoologie générale et appliquée, Faculté Universitaire des Sciences
agronimiques, Passage des Déportés, 2 B-5030 Gembloux (Belgium).
[1]School of Biological Sciences, University of East Anglia, Norwich NR4 7TJ.

Abstract

Coccinellids as well as other specific predators of aphids are often considered as
potential agents of biological control. Their efficiency depends on their ability to
show a strong numerical response to aphid infestations. The study of their foraging
strategies indicates that this is unlikely. Egg cannibalism is a major factor
constraining the numerical response of *Adalia bipunctata*. They lay only a few eggs
at the beginning of the development of aphid colonies. In addition, males eat few
aphids compared with females. Therefore the ability of natural populations of
ladybird beetles and other predators, which behave similarly, to control aphid
abundance is open to question.

Key words: coccinellids, aphids, biological control, foraging behaviour.

Introduction

In temperate countries of Europe, crops are usually attacked by indigenous
aphids plus a few exotic species introduced a long time ago which are now
completely acclimatized (Hill 1987). These pests are generally attacked by
several species of indigenous parasitoids and predators, a situation suggesting
that biological strategies of pest control are feasible. As aphids coexist with
guilds of aphidophagous natural enemies, classical biological control is not
necessary. It seems wiser to avoid the possible side effects of introducing exotic
species (Pimm 1991, Takagi & Hirose 1994, but see Mills 1994) by exploiting
indigenous natural enemies. In theory, crops can be protected by the mass release
of indigenous beneficial insects or by the enhancement of natural populations of

Arthropod natural enemies in arable land · III *The individual, the population and the community*
W. Powell (ed.). *Acta Jutlandica* vol. 72:2 1997, pp. 11-19
© Aarhus University Press, Denmark, ISBN 87 7288 673 0

antagonists by manipulating agro-ecosystems (Begon et al. 1990, Raupp et al. 1994).

Economic constraints favour strategies of biological control that are comparable to chemical control in terms of cost, ease of application and efficiency. In the case of aphidophagous predators, mass releases are very ineffective compared with pesticides. Therefore the only option available is the enhancement of natural populations of beneficial insects. This approach is ecologically sound. Predators are usually larger than their prey, have a longer generation time and a rate of increase which is an order of magnitude lower than aphids. Their rate of population increase is therefore not their major attribute. Natural populations of coccinellids can therefore only control a prey population if they show a strong numerical response. That is possible if large numbers of adults are attracted and arrested to patches of aphids. Reproduction should occur there followed by a massive migration of newborn insects to adjacent infested fields (Sabelis 1992).

In the last decade, habitat manipulations have often been practised. In several cases, the density of generalist and specialist predators increased in response to the creation of refuges, shelters or vegetation strips providing alternative sources of food. Nevertheless, there does not appear to be a relationship between the abundance of natural enemies and a reduction in the incidence of aphids (van Emden 1990, Wratten & Powell 1991, Hickman & Wratten 1994). This is puzzling. There has, however, been an attempt to understand why aphidophagous ladybirds are so ineffective. Contrary to expectation (Sabelis 1992), their numerical response is restricted to a narrow range of low aphid densities (Wright & Laing 1980, Mills 1982) which can be explained if satiation restricts the numerical response (Mills 1982). Mills also observed that the response of aphidophagous ladybirds to aphids is further reduced by density-dependent egg cannibalism. A simulation model of the interaction between aphids and ladybirds (Kindlmann & Dixon 1993), which takes the minimum aphid density requirements of the first instar larvae of ladybirds and the risk of cannibalism into account, indicates that the best strategy is for the ladybird to lay a few eggs at the beginning of the development of aphid colonies. If they lay their eggs later, the larvae will not mature before the prey becomes scarce. In addition, if many eggs are laid, the larvae reduce the rate of increase of the aphid colony and cause an earlier decline in prey abundance. In this case, the larvae resort to cannibalism to survive. This results in the production of a few small adults which are unlikely to overwinter successfully and have a low potential fecundity. That is, if the ladybirds have to maximise their fitness they should lay a few eggs early in the development of an aphid colony.

Foraging behaviour of ladybird beetles

It should not be forgotten that populations of coccinellids generally consist of equal numbers of males and females. The vast majority of papers devoted to ladybird beetles assume that both sexes have the same foraging strategy (Brown 1972, Karner & Manglitz 1985, Kareiva & Sahakian 1990, Grevstad & Kepletka 1992, Gutierrez et al. 1981, 1990).

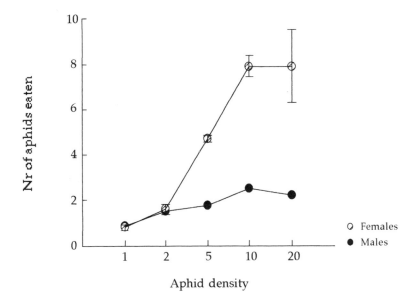

Fig. 1. The average number of aphids eaten daily by females and males of *Adalia bipunctata* at five prey densities (aphids / day / cm^2). (From Hemptinne et al. 1996)

However, males of *Adalia bipunctata* (L.) are less voracious and show a much weaker functional response to increasing aphid density compared with females (Fig. 1) (Hemptinne et al. 1996). In addition, their pattern of activity is not related to the availability of prey. Males in Petri dishes with 20 aphids are slightly more active than those kept with one aphid but, in both cases, they spend the same proportion of time in area restricted and intensive search (Fig. 2). In contrast, females do not react to the presence of more aphids by being more active but they reduce their extensive search and indulge more often in area restricted search (Fig. 2) (Hemptinne et al. 1992). These results indicate that males have little interest in aphids and that their relatively low energy demand may partly account for the weak numerical response shown by ladybirds in the field towards aggregations of aphids (i.e. Karner & Manglitz 1985, Ives et al. 1993).

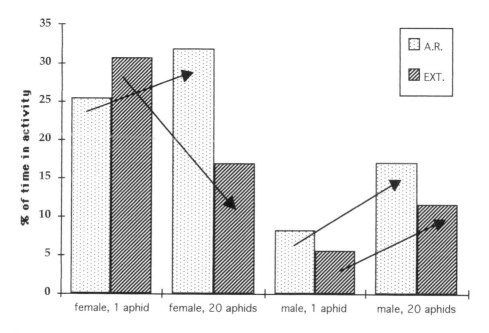

Fig. 2. Proportion of time devoted to area restricted search (A.R.) or extensive search (EXT.) by female or male *Adalia bipunctata* kept singly and fed with 1 or 20 aphids in a Petri dish for 24 h. (From Hemptinne et al. 1996)

What evidence is there that ladybird females forage optimally? In Belgium, *A. bipunctata* lay eggs on apple trees, nettles and wheat in that order as populations of aphids develop on these plants (Hemptinne & Dixon 1997). In these different habitats, most of the eggs are laid over a short period of time before the peak of each respective aphid population. That is, female ladybirds behave in accordance with the predictions of the model of Kindlmann & Dixon (1993). What causes the adults to cease laying eggs when aphids are still increasing in abundance? It is not due to a change in the quality of the host-plant of their prey, or in the age structure of aphid colonies (Hemptinne et al. 1995). It appears to be due to their reluctance to lay eggs in the presence of conspecific larvae (Fig. 3). The females respond to the tracks left by the larvae, which appear to act as an "oviposition inhibiting pheromone" (Doumbia et al. 1997). They are also more active (Fig. 4). Although there is a large difference between laboratory

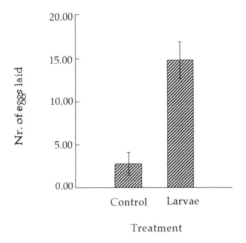

Fig. 3. Number of eggs laid in 3 h by female *Adalia bipunctata* kept singly or with 10 conspecific fourth instar larvae. There are 21 replicates for the two treatments which are significantly different (t = 4.80; 41 df; P = 0.001).

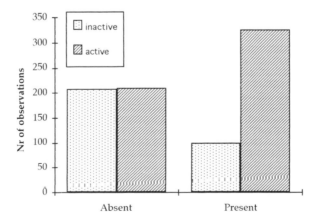

Fig. 4. Level of activity of female *Adalia bipunctata* kept singly or with a conspecific fourth instar larva. Levels of activity are either "active" or "inactive". (From Hemptinne et al. 1996).

and field conditions, this higher activity indicates that they are also more likely to leave patches of aphids already occupied by conspecific larvae. It is likely that the adaptive significance of this behaviour is that it reduces the incidence of cannibalism. Consequently, fewer eggs are laid, which restricts the effectiveness of *A. bipunctata* as biocontrol agents.

Conclusions

Among the strategies of biological control used against aphids in temperate countries, manipulations of agro-ecosystems to encourage populations of indigenous natural enemies are favoured for economical and ecological reasons. There are indeed many applications in orchards, and in cereal and vegetable fields. The results, however, are disappointing: although predators and parasitoids are sometimes more abundant in manipulated fields, their increased presence is not a guarantee of fewer aphids. An explanation of this can be obtained from our understanding of the foraging behaviour of ladybird beetles.

A. *bipunctata*, like most other aphidophagous predators, are larger than aphids. Based on their size (Fenchel 1974), their rate of population increase should lie between 0.01 and 0.07 (Ives 1981, Medvey 1988, Mills 1979, Tenhumberg & Poehling 1991). However, aphids have rates of increase in the range of 0.2-0.5 (Dixon 1985). In addition, the average predation rate of ladybird larvae (Barlow & Dixon 1980, Wratten 1973), is such that it is likely that an aphid population can only be suppressed when the prey/predator ratio ranges between 10/1 and 45/1 (Diekmann et al. 1988, 1989). This implies that ladybirds need to show a strong combined numerical response to aphids if they are to reduce their abundance.

A simulation model of the interactions between aphids and ladybirds (Kindlmann & Dixon 1993) indicates that in order to maximize their fitness, female ladybirds should lay a few eggs during the early development phase of aphid colonies. Field and laboratory experiments support the model's predictions (Hemptinne et al. 1992, Hemptinne & Dixon 1997). In addition, several species of hoverflies and one chrysopid species behave similarly (Hemptinne et al. 1994, Kan 1988a, b, Ruzicka 1994). All these predators lay a few eggs in each aphid colony, then cease laying eggs, presumably to avoid the risk of egg cannibalism. Therefore, the numerical response of natural populations of these predators to aphids is too weak for them to be effective biological control agents.

Modelling of the interactions of predators with aphids in biological control studies should take the oviposition behaviour of the predators into account. Finally, as natural populations of aphidophagous predators appear to be ineffective in biological control, more efforts should be devoted to other ways of reducing the rate of increase of aphids.

References

Barlow, N.D. & Dixon, A.F.G. 1980. *Simulation of lime aphid population dynamics*. Podoc, Wageningen.

Begon, M., Harper, J.L. & Townsend, C.R. 1990. *Ecology. Individuals, populations and communities*. Blackwell Scientific Publications, Oxford.

Brown, H.D. 1972. Predacious behaviour of four species of Coccinellidae (Coleoptera) associated with the wheat aphids, *Schizaphis graminum* (Rondani), in South Africa. *Trans. R. Ent. Soc. London* 124: 21-36.

Diekmann, O., Metz, J.A.J. & Sabelis, M.W. 1988. Mathematical models of predator-prey-plant interactions in a patchy environment. *Exper. Appl. Acarol.* 5: 319-342.

Diekmann, O., Metz, J.A.J. & Sabelis, M.W. 1989. Reflections and calculations on a prey-predator-patch problem. *Acta Applicandae Mathematicae* 14: 23-25.

Dixon, A.F.G. 1985. *Aphid ecology*. Blackie, Glasgow.

Doumbia, M., Hemptinne, J.-L. & Dixon, A.F.G. 1997. Assessment of patch quality by ladybirds: role of larval tracks. *Oecologia (Berl.)* in press.

Fenchel, T. 1974. Intrinsic rate of natural increase: the relationship with body size. *Oecologia (Berl.)* 14: 317-326.

Grevstad, F.S. & Kepletka, B.W. 1992. The influence of plant architecture on the foraging efficiencies of a suite of ladybird beetles on aphids. *Oecologia (Berl.)* 92: 399-404.

Gutierrez, A.P., Baumgaertner, J.U. & Hagen, K.S. 1981. A conceptual model for growth, development and reproduction in the ladybird beetle, *Hippodamia convergens* (Coleoptera; Coccinellidae).*Can. Ent.* 113: 21-33.

Gutierrez, A.P., Hagen, K.S. & Ellis, C.K. 1990. Evaluating the impact of natural enemies: a multitrophic perspective. In: M. Mackauer, L.E. Ehler & J. Roland (eds.) *Critical issues in biological control* pp 81-109, Intercept, Wimborne.

Hemptinne, J.-L. & Dixon, A.F.G. 1997. Are aphidophagous ladybirds (Coccinellidae) prudent predators? *Biological Agriculture & Horticulture*, in press.

Hemptinne, J.-L., Dixon, A.F.G. & Coffin, J. 1992. Attack strategy of ladybird beetles (Coccinellidae): factors shaping their numerical response. *Oecologia (Berl.)* 90: 238-245.

Hemptinne, J.-L., Doucet, J.-L. & Gaspar, C 1994. How do ladybirds and syrphids respond to aphids in the field? *Bull. IOBC/WPRS* 17: 101-111.

Hemptinne, J.-L., Doumbia, M. & Gaspar, C. 1995. The reproductive strategy of predators is a major constraint to the implementation of biological control in the field. *Meded. Fac. Landbouww. Univ. Gent* 60/3a: 735-741.

Hemptinne, J.-L., Dixon, A.F.G. & Lognay, G. 1996. Searching behaviour and mate recognition by males of the two-spot ladybird beetle, *Adalia bipunctata* (L.). *Ecol. Entomol.* 21: 165-170.

Hickman, J.M. & Wratten, S.D. 1994. Use of *Phacelia tanacetifolia* (Hydrophyllaceae) as a pollen resource to enhance hoverfly (Diptera: Syrphidae) populations in sweetcorn fields. *Bull. IOBC/WPRS* 17: 156-167.

Hill, D.S. 1987. *Agricultural insect pests of temperate regions and their control.* Cambridge University Press.

Ives, A.R., Kareiva, P. & Perry, R. 1993. Response of a predator to variation in prey density at three hierarchical scales; lady beetles feeding on aphids. *Ecology* 74: 1929-1938.

Ives, P.M. 1981. Feeding and egg production of two species of coccinellids in the laboratory. *Can. Ent.* 113: 999-1005.

Kan, E. 1988a. Assessment of aphid colonies by hoverflies. I. Maple aphids and *Episyrphus balteatus* (de Geer) (Diptera: Syrphidae). *J. Ethology* 6: 39-48.

Kan, E. 1988b. Assessment of aphid colonies by hoverflies. II. Pea aphids and three syrphid species: *Betasyrphus serarius* (Wiedemann), *Metasyrphus frequens* Matsumara and *Syrphus vitripennis* (Meigen) (Diptera: Syrphidae). *J. Ethology* 6: 135-142.

Kareiva, P. & Sahakian, R. 1990. Tritrophic effects of a simple architectural mutation in pea plants. *Nature* 345: 433-434.

Karner, M.A., & Manglitz, G.R. 1985. Effects of temperature and alfalfa cultivar on pea aphid (Homoptera: Aphididae) fecundity and feeding activity of convergent Lady beetle (Coleoptera: Coccinellidae). *J. Kansas Entomol. Soc.* 58: 131-136.

Kindlmann, P. & Dixon, A.F.G. 1993. Optimal foraging in ladybird beetles (Coleoptera: Coccinellidae) and its consequences for their use in biological control. *European J. Entomol.* 90: 443-450.

Medvey, M. 1988. On the rearing of *Episyrphus balteatus* (de Geer) (Diptera: Syrphidae) in the laboratory. In: E. Niemczyk & A.F.G. Dixon (eds.) *Ecology and effectiveness of aphidophaga* pp. 61-63. SPB Academic Publishing, The Hague.

Mills, N.J. 1979. *Adalia bipunctata* (L.) as a generalist predator of aphids. PhD thesis, University of East Anglia.

Mills, N.J. 1982. Voracity, cannibalism and coccinellid predation. *Ann. Appl. Biol.* 101: 144-48.

Mills, N.J. 1994. The structure and complexity of parasitoid communities in relation to biological control. In: B.A. Hawkins & W. Sheehan (eds.) *Parasitoid community ecology.* Oxford Science Publications, pp. 397-417.

Pimm, S.L. 1991. *The balance of nature? Ecological issues in the conservation of species and communities.* The University of Chicago Press.

Raupp, M.J., Hardin, M.R., Braxton, S.M. & Bull, B.B. 1994. Augmentative releases for aphid control on landscape plants. J. Arboriculture 20: 241-49.

Ruzicka, Z. 1994. Oviposition-deterring pheromone in *Chrysopa oculata* (Neuroptera: Chrysopidae). *European J. Entomol.* 91: 361-70.

Sabelis, M.W. 1992. Predatory arthropods. In: M.J. Crawley (ed.) *Natural enemies. The population biology of predators, parasites and diseases.* Blackwell Scientific Publications, Oxford, pp. 225-64.

Takagi, M. & Hirose, Y. 1994. Building parasitoid communities: the complementary role of two introduced parasitoid species in a case of successful biological control. In: B.A. Hawkins & W. Sheehan (eds.) *Parasitoid community ecology.* Oxford Science Publications, pp. 437-48.

Tenhumberg, B. & Poehling, H.-M. 1991. Studies on the efficiency of syrphid larvae as predators of aphids on winter wheat. In: L. Polgar, R.J. Chambers, A.F.G. Dixon & I. Hodek (eds.) *Behaviour and impact of aphidophaga* pp. 281-88. SPB Academic Publishing, The Hague.

van Emden, H.F. 1990. Plant diversity and natural enemy efficiency in agroecosystems. In: M. Mackauer, L.E. Ehler & J. Roland (eds.) *Critical issues in biological control* pp. 63-80. Intercept, Wimborne.

Wratten, S.D. & Powell, W. 1991. Cereal aphids and their natural enemies. In: L.G. Firbanks, N. Carter, J.F. Darbyshire & G.R. Potts (eds.) *The ecology of temperate cereal fields* pp. 233-57. Blackwell Scientific Publications, Oxford.

Wratten, S.D. 1973. The effectiveness of the coccinellid beetle *Adalia bipunctata* (L.) as a predator of the lime aphid *Eucallipterus tiliae* (L.). *J. Anim. Ecol.* 42: 785-802.

Wright, E.D. & Laing, J.E. 1980. Numerical response of coccinellids to aphids in corn in southern Ontario. *Can. Ent.* 112: 977-88.

A semi-field method to assess the effect of dimethoate on the density and mobility of ground beetles (Carabidae)

P.J. Kennedy & N.P. Randall

Entomology & Nematology Department, IACR-Rothamsted,
Harpenden, Herts. AL5 2JQ, UK.

Abstract

The feasibility of using mark-recapture techniques to assess the side-effects of dimethoate on the density and mobility of *Harpalus rufipes, Pterostichus cupreus, P. madidus* and *P. melanarius* was investigated in semi-field enclosures. The distribution and frequency of recaptures of individually-marked ground beetles were used to estimate mortality, recruitment, rate of displacement and displacement direction. The technique, with adequate replication, provides a useful semi-field method to study acute toxicity and sub-lethal effects of pesticides on such mobile, medium-to-large sized epigeal arthropods, which typically occur at low densities.

Key words: Carabidae, pitfall traps, pesticides, mark-recapture, absolute density estimates, displacement rate, displacement direction, semi-field enclosures.

Introduction

European governments are increasingly promoting responsible and efficient use of pesticides as part of integrated pest control strategies compatible with sustainable agriculture. To this end, pesticides should only be used when necessary, without recourse to excessive amounts, and alternative strategies should be adopted where possible. To promote natural pest control and to minimize the impact of agricultural practices on the environment, pesticide registration processes now take account of side-effects of pesticides on beneficial arthropods but standard methods to assess these effects in the field are still in their infancy. Any quantitative assessment of the impact of pesticides on the population dynamics of beneficial arthropods needs to be based on reliable estimates of population density and dispersal.

Arthropod natural enemies in arable land · III The individual, the population and the community
W. Powell (ed.). *Acta Jutlandica* vol. 72:2 1997, pp. 21-37
© Aarhus University Press, Denmark, ISBN 87 7288 673 0

Field studies of ground beetles, which form part of the natural enemy complex suppressing pest outbreaks, are usually based on data from pitfall trap catches. Pitfall traps provide a relatively cheap and efficient means for catching such mobile and cryptic epigeal arthropods, but their catch is influenced by the beetles' density, catchability and activity, and can only provide a relative measure of abundance. Sublethal effects of pesticides on the mobility of beneficial arthropods have been recorded in the literature (Moosebeckhofer 1983, Chiverton 1984, Heneghan 1992, Wiles & Jepson 1994) and may influence the subsequent pitfall trap catch as well as the long-term fitness of populations surviving pesticide applications. For a clear understanding of the impact of pesticides on population processes there is, therefore, a need to distinguish between pesticide-induced changes in density and mobility.

Absolute density estimates of ground beetles have been obtained by surface searching within quadrats, soil sampling, vacuum sampling, soil flooding, removal trapping and mark-recapture methods. The majority of these methods are laborious or biased towards less mobile and diurnal species. Alternatively, mark-recapture methods, when combined with live pitfall trapping, enable both diurnal and nocturnal species to be assessed, although they are biased towards large (for ease of marking) and mobile species.

Mark-recapture methods can be used to estimate both population densities and mobility. The ratio of marked to unmarked individuals in traps, after release of marked individuals into a population, can be used to estimate population size and, if the sampling area can be defined, population density. Mark-recapture studies performed within enclosures, over relatively short periods, are both geographically and demographically defined. This simplifies the estimation of population density and permits the use of relatively robust mark-recapture models (Loreau 1984). In addition, the number of recaptures over a period of time and their distance from a release site can be used to calculate a probability of recapture and be used as an indicator of displacement rate. Releasing marked beetles at the centre of a circular enclosure, with traps set along a barrier around its perimeter, sets the displacement distance to the radius of the enclosure. Barriers increase the sampling efficiency of pitfall traps (Durkis & Reeves 1982) and probably reduce the time to capture, following a beetle's displacement to a barrier, to a minimal period. Nevertheless, barriers may alter behaviour by restraining dispersal and by affecting the microclimate within enclosures. Enclosures must be large enough to mimimize the influence of barriers on behaviour, enclose sufficient individuals to obtain reliable density and probability of recapture estimates, contain sufficient resources to support these individuals during the period of study and small enough to ensure high rates of recapture.

This paper describes a mark-recapture study performed within circular enclosures to investigate lethal and sub-lethal effects of the organophosphorus insecticide dimethoate on ground beetles, at natural population densities. The study was intended as a preliminary investigation to demonstrate the feasiblity of the technique to determine pesticide-induced changes in population density (mortality and recruitment) and displacement (rate and direction).

Materials and Methods

The mark-recapture study was conducted, in June and July 1995, within three circular enclosures in a 5.1 ha field of spring wheat at IACR-Rothamsted, UK. Each enclosure was 10 m in diameter and, between mid-May and August, had a boundary formed by a 0.3 m high polythene barrier, buried to an equal depth and

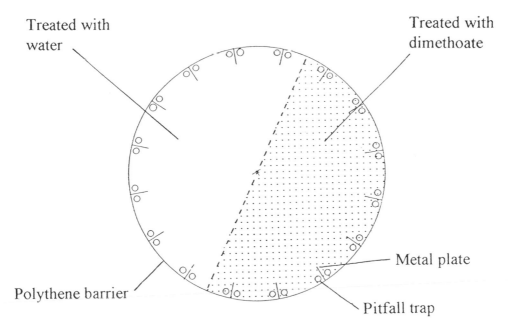

Fig. 1. Diagram of enclosure C, illustrating the areas treated with water (open) and with dimethoate (dotted).

supported by nylon guy ropes attached to wooden posts. The centres of the three enclosures were arranged in a line, separated by 20 m, at least 35 m from the field edge. A total of 32 pitfall traps were placed in pairs, along the inside of each barrier, at equidistant intervals (Fig. 1). Each pitfall trap consisted of a 60 mm diameter white plastic beaker sunk into the ground, inside a section of drainpipe, with its upper rim flush with the soil surface and covered by a raincover made from an inverted flower-pot saucer. Metal plates, 75 mm high and 500 mm long, radiating from the barrier towards the centre of the enclosures, were pushed into the ground between pairs of traps to facilitate the capture of ground beetles in pitfall traps near their point of encounter with the barrier.

Pitfall traps were opened on 5 June and checked daily until 28 July 1995. All medium-to-large ground beetles caught were identified, sexed and given an individual-based mark by scratching their elytra with coded values similar to the method described by Thomas (1995). Ground beetles were marked and released in the centre of the enclosures within which they were caught, on the same morning that they were collected. No additional beetles were added to the enclosures, nor were beetles 'stock-piled' for release before the study. This avoided changes in natural population densities which could lead to subsequent changes in behaviour of individuals. Recaptures were re-released within each enclosure and all recapture data (single and multiple recaptures) were considered in the analyses. Only data of the more abundant ground beetles, *Harpalus rufipes, Pterostichus cupreus, Pterostichus madidus* and *Pterostichus melanarius*, are presented in this paper.

On 5 July 1995, enclosures were sprayed either completely with water (enclosure A) or dimethoate (enclosure B), or had one half sprayed with water and the other half sprayed with dimethoate (enclosure C). Dimethoate was applied at the manufacturer's recommended field rate (840 ml in 200 l water/ha; 336 g a.i./ha) and water was applied at the same rate (200 l/ha). No other insecticides were applied to the enclosures or to a 20 m surround. All other agrochemical inputs to the enclosures and the field followed normal farm practice.

Densities, and their standard errors, were estimated using a weighted mean Petersen estimate (Begon 1979):

$$N = \frac{\Sigma \, M_i n_i}{(\Sigma m_i)+1} \qquad SE_N = N \sqrt{\frac{1}{\Sigma m_i+1}+\frac{2}{(\Sigma m_i+1)^2}+\frac{6}{(\Sigma m_i+1)^3}}$$

Where N is the weighted mean estimate of population size, SE_N is its standard error, M_i is the total number of marked individuals at time i, n_i is the total number of individuals caught at time i, and m_i is the number of marked individuals in that catch. Separate population estimates were derived for the four-week pre-treatment and the four-week post-treatment periods. For the post-treatment period, separate estimates were calculated for the subpopulation known to have been present before treatments were applied (marked before 5 July; 'survivors') and the subpopulation marked since treatments were applied ('new recruits'). The significance of a difference between density estimates was calculated using Bartlett's pooled estimate of variance and Student's t-test.

Mean individual daily recapture rates were determined assuming that the probability of recapture of an individual on any given day follows a geometric distribution. The following were found to be adequate estimates of the mean

$$P(r) = \frac{R}{D} \qquad SE_{P(r)} = \sqrt{\frac{R(D-R)}{D^3}}$$

daily probability of recapture and its standard error:

Where $P(r)$ is the mean individual daily recapture rate, $SE_{P(r)}$ is its standard error, R is the total number of recaptures within a sampling period and D is the total number of days available for recapture after initial release of individuals within a sampling period. Differences in recapture rates were tested for statistical significance using Bartlett's pooled estimate of variance and Student's t-test.

Displacement directions were assessed by analysing the distribution of recaptures around the perimeter of enclosures, grouped into 16 directions (pairs of pitfall traps per direction; 22.5° intervals), using circular statistics (Batschelet 1981, Fisher 1993). A chi-squared test was performed to test for a uniform distribution of recaptures against any alternative. For validity, the test requires the mean number of recaptures per direction interval to exceed 2. When this was not the case, the number of direction intervals was reduced by pooling data from pairs of intervals until this rule was satisfied. A Rayleigh test was performed to test for uniformity against a unimodal alternative. The Rayleigh is a more specific test than the chi-squared and is, consequently, more sensitive to the presence of a single mode in the underlying distribution. When raw data plots suggested a bipolar distribution, displacement directions were doubled and reduced to multiples modulo 360° prior to analysis by Rayleigh.

Results

Density and mortality

A total of 214 individuals of the four species were marked and released during the eight week duration of the study. Pitfall trap catches were initially composed of only unmarked individuals but, as the study progressed, contained an increasing proportion of marked individuals following the continual re-release of marked individuals to the enclosures. *Pterostichus melanarius* dominated, forming 43 % of individuals marked, followed by *Harpalus rufipes* (32 %), *P. madidus* (16 %) and *P. cupreus* (9 %). The same rank order was also obtained from mean pre-treatment density estimates for all enclosures: *P. melanarius* was present at a density (± SE) of 0.181 ± 0.040 individuals m^{-2}, *H. rufipes* at 0.178 ± 0.048 m^{-2}, *P. madidus* at 0.114 ± 0.076 m^{-2} and *P. cupreus* at 0.039 ± 0.020m^{-2}. Nearly all individuals (at least 73 %) estimated to have been present, either pre-treatment or post-treatment, were marked and released within each four-week sampling period.

Density estimates, calculated using the weighted mean Petersen method, varied greatly between species, enclosures and sampling periods (Fig. 2). Significant (P < 0.05) pre-treatment differences between enclosures were observed for all species. Post-treatment 'survivor' density estimates were, with one exception, lower than pre-treatment estimates and the difference between the

Table 1. Estimated % mortality of four carabid species, between the pre-treatment and post-treatment sampling periods, based on the relative difference in population size before and after treatment of individuals marked before 5 July 1995. (Enclosure A was treated with water, enclosure B with dimethoate and enclosure C was treated half with water and half with dimethoate).

	% mortality in enclosure		
	A	B	C
H. rufipes	24	36	53
P. cupreus	—	100	20
P. melanarius	− 21 *	21	56
P. madidus	31	50	53

* A negative mortality index indicates a higher density estimate for 'survivors' than for the pre-treatment population.

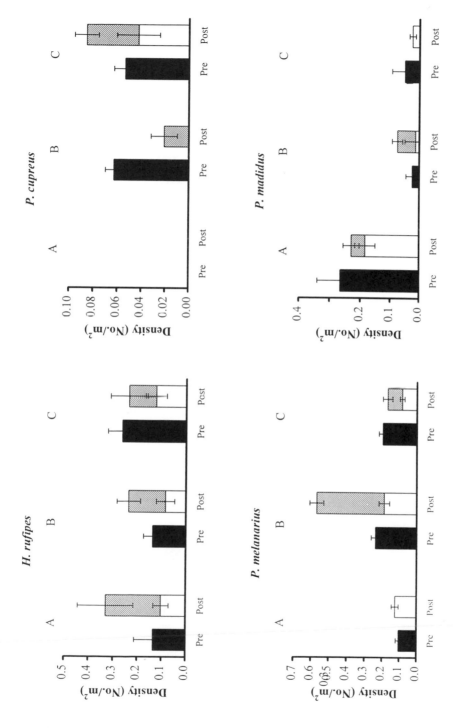

Fig. 2. Estimated densities, and their standard errors, of pre-treatment (filled), post-treatment 'survivor' (open) and post-treatment 'new recruit' (hatched) populations of four carabid species in circular enclosures. Enclosure A was treated with water, enclosure B was treated with dimethoate and enclosure C was treated half with water and half with dimethoate.

(a) **Pre-treatment**

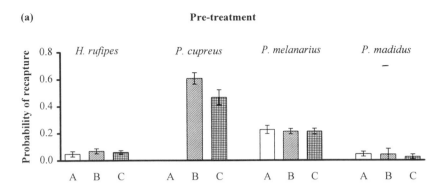

(b) **Post-treatment ('survivors')**

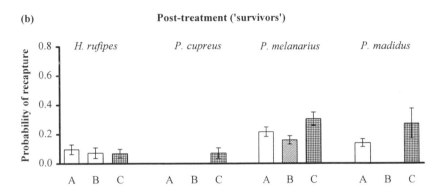

(c) **Post-treatment ('new recruits')**

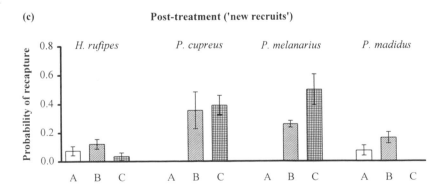

Fig. 3. Mean daily probability of recapture of individuals from (a) pre-treatment, (b) post-treatment 'survivor' and (c) 'new recruit' populations of four carabid species in enclosures A (open), B (hatched) and C (cross-hatched). Treatments applied as described in Fig. 2.

two was used as an index of mortality between the two sampling periods (Table 1). All species marked in the water-treated enclosure A had a lower mortality than in the dimethoate-treated enclosure B. Comparisons between pre-treatment and post-treatment 'survivor' population densities revealed no significant differences in enclosure A (water) but significantly smaller 'survivor' populations of *P. cupreus* in enclosure B (dimethoate), and of *P. melanarius* and *H. rufipes* in enclosure C (dimethoate and water), compared with pre-treatment populations. Recaptures of ground beetles marked after treatments were applied were used to estimate recruitment. Significantly more ($P < 0.05$) 'new recruits' of *P. melanarius* were estimated in enclosure B than in either enclosure A or C. Similarly, more 'new recruits' of *P. madidus* were estimated in enclosure B than in enclosure C ($P < 0.01$), but the reverse was the case for *P. cupreus* ($P < 0.05$). No significant differences in recruitment between enclosures were determined for *H. rufipes*.

Probability of recapture

Pre-treatment recapture rates were generally consistent across enclosures but varied greatly between species (Fig. 3a). No significant differences could be determined between sexes and data were pooled for subsequent analyses. *Pterostichus cupreus* and *P. melanarius* were those recaught most frequently ($P(r) \pm SE_{P(r)} = 0.538 \pm 0.070$ and 0.218 ± 0.004, respectively) and *H. rufipes* and *P. madidus* were those least frequently recaught ($P(r) \pm SE_{P(r)} = 0.060 \pm 0.006$ and 0.038 ± 0.006, respectively). A significant difference between enclosures was only found with *P. cupreus* ($P(r)$ in enclosure B > $P(r)$ in enclosure C; $P < 0.01$). Although the probability of recapture is also influenced by catchability and mortality, it does provide a crude estimate of displacement rate and its inverse can be used to estimate the average time taken to displace 5 m. Thus, *P. cupreus, P. melanarius, H. rufipes* and *P. madidus* would take an average of 2, 5, 17 and 26 days, respectively, to displace 5 m.

Post-treatment recapture rates of 'survivors' varied between enclosures and differed from pre-treatment rates for some species (Fig. 3b). Both *P. melanarius* and *P. madidus* were recaught more frequently, post-treatment, in the water and dimethoate treated enclosure C than in either of the other two enclosures, and *P. madidus* was recaught less frequently in the dimethoate-treated enclosure B than in water-treated enclosure A. *H. rufipes* was recaught at the same low rate in all enclosures and *P. cupreus* 'survivors' were only recaught in enclosure C. Both *P. melanarius* and *P. madidus* were recaught more frequently post-

(a) *P. cupreus*

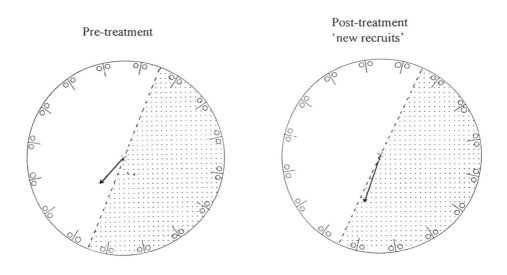

Pre-treatment

Post-treatment
'new recruits'

(b) *P. melanarius*

Post-treatment
'new recruits'

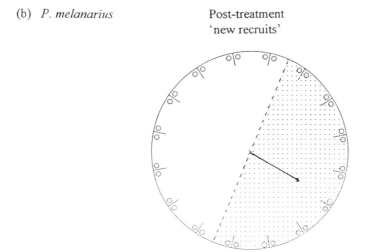

Fig. 4. Mean vector of displacement based on recaptures of (a) *P. cupreus* and (b) *P. melanarius* in enclosure C. The area of the enclosure treated with dimethoate is indicated by dotted shading.

Table 2. Analysis of the distribution of recaptures of *P. cupreus* and *P. melanarius* around the perimeter of enclosures during pre-treatment and post-treatment sampling periods. The mean direction is given as a compass direction. Chi-squared tests were performed to test for uniformity; the chi-squared value and its degrees of freedom are shown. Significant results, i.e. $P_{X2} < 0.05$, indicate a non-uniform distribution. The null hypothesis of the Rayleigh test is a uniform distribution. A P_Z value < 0.05 indicates a significant difference from the null hypothesis, i.e. a unimodal distribution. Enclosure A was treated with water, enclosure B was treated with dimethoate and enclosure C was treated half with water and half with dimethoate.

		Pre-treatment			Post-treatment 'Survivors'			Post-treatment 'New Recruits'		
		A	B	C	A	B	C	A	B	C
P. cupreus	No. recaptures	—	79	37	—	—	11	—	5	20
	Mean direction	—	100°	223°	—	—	188°	—	19°	199°
	Chi-squared, X^2	—	25.62	23.14	—	—	6.09	—	—	20.00
	d.f.	—	15	15	—	—	3	—	—	7
	P_{X2}	—	<0.05	>0.05	—	—	>0.05	—	—	<0.01
	Rayleigh, Z	—	2.60	4.36	—	—	0.86	—	—	5.25
	P_Z	—	0.076	0.017	—	—	0.413	—	—	0.017
P. melanarius	No. recaptures	44	94	83	50	51	43	—	98	11
	Mean direction	124°	112°	126°	325°	344°	22°	—	168°	120°
	Chi-squared, X^2	21.55	12.22	28.23	15.56	10.88	34.77	—	25.14	6.82
	d.f.	15	15	15	15	15	15	—	15	3
	P_{X2}	>0.05	>0.05	<0.05	>0.05	>0.05	<0.01	—	<0.005	>0.05
	Rayleigh, Z	0.85	1.15	0.02	0.34	0.33	0.02	—	0.26	3.49
	P	0.426	0.315	0.978	0.712	0.720	0.977	—	0.773	0.048

treatment than pre-treatment in enclosure C (P < 0.05 and p < 0.001, respectively) and less frequently post-treatment than pre-treatment in enclosure B (P < 0.001 for *P. madidus*).

Post-treatment recapture rates of 'new recruits' were in most cases greater than those of 'survivors', but differences between enclosures were inconsistent between species (Fig. 3c).

Direction of displacement

The only distributions of recapture that were significantly (P<0.05) concentrated around a single mode, and therefore indicated directed displacement, were found in enclosure A (Table 2). Half of this enclosure was sprayed with dimethoate and the other half with water. 'New recruits' of *P. cupreus* displaced southwest, along the treatment boundary post-treatment, but pre-treatment recaptures of *P. cupreus* were also concentrated in this direction (Fig. 4a). 'New recruits' of *P. melanarius* displaced east, towards the dimethoate treated area, (Fig. 4b) but recaptures of 'survivors' and pre-treatment individuals were not concentrated in one direction. Recaptures of *H. rufipes* were uniformly distributed around the perimeter of the enclosure, while *P. madidus* were recaught in insufficient numbers for analysis.

In either of the two other enclosures, the distributions of recapture of all species were not significantly different from uniform distributions whenever recaptures were sufficiently large enough for analysis (Table 2). In enclosure C, *P. cupreus* did tend to displace east during the pre-treatment period, but this observation was not significant.

Discussion

This preliminary study demonstrates the feasibility of using mark-recapture techniques within enclosures to estimate, simultaneously, changes in carabid population densities, displacement rates and displacement direction, following pesticide applications. Lack of replication in this study, other than replication through time and by individuals, prevented the separation of treatment effects from effects of enclosure location. Nevertheless, it was possible to discern some probable effects of treatment when pre-treatment comparisons implied no significant effect of location and by comparison with side-effects of pesticides recorded in the literature. In addition, individually marked beetles caught post-treatment were separable into 'survivors' (individuals marked before spray application) and 'new recruits' (beetles marked since spray application), and this permitted levels of survival and recruitment to be estimated. Without this distinction, changes in density would provide underestimates of mortality.

Density estimates made using the mean weighted Petersen method are based on a number of assumptions common to all mark-recapture models (e.g. marks are permanent, marked and unmarked individuals do not differ in catchability or mortality; Begon 1979). In addition, the Petersen method assumes that the sampled population is closed, i.e. there are no births (or immigration) and no deaths (or emigration) within the population. Petersen's method can allow for either births or deaths, but not both. Enclosing populations using physical barriers reduces immigration and emigration, and conducting mark-recapture studies over short periods of time ensures that the effects of natural mortality remain negligible. Determining separate and independent density estimates of 'survivors' for the pre- and post-treatment periods avoided any bias caused by treatment induced mortality and, by definition of 'survivors' as recaptured individuals, any bias by recruitment to their density estimation. However, marked individuals were released at the centre of enclosures and, for some unknown length of time, their distribution would have differed from that of unmarked individuals. This may have led to unequal probabilities of capture immediately after release. Nevertheless, estimates of pre-treatment densities calculated by the weighted mean Petersen method were of the same order of magnitude as densities in the literature (Basedow 1991, Thiele 1977, Hawthorne 1995, Ulber & Wolf-Schwerin 1995, Lövei & Sunderland 1996) for medium-to-large ground beetles in arable habitats.

Significant pre-treatment differences between enclosures were found in the densities of virtually all species. Considering the proximity of these enclosures to one another, the scale of variation was unexpected. Differences in ground beetle abundance may have been induced by patchy crop germination, following a spring drought, and beetles' micro-climatic requirements prior to enclosure. The complete absence of *P. cupreus* in enclosure A (treated with water) made determination of potential treatment effects on this species particularly problematic, although the frequent recapture of individuals in enclosure B prior to dimethoate treatment and the sudden lack of recaptures thereafter imply either a treatment effect or a coincidental change in activity or density. Toxic effects of dimethoate on beneficial arthropods have been documented (Vickerman & Sunderland 1977, Thacker 1991) and dimethoate is frequently used as a toxic standard in laboratory (Lewis *et al.* 1992) and semi-field tests (Heimbach, Büchs

& Abel 1992) of pesticide side effects. The observed higher mortality of the remaining three species, in dimethoate treated enclosures compared to water-treated enclosures, would therefore agree with an effect of spray treatment on densities of ground beetles, but lack of replication prevents the separation of treatment effects from effect of enclosure location.

Recruitment to post-treatment adult populations were observed in all species in most enclosures. *Harpalus rufipes, P. madidus* and *P. melanarius* are autumn breeding species with larvae developing during the winter and teneral beetles emerging in June and July. *Pterostichus cupreus* is a spring breeding species with larvae developing during the summer and teneral beetles emerging in August and September. Consequently, recruitment was expected to the autumn breeding species but not to *P. cupreus*. Recruitment may have been over-estimated through some adults exhibiting low levels of activity pre-treatment and increased activity post-treatment.

Sublethal effects of pesticides can lead to alterations in behaviour of beneficial arthropods, disrupting mate location, foraging behaviour, dispersal and predator avoidance, and can therefore influence their long-term population dynamics (Croft & Brown 1975, Elzen 1989). Sub-lethal effects can be measured quantitatively soon after exposure through detailed analysis of locomotory behaviour (Little & Finger, 1990). In this study, the probability of recapture was used as a crude estimate of displacement rate. Displacement rate is influenced by both the speed and pattern of movement and, consequently, provides a useful summary statistic of mobility. But the probability of recapture is also influenced by the catchability of a species and its mortality. Further, catchability is influenced by the speed and pattern of movement, as individuals exhibiting 'directed movement' are less likely to avoid traps than individuals exhibiting 'random movement' (Paling 1992, Hawthorne 1995). The influence of treatment-induced acute mortality on the probability of recapture should have been negligible, since calculations of post-treatment recapture probabilities were restricted to recaptures following an initial post-treatment capture (usually a few days after treatment), but some delayed mortality may still have been included in the estimate.

Significant differences in probabilities of recapture of *Pterostichus* species were found between enclosures post-treatment despite the lack of any significant pre-treatment differences. Dimethoate treatment resulted in reduced recapture rates of 'survivors' when a whole enclosure was treated but elevated recapture rates when only half an enclosure was treated, relative to the water-treated control enclosure. In addition, 'new recruits' displaced more frequently towards the dimethoate-treated area than the water-treated area in the enclosure treated with both treatments. A higher probability of recapture in the dimethoate-treated half than in the water-treated half could have been a consequence of increased searching activity in the absence of suitable prey. 'Survivors' showed no such unimodal displacement. Sublethal effects of pesticides typically result in an increase in speed of movement for a few hours (Moosebeckhofer 1983, Heneghan 1994) followed by a reduction in speed of movement for a few days after treatment (Heneghan 1994, N. Randall & K. Taunton, unpublished data). Although a reduction in speed of carabid movement could explain the observed reduction in probability of recapture of 'survivors' during the four-week post-treatment period, further work will need to be undertaken to investigate the influence of dimethoate treatment on the duration and pattern of carabid movement and on the catchability of carabid species.

Mark-recapture techniques employed within enclosures, treated and untreated with pesticides, provide a useful means of determining the side-effects of pesticides on mortality, recruitment and displacement. Enclosures ensure that sampling areas are defined, immigration and emigration are reduced or eliminated, and standard errors of population estimates are reduced by increasing the probability of recapture. Studies of the effects of pesticides on mortality, recruitment and displacement provide an insight into the likelihood of subsequent population recovery. Such studies, combined with laboratory and field studies of the side-effects of pesticides as well as data on the population dynamics of individuals species, permit a greater understanding of the impact of chemical pest control on natural control processes.

Acknowledgements

We thank B. Hackett, K.L. Sykes & P.J. Tuck for their help erecting the polythene barriers, and J.N. Perry & G. Ross for their statistical advice. This research was financially supported by the Pesticides Safety Directorate of the MAFF. IACR receives grant-aided support from the Biotechnology and Biological Sciences Research Council of the United Kingdom.

References

Basedow, Th. 1991. Population density and biomass of epigeal predatory arthropods, natural enemies of insect pests, in winter wheat fields of areas with extremely different intensity of agricultural production. *J. Pl. Diseases & Protection* 98: 371-77.

Batschelet, E. 1981. *Circular Statistics in Biology.* Academic Press, London.

Begon, M. 1979. *Investigating animal abundance: capture-recapture for biologists.* Edward Arnold, London.

Croft, B.A. & Brown, A.W.A. 1975. Responses of arthropod natural enemies to insecticides. *Ann. Rev. Entomol.* 20: 285-335.

Chiverton, P.A. 1984. Pitfall-trap catches of the carabid beetle Pterostichus melanarius, in relation to gut contents and prey densities, in insecticide treated and untreated spring barley. *Entomol. Exp. Appl.* 36: 23-30.

Durkis, T.J. & Reeves, R.M. 1982. Barriers increase efficiency of pitfall traps. *Entomol. News* 93: 8-12.

Elzen, G.W. 1989. Sublethal effects of pesticides on beneficial parasitoids. In: Jepson, P. (ed.) *Pesticides and Non-target Invertebrates* pp.129-50. Wimborne.

Fisher, N.I. 1993. *Statistical Analysis of Circular Data.* Cambridge University Press, Cambridge.

Hawthorne, A. 1995. Validation of the use of pitfall traps to the study of carabid populations in cereal field headlands. In: Toft, S. & Riedel, W. (eds.) *Arthropod natural enemies in arable land · 1, Density, spatial heterogeneity and dispersal. Acta Jutl.* 70 (2): 61-75.

Heimbach, U., Büchs, W. & Abel, C. 1992. A semi-field method close to field conditions to test effects of pesticides on *Poecilus cupreus* L. (Coleoptera, Carabidae). *Bull. IOBC/WPRS* 15: 159-65.

Heneghan, P.A. 1992. Assessing the effects on an insecticide on the activity of predatory ground beetles. *Aspects of Applied Biology 31, Interpretation of Pesticide Effects on Beneficial Arthropods,* 113-19.

Heneghan, P.A. 1994. Side-effects of synthetic pyrethroid insecticides on the dispersal activity of predatory Coleoptera, with particular reference to the Carabidae. PhD thesis, University of Southampton.

Lewis, G.B., Brown, R.A., Edwards, P.J. & Canning, L.C. 1992. A laboratory test method for assessing the toxicity of pesticides to the ground beetle *Pterostichus melanarius* (L.) (Carabidae, Coleoptera). *Bull. IOBC/WPRS* 15: 110-15.

Little, E.E. & Finger, S.E. 1990. Swimming behaviour as an indicator of sublethal toxicity in fish. *Environ. Toxicol. Chem.* 9: 13-19.

Loreau, M. 1984. Population density and biomass of Carabidae (Coleoptera) in a forest community. *Pedobiologia* 27: 269-78.

Lövei, G.L. & Sunderland, K.D. 1996. Ecology and behaviour of ground beetles (Coleoptera: Carabidae). *Ann. Rev. Entomol.* 41: 231-56.

Moosebeckhofer, R. 1983. Laboruntersuchungen über den Einfluß von Diazinon, Carbofuran und Chlorfenvinphos auf die Laufaktivität von *Poecilus cupreus* L. (Col., Carabidae). *Z. angew. Entomol.* 95: 15-21.

Paling, N.J. 1992. *Studies on activity and population density of Pterostichus madidus (F.) (Coleoptera: Carabidae) by video recordings.* PhD thesis, Cranfield Institute of Technology.

Thacker, J.R.M. 1991. Integrating laboratory, semi-field and field collected data on the effects of pesticides on Coccinella septempunctata in U.K. In: Polgár, L., Chambers, R.J., Dixon, A.F.G. & Hodek, I. (eds.) *Behaviour and Impact of Aphidophaga.*

The Hague, pp. 337-345.

Thiele, H.-U. 1977. *Carabid Beetles in Their Environments*. Springer Verlag, Berlin.

Thomas, C.F.G. 1995. A rapid method for handling and marking carabids in the field. In: Toft, S. & Riedel, W. (eds.) *Arthropod natural enemies in arable land · 1, Density, spatial heterogeneity and dispersal. Acta Jutlandica* 70 (2): 57-59.

Ulber, B. & Wolf-Schwerin, G. 1995. A comparison of pitfall trap catches and absolute density estimates of carabid beetles in oilseed rape field. In: Toft, S. & Riedel, W. (eds.) *Arthropod natural enemies in arable land · 1, Density, spatial heterogeneity and dispersal. Acta Jutlandica* 70 (2): 77-86.

Vickerman, G.P. & Sunderland, K.D. 1977. Some effects of dimethoate on arthropods in winter wheat. *J. Appl. Ecol.* 14: 767-77.

Wiles, J.A. & Jepson, P.C. 1994. Sub-lethal effects of deltamethrin on the within-crop behaviour and distribution of *Coccinella septempunctata. Entomol. Exp. Appl.* 72: 33-45.

Movement patterns of *Pterostichus cupreus* and *Nebria brevicollis* in habitat patches with different vegetation structures

L. Winstone, E.J.P. Marshall andG.M. Arnold

IACR-Long Ashton Research Station, Department of Agricultural Sciences, University of Bristol, Long Ashton, Bristol, BS18 9AF, UK

Abstract
The movement patterns of the carabid beetles *Pterostichus cupreus* and *Nebria brevicollis* were observed in three different vegetation structures; a grass ley, a cultivated area and the boundary between them. Ten individuals of each species were released from a central point and their movement patterns mapped. *Nebria brevicollis* and female *P.cupreus* showed a preference for grass habitat and only male *P.cupreus* crossed the boundary into the cultivated, open area.

Key words: carabidae, *Pterostichus cupreus, Nebria brevicollis, movement pattern, vegetation structure.*

Introduction

In animal species, dispersal may influence competition for food (Hassell & Southwood 1978) or mates. It may also reduce the risk of inbreeding among related individuals and extinction in ephemeral habitats as part of metapopulation dynamics (Perry & Gonzalez-Andujar 1993). The survival of metapopulations may depend upon the dispersal processes of individual species and the habitat quality (Opdam 1990). Fundamental to the development and success of integrated pest management systems is an understanding of the invasive potential of natural predators, such as carabids, into crop habitats, and the effects of landscape features on the dispersal processes of such predators. Both large- and small-scale features may impede, modify or even facilitate dispersal processes of carabids and, thus, influence their biocontrol potential.

Many studies of dispersal take account of large-scale features, such as hedges and other habitat boundaries (Fry 1994, Jepson 1994). However, few

Arthropod natural enemies in arable land · III The individual, the population and the community
W. Powell (ed.). *Acta Jutlandica* vol. 72:2 1997, pp. 39-46
© Aarhus University Press, Denmark, ISBN 87 7288 673 0

address the impact of small-scale features, notably soil topography and vegetation structure. The structure and diversity of vegetation ground cover can influence habitat favourability and affect the activity of ground predators (Carcamo & Spence 1994). For example, vegetation structure will affect colonisation (Burel & Baudry 1990) into crop or overwintering habitats if it impedes carabid movement (Greenslade 1964). Only mobile natural enemies can regulate actual or potential pests (Kruess & Tscharntke 1994).

Turchin et al. (1991) suggested that direct observations on carabid movement behaviour, with recording of movement paths, have more advantages than traditional mark-recapture techniques in studies of carabid movement. In such techniques, the movements of the majority of the population are not detected and they may act very differently to the small proportion of organisms that are recaptured. Thus, actual dispersal rates can be underestimated. Furthermore, by recording the actual movement tracks and the environment through which insects pass, environmental cues that affect insect movement may be recorded more accurately.

By understanding the factors that affect dispersal we may be able to predict those species able to re-establish populations (Bunce & Howard 1990) after local extinction and more accurately model local population dynamics. We may also create suitable habitats for natural predators in areas where they are required and, thus, enhance methods of integrated pest management.

Materials and methods

The movement patterns of ten *Pterostichus cupreus* (five males and five females) and ten *Nebria brevicollis* were observed in the field in each of three areas: grass ley (area 1), a 27 x 27 m cultivated patch (area 2) containing mostly annual plants, and a habitat boundary between them (area 3). The study was undertaken during May and June. Each beetle was marked using a metallic pen (Edding 780 paint marker) so that the beetles could be seen easily (Griffiths 1983) without approaching too close and, thus, affecting their behaviour. The beetles were then placed individually in the centre of a 4 m^2 section (grid) in each area within a release-ring (diameter 7 cm). This ring was used to avoid immediate high activity directly after release as a response to handling. Once activity had subsided, the release-ring was removed and each beetle was observed for 10 minutes or until it left the grid. A note was made of each beetle's track using cocktail sticks to mark points crossed and therefore give coordinates. Path length, number of stops made and total time moving were also noted. Individual plants in each area were then mapped onto graph paper and digitized into Freelance Graphics using a Summasketch board.

The movement data were analysed using analysis of variance (ANOVA) to look at possible differences between the two species under study. The sexes of *P. cupreus* were also compared; this species was easy to sex in the field. Any effect of the different habitats on beetle movement was observed by studying the total distance travelled (analysed with a log transformation), the displacement ratio (total distance/direct distance) which would show the degree of convolution in the walking pattern, the time spent travelling in seconds during the ten minute period of observation, and the number of stops made by each beetle (analysed with a square-root transformation).

Results

Analysis of variance of movement measurements of the two beetle species in the three habitats indicated there were no significant interactions between habitat and species. Further, there were no significant differences in movement measurements between the grass ley, the cultivated area or the boundary between them (Table 1). However, there were some differences in the movement patterns between the two species. Whilst there was no significant difference in total distance travelled by these species (Table 2), *N. brevicollis* had a significantly ($p=0.025$) higher displacement ratio, indicating a more convoluted walking pattern than that of *P. cupreus*. Males of *P. cupreus* spent significantly more time on the move than females or than individuals of *N. brevicollis*. Similarly, male *P. cupreus* had significantly fewer stops than females or *N. brevicollis*.

Although the movement measures (Table 1) indicated there were no significant differences between the habitats, observations of beetle behaviour in the boundary arena revealed some habitat preferences. Within this area, the beetle release-ring was located on the boundary between the grass ley and the cultivated patch. Only *P. cupreus* males moved some distance into the cultivated area (Figure 1). *Nebria brevicollis* (and female *P. cupreus*) did not cross from the grass into the cultivated area (Figure 2).

Table 1. Effect of different habitats on beetle movement averaged over both species

	Area 1 Grass Ley	Area 2 Cultivated Patch	Area 3 Habitat Boundary	SED 51df
Total distance (cm) (log distance)	3.71	3.93	3.97	0.288
Back-transformed mean	40.9	50.9	53.0	NS
Displacement ratio	0.71	0.57	0.63	0.1321
Back-transformed mean	2.03	1.76	1.88	NS
Time travelling (s)	180	167	179	38.3 NS
Number of stops	1.69	1.78	1.85	0.155
Back-transformed mean	2.86	3.17	3.42	NS

NS = Not significant

Table 2. Movement of two beetle species averaged over three habitats (grass ley, cultivated patch and a habitat boundary between them)

	N.brevicollis	P.cupreus	P.cupreus (male)	P.cupreus (female)	SED 51df
Total distance (cm) (log distance)	3.81	3.93	-	-	0.235
Back-transformed mean (cm)	45.2	50.9			NS
Displacement ratio	0.76	0.51	-	-	0.1078
Back-transformed mean	2.14	1.67			
Time travelling (s)	140	-	288	133	38.3* 44.2**
Number of stops	1.99	-	1.24	1.86	0.155*
Back-transformed mean	3.98		1.54	3.46	0.179**

NS = Not significant.
* = SED when comparing species.   ** = SED when comparing sexes

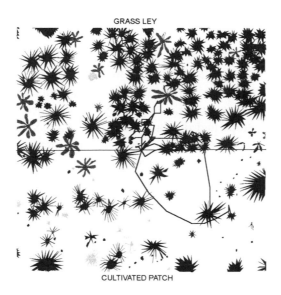

Fig. 1. Plant map of boundary (area3) showing movement of *P. cupreus*. The paths crossing the boundary are those of two males.

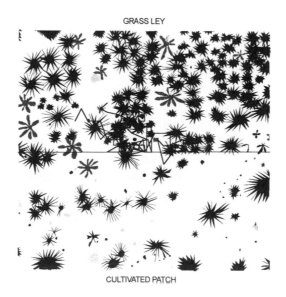

Fig. 2. Plant map of boundary (area 3) showing movement paths of *N. brevicollis*.

Discussion

The behaviour of two carabid beetle species was examined in a grass ley, a more open annually-cultivated area and on the border between these areas. The latter area was similar in structure to a field boundary adjacent to an arable field. The measures of beetle movement used, comprising total distance moved, time spent travelling, number of stops and displacement ratio, showed no significant effect of habitat or of any interaction between species and habitat. However, there were significant differences between species, with *P. cupreus* males stopping least and travelling for the longest time; *N. brevicollis* had a higher displacement ratio than *P. cupreus*.

Although movement measures indicated no significant difference in movement in the three areas, possibly resulting from the small numbers of beetles used, observation of the beetles indicated some preferences for the grass habitat. The exceptions were males of P. *cupreus*, a diurnal species (Thiele 1977), which moved into the cultivated area regardless of its open nature.

Pterostichus cupreus males may be more active and stop less often because they are searching for mates, and their reproductive period is during April and July (Zangger 1994). Males are characteristically the dispersers in animal populations (MacDonald & Smith 1990) and *P. cupreus* appears to be no exception. This behaviour has also been found in the larger scale mark-recapture studies of Rivard (1965).

The more convoluted movement pattern displayed by *N. brevicollis* may imply the species is in a favourable environment (Wallin & Ekbom 1988), indicating foraging behaviour by this species. The lack of any difference in movement measures in the open cultivated area may reflect the small sample size. *Nebria brevicollis* did not cross the boundary from grass in area 3. This species is nocturnal (Halsall & Wratten 1988) and, as the experiments were performed in daylight, it may explain its preference for denser vegetation cover. The grass ley may have been chosen as a possible aestivation site as this species does have an inactivity period in early summer (Greenslade 1964). However, it may not have preferred the cover in itself, but the composition of the vegetation in the grass ley, where *Lolium perenne* was the dominant grass species. During the course of study, *N.brevicollis* was found on the seed heads of this grass. It is not known, however, whether it was feeding on the seeds or animals within them.

Although known to eat aphids, *N. brevicollis* will also consume Dipteran larvae, collembola and small earthworms (Penney 1966). Due to its inactivity period in summer, it may be a less efficient aphid predator than some other species. *Pterostichus cupreus* is a voracious predator of aphids (Penney 1966) and is a particularly efficient predator of cereal aphids, due to its high energy demands (Wallin Ekbom 1994). As this species is also relatively unaffected by vegetation structure, it is likely to disperse throughout arable crops during the summer and may be useful in integrated pest management systems. *Nebria brevicollis*, in contrast, is likely to be limited to dense vegetation at arable field edges, at least during summer days; later in the year, it may be of greater importance for aphid predation.

Whilst the data presented are for a relatively small number of individuals, the technique of direct observation produced significant results. The methods can be applied to study beetle behaviour at habitat boundaries, which may affect patterns of colonisation or re-colonisation of adjacent fields. The technique may also be useful in identifying the environmental cues that affect the timing of movement into arable crops of species, such as *Agonum dorsale*, which over-winter in field margins (Coombes & Sotherton 1986).

Acknowledgements

IACR receives grant-aided support from the Biotechnology and Biological Sciences Research Council of the United Kingdom.

References

Bunce, R.G.H. & Howard, D.C. 1990. *Species Dispersal in Agricultural Habitats*. Belhaven Press, London and New York.

Burel, F. & Baudry, J. 1990. Hedgerow networks as habitats for forest species: implications for colonising abandoned agricultural land. In: Bunce, R.G.H. & Howard, D.C. (eds.) *Species Dispersal in Agricultural Habitats*, pp. 238-55. Belhaven Press, London.

Carcamo, H.A. & Spence, J.R. 1994. Crop type effects on the activity and distribution of ground beetles (Coleoptera: Carabidae). *Environ. Entomol.* 23: 684-92.

Coombes, D.S. & Sotherton, N.W. 1986. The dispersal and distribution of polyphagous predatory Coleoptera in cereals. *Ann. Appl. Biol.* 108: 461-74.

Fry, G.L.A. 1994. The role of field margins in the landscape. *BCPC Monograph No.58: Field Margins: Integrating Agriculture and Conservation*, pp. 31-40.

Greenslade, P.J.M. 1964. The distribution, dispersal and size of a population of *Nebria brevicollis* (F), with comparative studies of three other carabidae. *J. Anim. Ecol.* 33: 311-33.

Griffiths, E. 1983. The Feeding Ecology of the Carabid Beetle *Agonum dorsale* in Cereal Crops. PhD thesis, University of Southampton.

Hassell, M.P. & Southwood, T.R.E. 1978. Foraging strategies of insects. *Ann. Rev. Ecol. Syst.* 9: 75-98.

Halsall, N.B. & Wratten, S.D. 1988. The efficiency of pitfall trapping for polyphagous predatory Carabidae. *Ecol. Entomol.* 13: 293-99.

Jepson, P.C. 1994. Field margins as habitats, refuges and barriers of variable permeability to Carabidae. *BCPC Monograph No.58: Field Margins: Integrating Agriculture and Conservation*, pp. 67-76.

Kruess, A. & Tscharntke, T. 1994. Habitat fragmentation, species loss and biological control. *Science*, 264: 1581-84.

MacDonald, D.W. & Smith, H. 1990. Dispersal, dispersion and conservation in the agricultural ecosystem. In: Bunce, R.G.H. & Howard, D.C. (eds.) *Species Dispersal in Agricultural Habitats*, pp. 18-64. Belhaven Press, London.

Opdam, P. 1990. Dispersal in fragmented populations: the key to survival. In: Bunce, R.G.H. & Howard, D.C. (eds.) *Species Dispersal in Agricultural Habitats*, pp. 3-17. Belhaven Press, London.

Penney, M.M. 1966. Studies on certain aspects of the ecology of *Nebria brevicollis* (F) (Col: Carabidae). *J.Anim. Ecol.* 35: 505-12.

Perry, J.N. & Gonzalez-Andujar, J.L. 1993. Dispersal in a metapopulation neighbourhood model of an annual plant with a seedbank. *J. Ecol.* 81: 453-63.

Rivard, I. 1965. Dispersal of ground beetles (Coleoptera: Carabidae) on soil surface. *Can. J. Zool.* 43: 465-73.

Thiele, H.-U. 1977. Carabid Beetles in their Environments. Springer Verlag, New York.

Turchin, P., Odendaal, F.J. & Rausher, M.D. 1991. Quantifying insect movement in the field. *Environ. Entomol.* 20: 955-63.

Wallin, H. & Ekbom, B.S. 1988. Movements of carabid beetles (Coleoptera: Carabidae) inhabiting cereal fields: a field tracing study. *Oecologia (Berl.)* 77: 39-43.

Wallin, H. & Ekbom, B.S. 1994. Influence of hunger level and prey densities on movement patterns in three species of *Pterostichus* beetles (Coleoptera: Carabidae). *Environ. Entomol.* 23: 1171-81.

Zangger, A. 1994. The positive influence of strip-management on carabid beetles in a cereal field: accessibility of food and reproduction in *Poecilus cupreus*. In: Desender, K., Dufrene, M., Loreau, M., Luff, M.L. & Maelfait, J.-P. (eds.) *Carabid Beetles: Ecology and Evolution*, pp. 469-472. Kluwer Academic Publishers, Dordrecht.

Spatial heterogeneity and predator searching behaviour - can carabids detect patches of their aphid prey ?

L. Winder[1,2], S. D. Wratten[3] and N. Carter[2]

[1] School of Conservation Sciences, Bournemouth University, Poole, Dorset, U.K.
[2]IACR-Rothamsted Experimental Station, Harpenden, Herts, U.K.
[3]Department of Entomology and Animal Ecology, Lincoln University, New Zealand.

Abstract

Models were constructed which described the distribution of the grain aphid *Sitobion avenae* within a cereal crop and a ground predator, the carabid *Agonum dorsale* searched for these prey. The model allowed varying levels of prey aggregations to be generated in order to evaluate the response of the predator to prey patches. The model indicated that the predator was unable to detect aphid patches at aphid densities of between 0.1 and 5.0 aphids shoot^{-1} and that a simple analytical model could predict consumption rate accurately. If the model parameters were manipulated, a patch response could be detected. It is suggested that field trials showing patch responses are due to the artificial aphid distributions used in such experiments.

Key words: Analytical model, simulation, carabid beetle, *Agonum dorsale*, cereal aphid, *Sitobion avenae*, spatial distribution, predation.

Introduction

The role of ground beetles as aphid predators has been widely reported (Edwards *et al.* 1979, Sunderland & Vickerman 1980). Ground beetles search predominantly at ground level and most are unable to climb plants. This predatory guild must rely on aphids falling to the ground from the crop canopy in order to predate them. Predatory beetles may exhibit searching behaviour which enhances predation when aggregations of prey are encountered. Following prey capture, behavioural changes include an increase in turning rate (klinokinesis) and reduction in speed of movement (orthokinesis), the predator reverting to the pre-feeding searching pattern if such behaviour is not reinforced by the capture of

Arthropod natural enemies in arable land · III The individual, the population and the community
W. Powell (ed.). *Acta Jutlandica* vol. 72:2 1997, pp. 47-62
© Aarhus University Press, Denmark, ISBN 87 7288 673 0

a further prey item. This searching strategy has been observed in, for example, the carabid *Pterostichus coerulescens* L. (Mols 1986), the housefly *Musca domestica* L. (Murdie & Hassel 1973) and the predatory mite *Phytoselius persimilis* Athias-Herriot (Sabelis 1981). This ability to search patches where prey is plentiful has been identified as a parameter which could be used to rank the efficacy of predators as biological control agents (Putman and Wratten 1984).

Cereal aphids are distributed heterogeneously in cereal fields (Rabbinge & Mantel 1981). Aphids, may, therefore be aggregated in a way which could lead to the behaviour described above enhancing consumption rates. Field studies support this hypothesis, showing that ground beetles may aggregate in artificially created patches, albeit at high aphid densities (Bryan and Wratten 1984, Halsall and Wratten 1988).

This paper investigates whether the patch response behaviour of *Agonum dorsale* Pont. (Coleoptera: Carabidae) influences prey consumption rates when realistic distributions and densities of aphid prey are encountered. This is achieved in two ways; by simulation and by a simple analytical type model.

Predatory behaviour

A. dorsale searches for prey predominantly on the ground, consuming aphids (Sunderland *et al.* 1987) which have fallen from plants (or alternative prey such as Collembola due to its polyphagy). Its searching behaviour is characterised by longer periods of time being spent in patches where the rate of encounter with prey is highest (Griffiths 1983, Griffiths *et al.* 1985, Halsall & Wratten 1988). Its behaviour has been recorded in detail and quantified in ways which is suitable for the construction of models (Griffiths 1983, N. Halsall personal communication). These are summarised in Table 1.

Prey distribution and behaviour

Aphids are distributed in cereal crops in a heterogeneous spatial pattern. This distribution can be defined by three components:

- the number of aphids on each shoot;
- the location of each shoot;

- the position of aphids from each colony (aphids may be on shoots or on the ground).

Table 1. Values of parameters describing the searching behaviour of *A. dorsale*. Capture efficiency defined as the proportion of encounters leading to prey consumption, and handling time, the time taken to consume a captured prey item. The activity period was recorded by Griffiths (1983), other parameters recorded by Halsall (personal communication).

Behaviour	Symbol	Value
Activity period 24h^{-1}	P	6 hours
Mean walking speed prior to prey capture	V	0.024 m s^{-1}
Mean turning rate prior to prey capture	T	32.5 ^{o} s^{-1}
Mean walking speed post prey capture	v	0.0068 m s^{-1}
Mean turning rate post prey capture	t	32.5 ^{o} s^{-1}
Period of tortuous walk post prey capture	p	0.05 h
Detection distance	D	0.001m
Capture efficiency	E	60 %
Handling time	H	0.05 h

The number of aphids on shoots may be modelled descriptively using the negative binomial probability distribution, defined by two parameters, the mean number of insects sampling unit^{-1} (μ) and the dispersion parameter, k (Bliss and Owen 1958, Harcourt 1965, Shiyomi & Nakamura 1964, Elliott 1977). The distribution of *Sitobion avenae* (F.), the commonest cereal aphid in the UK may be described in this way in order to define the probability of finding 0, 1, 2,....,n aphids on a shoot for given values of μ and k (Rabbinge & Mantel 1981). The relationship between variance and mean obeys a power law;

$$s^2 = a\mu^b$$

which is related to the negative binomial distribution (Taylor *et al.* 1979);

$$k^{-1} = a\mu^{b-2} - \mu^{-1}$$

In the simulation described below, the distribution of aphids between shoots is described, using data collected at Leckford, Hampshire and at Rothamsted Experimental Station, Hertfordshire in 1987 and 1988 respectively (Winder 1990a), by generating a regression equation from field counts of aphid densities to estimate the constants a and b:

Leckford: $\text{Logs}^2 = 0.99\log\mu + 0.74$ P<0.01   r=0.94   n=11

Rothamsted: $\text{Logs}^2 = 1.67\log\mu + 0.71$ P<0.01   r=0.95   n=12

This allows the estimation of k for any given aphid density which allows the generation of a negative binomial distribution. Little data is currently available for the spatial location of shoots infested with aphids within a crop and this distribution is modelled in the simulation described below.

The position of aphids from individual colonies may be described by measuring the rates at which aphids leave their host plants and subsequently return. The rate of aphid fall-off (F, number of aphids falling day^{-1}) at a given aphid density (μ, aphids shoot^{-1}) may be described by empirically derived regression equations and these rates were measures as: $\log F = 0.85\log \mu - 0.84$ and $\log F = 0.826\log\mu - 0.572$ for the 1987 and 1988 data respectively (Winder 1990b). The average time living aphids reside on the ground (R, hours) prior to returning to the crop canopy has been measured as 0.076 hours (Winder *et al.* 1994).

Description of analytical model

This model assumes that searching is a random process and that the patch response to prey capture does not occur. If this assumption is made, then the consumption rate of a predator may be estimated using equations similar to Skellam (1958) and Mols (1993). The number of aphids consumed day^{-1} (C) may be calculated as a product of the area searched by a predator (S, m^{-2}) and the availability of aphids within that searched area (A, number of aphids). Hence, S may be calculated from:

$$S = (3600.P).2D.V$$

and A from:

$$A = \frac{500.F.E.R}{24}$$

where a shoot density of 500 m^{-2} is assumed. These equations were used to estimate the number of aphids consumed day^{-1} by an individual *A. dorsale* searching at aphid densities ranging from 0.1 to 5.0 aphids shoot^{-1}.

Description of the simulation model

A stochastic model, written in FORTRAN 77, simulated the searching behaviour of *A. dorsale* within a 1 m x 0.96 m arena where predation of aphids was possible on the "ground". The model generated aphid distributions at chosen aphid densities, levels of aggregation and residence times. The model was used to investigate the effect of:

- aphid density;
- living aphids returning to the crop canopy after a residence time compared to aphids remaining on the ground;
- the level of aphid aggregation;

on the consumption rate of an individual *A. dorsale*. Data from both 1987 and 1988 were used in the simulations. The model has three components; aphid distribution within the crop canopy, aphids on the ground and predator searching model.

Aphid distribution within the crop canopy
Using Taylor's equations described above, k for a given aphid density (μ) may be estimated and the probability of 0, 1, 2,....,n aphids being present on a shoot calculated. The sub-model generated numbers of aphids shoot^{-1} by sampling randomly from the negative binomial distribution and was repeated until 480 values had been generated. These values were ranked in descending order and then assigned to a shoot within an area of 1 m x 0.96 m giving an overall shoot density of 500 m^{-2} with forty-eight shoots arranged in rows 0.1 m apart to simulate field spacing. The number of aphids assigned to each shoot was determined by an aggregation parameter used to generate different levels of aggregation (J. Perry personal communication). The arena was divided into 20 sub-areas. Aphid numbers, derived from the negative binomial distribution described above were assigned to the sub-areas. By assigning the highest aphid numbers to a single sub-unit, high levels of aggregation could be generated and conversely, low levels of aggregation were obtained by assigning aphid numbers

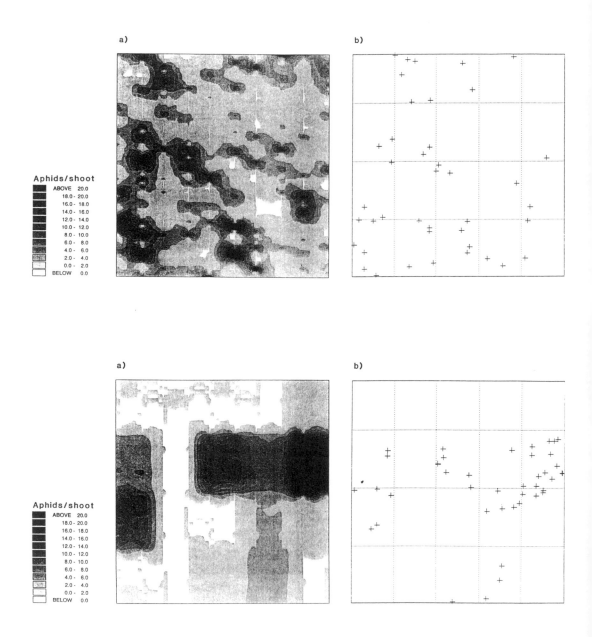

Fig. 1. Example of (a) the distribution of *S. avenae* within the crop canopy and (b) the position of aphids which fell to the ground during a simulation of six hours' duration at low and high levels of aggregation.

randomly to all sub-areas (Fig. 1). Intermediate levels of aggregation could also be generated. The aggregation parameter could be varied between 0.05 and 1, the former representing the lowest aggregation and the latter the highest level of aggregation within the model respectively. Aphid densities were varied between 0.1 and 5 aphids shoot^{-1}.

Aphids on the ground

The second component of the model simulated aphids falling to the surface of the arena. The number of aphids falling per day was calculated from the aphid density μ using the empirical relationships described above. It was assumed that the times at which aphids fell occurred at random. The aphid which fell was chosen at random and was assumed to land within 5 cm of the base of the plant from which it originated. Each aphid then remained at that position for the residence time, R, after which it returned to the plant from which it fell unless it had been consumed. The value of R was set either to represent the residence time of a living aphid able to re-climb or to represent aphids unable to return to the crop canopy. Spatial distribution of aphids on the ground was therefore dependent on that present within the crop canopy (Fig. 1).

Predator searching model

At the start of each simulation a predator was positioned randomly within the arena. The predator searched for its activity period, P. Its behaviour was defined by the experimental observations described above for *A. dorsale* (Table 1). At each iteration of 1 sec, the predator moved and turned at rates determined by the values of V and T or v and t, dependent on whether an aphid had been consumed. If an aphid had not been consumed within the previous p hours, values were V and T; if a prey had been consumed values were v and t. If the predator was outside the boundary of the arena it was repositioned to re-enter at the next iteration (Fig. 2). As the model was stochastic in nature, repeated simulations were conducted and mean results calculated.

At each iteration the locations of the predator and aphids were compared. If the distance was within the detection distance an aphid was located, and subsequent consumption was dependent on capture efficiency. Satiation was not included in the model as consumption rates were generally so low that it did not occur.

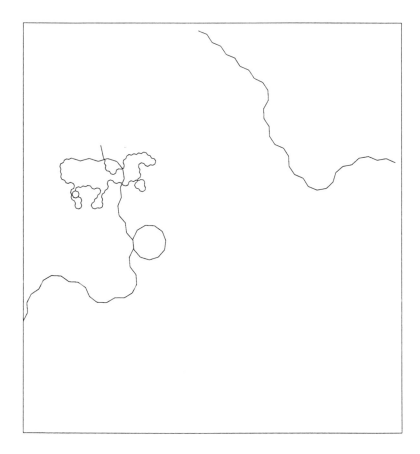

Fig. 2. Simulated searching pattern (to scale) of *A. dorsale* within 1 m x 0.96 m arena, showing normal and tortuous search patterns.

Results

Consumption rates calculated by the analytical and simulation models were similar and increased with aphid density (Table 2).

A comparison of rates of consumption of aphids when they had a residence time representing that of an aphid which remained permanently on the ground once it had fallen, demonstrated that at a high level of aggregation, the consumption rate of *A. dorsale* was measurably higher than at a low level of aggregation. This effect was more evident at relatively high aphid densities (Fig. 3). When parameters of the model were set for those representing living aphids the converse was true, being independent of level of aggregation (Fig. 4). Consumption rate was dependent solely upon aphid density.

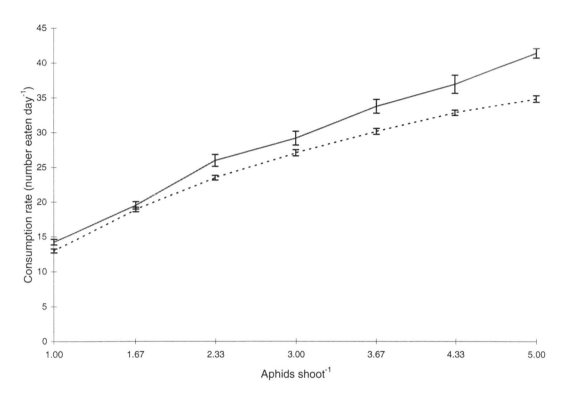

Fig. 3. Mean number of aphids eaten day (± 1 s.e.) under conditions of low (_ _ _ _) and high (_____) levels of aphid aggregation with *S.avenae* remaining on ground once they had fallen.

Sensitivity analysis (by varying parameters by ± 20%) of the components of the searching behaviour model indicated that detection distance, speed at which the predator searched, capture efficiency and residence time affected the consumption rate. Handling time, turning rate and duration of tortuous walk did not have a measurable effect. Sensitivity analysis indicated that parameters related to the *post* prey capture behaviour had little effect on consumption rate, whilst parameters which influenced behaviour *before* an aphid was encountered did influence consumption.

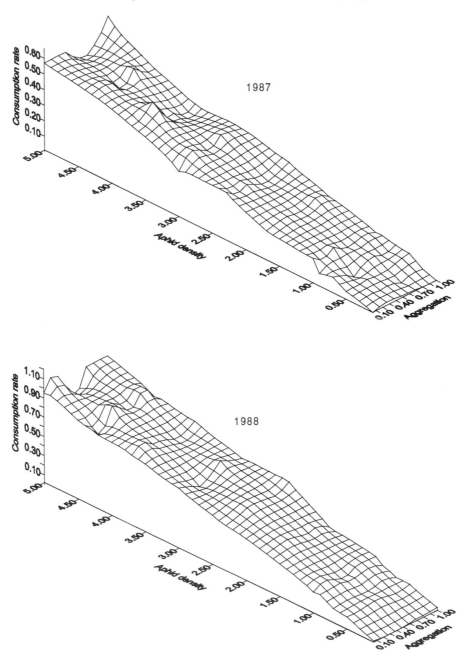

Fig. 4. Mean numbers of eaten day^{-1} at a range of aphid densities and levels of aggregation for simulations from 1987 and 1988.

Table 2. Consumption rate (number eaten day^{-1}) of live aphids by *A.dorsale* estimated by analytical and simulation models described in text. Simulations means ±1 s.e.

Aphid density	1987		1988	
	Analytical	Simulation	Analytical	Simulation
0.1	0.02	0.019 ± 0.003	0.039	0.031 ± 0.005
0.3	0.051	0.047 ± 0.002	0.097	0.090 ± 0.004
0.5	0.079	0.0072 ± 0.007	0.148	0.154 ± 0.006
0.7	0.105	0.091 ± 0.009	0.195	0.176 ± 0.006
0.9	0.129	0.123 ± 0.006	0.24	0.246 ± 0.006
1.0	0.142	0.141 ± 0.008	0.263	0.266 ± 0.012
1.67	0.219	0.214 ± 0.008	0.401	0.402 ± 0.013
2.33	0.29	0.282 ± 0.012	0.527	0.56 ± 0.016
3.00	0.36	0.366 ± 0.014	0.65	0.628 ± 0.019
3.67	0.427	0.388 ± 0.017	0.768	0.741 ± 0.02
4.33	0.492	0.510 ± 0.019	0.88	0.848 ± 0.094
5.0	0.556	0.546 ± 0.023	0.991	0.923 ± 0.052

Discussion

The simulation and analytical models described in this study provide a valuable insight into the importance of patch response behaviour for a ground predator searching at realistic prey densities and distributions. The analytical and simulation models gave near identical values demonstrating that predation rate could be estimated assuming simple random search. This implies that the predator does not benefit from its patch response behaviour under conditions set within the model. This was due primarily to the residence time of living aphids being relatively short, preventing any patches from developing to which the predator could respond. However, a clear patch response could be demonstrated when aphids did not return to the crop canopy. This difference in response has important implications for assessing the bio-control potential of such predators. It is possible that, rather than consuming live aphid prey, the predator is either consuming alternative prey or consuming moribund or dead aphids present on the ground. Winder *et al.* (1994) calculated that dead aphids were more apparent than live aphids to ground predators such as *A. dorsale* due to the short residence time of the latter. Mols (1986) constructed a simulation model of the carabid *P. coerulescens*, which shows a similar response to prey patches as *A. dorsale*, and observed that predation rates were higher at the same overall density when prey were aggregated and remained on the ground.

Table 3. Numbers of *A. dorsale* caught in pitfall traps in patch and control areas in field trials constructed by Bryan & Wratten (1984). * indicates a significant difference between patch and control.

	Patch	Control	Patch/control
1982			
7 to 14 June	192	34	5.6*
14 to 21 June	66	44	1.5
21 to 28 June	77	33	2.3
28 June to 5 July	35	58	0.6
5 to 12 July	21	17	1.2
1983			
31 May to 6 June	174	48	3.6*
6 to 13 June	268	44	6.1*
13 to 20 June	65	21	3.1*
20 to 27 June	49	27	1.8
27 June to 4 July	51	49	1.0
4 to 11 July	14	16	0.9

Bryan & Wratten (1984) demonstrated that *A. dorsale* can respond to prey patches generated artificially in the field (Table 3). However, in order to do this, aphid densities were manipulated so that there were large differences in aphid density between control and patch areas; aphid densities in the patches were up to 707.5 times higher when compared to control plots (Table 4). The measured response may be an artefact of the artificially high densities. Halsall & Wratten (1988) also demonstrated a patch response by *A. dorsale* to aphid prey; in this case aphid prey were dead and therefore available continuously, which concurs with the results of the simulation described above.

Van der Valk (1987) suggested that aphid populations on the soil would be aggregated for very short periods because they are likely to disperse as soon as they reach the ground. The short residence times measured by Winder *et al.* (1994) support this, although experiments were done at 13°C or 18°C and the response at other temperatures is not known. Aggregations of live aphids are therefore perhaps unlikely to occur or to exist for periods too short for predators to respond to them.

Other polyphagous predators which are known to climb, such as *Tachyporus* spp. (Staphylinidae), may have a stronger response to aphid aggregations because they may exploit the aphid population present within the crop canopy as well as that present on the ground. Fraser (1982) studied the aggregative response of Linyphiid spiders and concluded that they would not respond to small aggregations of prey because their patch response was too slow and pointed out that polyphagous predators could respond to a particular prey type only if it was the most important source of food. Other natural enemies such as parasitoids are able to respond to patches of aphids at natural field densities (Feng & Nowierski 1992).

Table 4. Mean number of aphids shoot^{-1} in control and artificially created patch areas in trials conducted by Bryan & Wratten (1984). * indicates a significant difference between patch and control.

	Patch	Control	Patch/control
1982			
7 June	57.8	0.5	115.6*
14 June	0.8	0.8	1
21 June	0.4	1.3	0.3*
28 June	0.5	1.1	0.45*
5 July	0.7	0.9	0.78
12 July	0.7	0.6	1.2
1983			
31 May	32.5	0	_*
6 June	28.3	0.04	707.5*
13 June	48.0	0	_*
20 June	38.8	0.3	129.3*
27 June	1.1	1.3	0.85
4 July	1.1	1.4	0.79
11 July	0.7	1.3	0.54

The model was run at a fixed spatial scale and it is possible that patch responses were not measurable. Heads & Lawton (1983) studied the effect of spatial scale on the detection of the aggregative response of the parasitoid *Chrysocharis gemma* Wlk. to the holly leaf miner *Phytomyza illicis* Curtis. The parasitoid appeared to aggregate strongly in regions of high host density. The effect was most obvious at the smallest sampling scale (0.03 m^2) and became progressively weaker at larger spatial scales. Aggregative response could not be detected at a spatial scale of 1 m^2. However, in the case of *P. coerulescens*, Mols (1986) concluded that it was not adapted especially to any distinct prey cluster size.

Sensitivity analysis demonstrated that consumption rate was dependent on parameters which affected aphid availability or total area searched. Parameters relating to behaviour after a prey item was located did not influence consumption rate and was a consequence of the behaviour being inappropriate given the aphid distributions generated. In this case, factors which affected the response of the predator to aphid aggregations were not important, although they may be important for other prey items which may form aggregations to which the predator could respond. Mols (1986) demonstrated that when aphids were randomly distributed, tortuous walk reduced the rate of prey capture due to the slower walking speed reducing the total area searched. If the aphid distributions are effectively random, then this searching strategy is disadvantageous. As *A. dorsale* is a polyphagous predator, the tortuous walking strategy may act to reduce the rate at which aphids are consumed, firstly because time is wasted when searching at a slower speed and secondly because the predator may be responding to aggregations of alternative prey items or dead aphid prey (Winder *et al.* 1994), thus reducing the likelihood of live aphids being eaten. Collembola, for example, are ground dwelling and may provide aggregations of prey which are continuously on the ground and *A. dorsale* could respond to these.

Acknowledgements

The authors would like to thank Joe Perry for help in constructing the simulation model and N. Halsall for providing data on the searching behaviour of *A. dorsale*. The authors would also like to thank Peter Mols for comment during the preparation of this manuscript.

References

Bliss, C.I & Owen, A.R.G. 1958. Negative binomial distribution with a common k. *Biometrica* 45: 37-58.

Bryan, K.M. & Wratten, S.D. 1984. The responses of polyphagous predators to prey spatial heterogeneity: aggregation by carabid and staphylinid beetles to their cereal aphid prey. *Ecol. Entomol.* 9: 251-59.

Edwards, C.A., Sunderland, K.D. & George, K.S. 1979. Studies on polyphagous predators of cereal aphids. *J. Appl. Ecol.* 16: 811-23.

Elliott, J.M. 1977. *Some Methods for the Statistical Analysis of Samples of Benthic Invertebrates*. Freshwater Biological Association.

Feng, M.G. & Nowierski, R.M. 1992. Spatial patterns and sampling plans for cereal aphids (Hom: Aphididae) killed by entomopthoran fungi and hymenopterous parasitoids in spring wheat. *Entomophaga* 37: 265-75.

Fraser, A.M. 1982. The role of spiders in determining cereal aphid numbers. PhD thesis, University of East Anglia.

Griffiths, E. 1983. Feeding ecology of the carabid *Agonum dorsale* on cereals. PhD thesis, University of Southampton.

Griffiths, E., Wratten, S.D. & Vickerman, G.P. 1985. Foraging by the carabid *Agonum dorsale* in the field. *Ecol. Entomol.* 10: 181-89.

Halsall, N.B. & Wratten, S.D. 1988. Video recordings of aphid predation in a wheat crop. *Asp. Appl. Biol.* 17: 277-80.

Harcourt, D.G. 1965. Spatial patter of the cabbage looper, *Trichoplusia ni* on crucifers. *Ann. Entomol. Soc. Amer.* 58: 89-94.

Heads, P.A. & Lawton, J.H. 1983. Studies on the natural enemy complex of the holly leaf miner: the effects of scale on the detection of aggregative responses and the implications for biological control. *Oikos* 40: 267-76.

Mols, P.J.M. 1986. Predation in the carabid beetle *Pterostichus coerulescens* (L.). In: den Boer, P.J., Grum, L. & Szyszko, J. (eds.) *Feeding behaviour and accessibility of food for carabid beetles.* pp. 49-58. Report of the fifth meeting of European Carabidologists. Warsaw Agricultural University SGGW-AR.

Mols, P.J.M. 1993. Foraging behaviour of the carabid beetles *Pterostichus coerulescens* L. at different densities and distributions of the prey. Part II. Wageningen Agricultural University Papers 93-5.

Murdie, G. & Hassell, M.P. 1973. Food distribution, searching success and predator-prey models. In: Hiorns, R.W. (ed.) *The mathematical Theory of the Dynamics of Biological Populations* pp. 87-101. Academic Press.

Putman, R.J. & Wratten, S.D. 1984. *Principles of Ecology.* Croom Helm.

Rabbinge, R. & Mantel, W.P. 1981. Monitoring for cereal aphids in winter wheat. *Netherlands J. Pl. Pathol.* 87: 25-29.

Sabelis, M.W. 1981. *Biological control of the two-spotted spider mites using phytoseiid predators, Part 1: Modelling the predator-prey interaction at the individual level.* Agricultural Research Report 910. Pudoc, Wageningen.

Shiyomi, M. & Nakamura, K. 1964. Experimental studies on the distribution of aphid counts. *Research into Population Research* 6: 79-87.

Skellam, J.G. 1958. Random dispersal in theoretical populations. *Biometrica* 38: 196-218.

Sunderland, K.D., Crook, N.E., Stacey, D.L. & Fuller, B.T. 1987. A study of feeding by polyphagous predators on cereal aphids using ELISA and gut dissection. *J. Appl. Ecol.* 24: 907-34.

Sunderland, K.D. & Vickerman, G.P. 1980. Aphid feeding by some polyphagous predators in relation to aphid densities in cereal fields. *J. Appl. Ecol.* 17: 389-96.

Taylor, L.R., Woiwood, I.P. & Perry, J.N. 1979. The negative binomial as a dynamic ecological model for aggregation, and the density dependence of k. *J. Appl. Ecol.* 48: 289-304.

van der Valk 1987. Kevers tussen het graan. Een simulatiestudie naar het effekt van *Pterostichus cupreus* op de graanius *Sitobion avenae.* Doctoraalverslag Theoretische Teeltkunde Landbouwuniversiteit, Wageningen.

Winder, L. 1990a. Modelling the effects of polyphagous predators on the population dynamics of the grain aphid *Sitobion avenae* (F.). PhD thesis, University of Southampton.

Winder, L. 1990b. Predation of the cereal aphid Sitobion avenae by polyphagous predators on the ground. *Ecol. Entomol.* 15: 105-10.

Winder, L., Hirst, D.J., Carter, N., Wratten, S.D. & Sopp, P.I. 1994. Estimating predation of the grain aphid *Sitobion avenae* by polyphagous predators. *J. Appl. Ecol.* 31: 1-12.

THE POPULATION

The construction of a simulation model of the population dynamics of *Lepthyphantes tenuis* (Araneae: Linyphiidae) in an agroecosystem.

C. J. Topping

National Environmental Institute, Kalø, Grenåvej 12, DK-8410 Rønde, Denmark
Previous Address: S.A.C., 581 King Street, Aberdeen, U.K.

Abstract

Details of the construction of a landscape simulation model for the linyphiid spider *Lepthyphantes tenuis* are presented. The model is based on a structured metapopulation model and simulates the population dynamics of *L. tenuis* on a landscape with a resolution of 50 m. Different agricultural operations and crops are simulated largely by their influence on spider mortality. Crop rotations, different husbandry and set-aside areas are all incorporated into the model. Dispersal from each sub-population is based on a simple diffusion process. The model has been used to investigate the potential for augmenting spiders by changing patterns of land-use and crop husbandry and has been used to simulate arable and grazing systems. Patterns generated by the model are largely dependent upon land management during the winter and early spring and on the growth rate of the spider population during summer. It is suggested that such autecology based models might be useful in the design of programmes aimed at the modelling of a much wider range of species for investigating the interactions between management and biodiversity as well as for planning purposes.

Key words: simulation model, population dynamics, spider, *Lepthyphantes tenuis*, landscape, dispersal.

Introduction

Much attention has been focused on the beneficial action of polyphagous predators, such as spiders, in agroecosystems. In many of these systems spiders are abundant and are therefore potentially important pest control organisms (Riechert & Lockley 1984, Nyffeler & Benz 1988). *Lepthyphantes tenuis* is one species which is particularly common in cereal crops (e.g. Sunderland 1991,

Arthropod natural enemies in arable land · III *The individual, the population and the community*
W. Powell (ed.). *Acta Jutlandica* vol. 72:2 1997, pp. 65-77
© Aarhus University Press, Denmark, ISBN 87 7288 673 0

1996). It occurs in a wide range of habitats and has a very large geographical range, including most of Europe and New Zealand, North Africa and North and South America (Sunderland 1995). It has been shown that this species, acting in concert with other polyphagous predators, can kill a potentially economically significant proportion of aphids present in cereal crops both by direct feeding, and through the large area of web present in the field which acts as a lethal trap for small aphids (Fraser 1982, Sunderland et al. 1986).

The population dynamics of this species has been investigated by Topping & Sunderland (unpublished data). In southern England, adults can be found throughout the year. Eggs are laid in egg-sacs containing approximately 21 eggs from early spring through to November. The eggs hatch to produce first instar spiderlings which remain in the egg-sac until they moult to the second instar. There are a further five moults before becoming adult (Sunderland 1995). In southern England, there are probably two to three generations per year. This species makes considerable use of aerial dispersal (ballooning). All mobile stages engage in ballooning to some extent but it is the adults which are most aerially active (Topping & Sunderland, unpublished data). Factors such as the suitability of the local conditions may affect ballooning in most life stages but adult females balloon whenever the meteorological conditions are suitable (rising air (Humphrey 1987), windspeed <3 m sec^{-1} (Richter 1970, van Wingerden & Vugts 1974, Greenstone 1990)).

If we want to promote the use of beneficials in agroecosystems we must first understand the factors controlling population fluctuations. Once this has been determined, this information could be used to help manage the system to promote beneficial populations in the crops. This is an extremely difficult task. Determining the details of the population processes is only part of the problem because any organism which is useful as a pest control agent is likely to be abundant and therefore, in an ephemeral habitat such as that provided by arable agriculture, dispersive (Southwood 1962, Den Boer 1977, Ebenhard 1991). A dispersive ability which involves the mixing of populations from different areas, containing different habitats and different local conditions results in the need to consider the population dynamics on a landscape scale. This is certainly the case with linyphiid spiders, which will regularly disperse over distances from 1 to 6 km in a day (Thomas 1993). If we need to consider the complexities of spatial interactions on a large scale then the only way to integrate the information required is by using a computer model.

Two basic approaches to modelling the effects of management on a landscape scale can be adopted. One is to try to correlate observed patterns of distribution with a range of potential causal factors (e.g. Sparks et al. 1994) the other is to work from the population dynamics and extrapolate this behaviour onto a landscape simulation. The second approach can be used to simulate the dynamics of populations based on modelling the behaviour of individuals (e.g. Booij 1996). Alternatively, it can be based on integrating the dynamics of a series of linked sub-populations which are mapped onto the simulation. The individual approach works well when individuals are large and population density is low when compared to the spatial resolution of the model. In the case of *L. tenuis*, population density can be very high and dispersal distances large, so the sub-population based approach was adopted in order to obtain an adequate representation of distribution of densities across the landscape.

The resulting model has been used to investigate the potential effects of changing management and area of set-aside in the arable landscape (Topping & Sunderland 1994a) and well as the possibility of altering *L. tenuis* densities by altering field sizes and crop management (Topping 1997). The aim of this paper is to describe the details of the construction of the model, which have not been documented previously.

Methods

Model Overview
The present landscape simulation model is an extension of a preliminary metapopulation model of Topping & Sunderland (1994b). The present model is a structured metapopulation model, considering populations occurring in a square matrix of up to 400,000 sub-population units. Each sub-population is positioned on a simulated landscape and is represented as occupying a discrete patch of habitat covering a square area of 50 x 50 m^2. Habitats are designated by a unique set of parameters which affect the population processes and spatial dynamics of the organisms. Sub-populations interact by means of dispersal. As the model runs, population processes are simulated on a monthly basis within a yearly cycle. Hence all population dynamics parameters used for input are given 12 values, one for each month.

Population Processes
The relationships between the various population processes and the life-history of the spider are summarised in Figure 1. Population dynamics data were available from a study of spiders in winter wheat (Sunderland & Topping 1992, Topping & Sunderland 1994a, 1996, Sunderland et al. 1996). Reproduction was measured as

a function of eggs per female spider per month and incorporated into the population dynamics processes by:

$$eggs_{t+1} = Density_{females} \times R_t,$$

where R_t is the reproductive rate per female for that month.

Development of eggs was found to be dependent on a day-degrees calculation, and so the month during which eggs would hatch could be calculated by using temperature data. Spiderlings were therefore introduced into the model population at the time when the eggs were predicted to hatch. Development of the spiders was also dependent upon temperature (Topping & Sunderland 1996) and hence the month during which the spiderlings became adult could also be predicted in the same way. However, mortality of the eggs and spiderlings must also be considered. This was incorporated as three mortality factors. The first factor represents the mortality of the eggs. This is based on estimates of parasitoid attack of egg-sacs in the field and is seasonally variable. A further small but constant mortality is assumed due to attack by other predators. This mortality is applied as a fixed proportion mortality of the eggs in the population per month. A similar mortality was applied for the spiderlings based on the daily fixed mortality estimate from Topping & Sunderland (1996). Hence at each month the number of juveniles was calculated to be:

$$N_{t+1} = B + (N_t(1 - m_j)),$$

where B is the number of juveniles hatching from eggs and m_j is the fixed proportion mortality for that month.

Mortality for the adults was on a fixed proportion basis per month below a critical density of 10 spiders m^{-2}. Above this density a higher proportion mortality was used. This density-related fixed proportion mortality gave a growth curve which approximated to the densities expected from field data (Topping & Sunderland 1994b).

Dispersal was modelled using a simple diffusion model by assuming that a proportion of the population would disperse from the field at each month. Dispersal was assumed to occur in all directions equally with individuals being distributed according to a simple model:

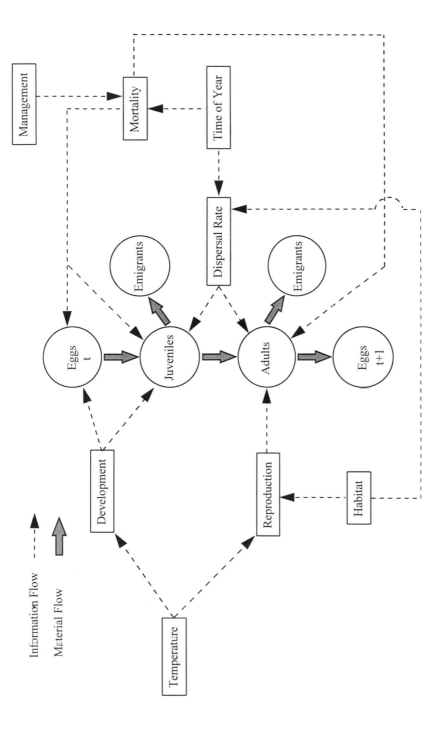

Fig. 1. The relationship between life-history and other factors considered in the population dynamics of a sub-population of a landscape scale simulation model for *Lepthyphantes tenuis*

$$N_s = \frac{N_e}{(D8d)},$$

where N_s is the number of emigrants arriving in sub-population s, N_e is the number of emigrants leaving the source sub-population, D is the maximum number of sub-populations which can be crossed by dispersal that month, d is the number of sub-populations between the source of immigrants and destination population (including the destination population). In pseudocode the algorithm is:

N_1, N_s : Floating point numbers;
$N_1 = N_e/D$;
for j =1 to D, increment j:
begin
  $N_s = N_1 / (j * 8)$;
  add N_s to the appropriate sub-population;
end;

This model does have the disadvantage that the spiders travelling diagonally, i.e. towards the corners of the square pattern described by this formula travel further than the ones travelling horizontally or vertically. The advantage of using this model is that it provides a very simple computation method for this phase of the simulation, and correlates well with the predicted dispersal pattern of more complex simulations (Thomas 1996).

When running, the model considers the population processes of births and deaths for each sub-population separately. Following this the emigrants are dispersed simultaneously from all populations following the above dispersal model. The number of emigrants from each sub-population depends on the size of the sub-population, the habitat type and the time of year. Data was available from 4 years of suction trap sampling and two years of other aerial trapping methods which was used to generate temporal dispersal rates (Topping & Sunderland 1994b, Topping & Sunderland 1995). Habitat related dispersal rates influenced the temporal rate only when habitats were clearly unsuitable for ballooning (e.g. water).

Practically, this approach takes considerable processing power. For each sub-population it is necessary to maintain a record for cohorts of eggs and spiderlings for each month of the year. Dispersal is also very time consuming because of the very large number of calculations required to disperse individuals long distances from each sub-population. For instance a dispersal distance of 1 km requires the distribution of emigrants from each of the 400,000 sub-populations to 1680 surrounding sub-population units, at total of 672 million calculations. Hence on a 486dx PC, running at a clock speed of 60 MHz a single run of 10 years simulation of a 10 x 10 km matrix will take approximately 48 hours.

Construction of the Landscape Data

Within the constraints of a resolution of 50 m, a relatively complex representation of a farm landscape was developed. Areas of farmland were mapped into 2500, m^2 (50 x 50 m) units by placing a grid over 1:25,000 Ordinance Survey maps and allocating each unit a habitat type. These habitat types were either recorded as agricultural fields, or other habitats as listed in Table 1. If the unit formed part of a field the whole field was designated by the use of a unique numeric code for that field. A field could then be delimited by the model by searching adjacent units until all units containing the same code were found. These units were then treated in unison when applying agricultural managements. All agricultural fields were initially allocated a crop. This was achieved using a probability table which listed the crops found in that landscape and the area covered. The probability of encountering a particular crop was calculated accordingly. Each crop allocation was divided into a series of sub-divisions based on a set of typical crop husbandries for that farming area. Again distribution of crop husbandry was based on estimates of the proportion of crop covered by each husbandry in the real landscape. In total, 44 crop and husbandry combinations were used (see Topping 1997 for further details) although the total number of crop/husbandry and other habitats which can be used is only limited by the availability of computer memory.

Table 1. Non-agricultural habitat designations used in the landscape simulation of *L. tenuis* population dynamics

Non-agricultural habitats

Semi-natural grassland
Woodland
Railway
Drains
Roads
Urban - built
Urban space - allotments, parks and gardens
Water

Crop rotation was simulated by fixing a series of 10 crop and husbandry patterns, which repeat over a six year period. This provided the facility to use any rotation ranging from continuous cropping of one type to complicated 6 year rotations, with different crops and husbandries.

"Set-aside" was also built into the model by allocating a percentage of the landscape to set-aside and by treating set-aside as a crop type. Rotational set-aside was moved each year whereas non-rotational set-aside was fixed throughout the simulation. A field used for rotational set-aside in one year was not allocated as set-aside the following year.

Each crop husbandry was simulated in terms of the mortalities expected as a result of the farming operations. Data on the mortality rate of farming operations in the crops to be simulated was not available and had to be estimated by analogy with other studies (e.g. Everts et al. 1989, Topping 1991, Pullen et al. 1992, Sunderland & Topping 1993). Hence, harvest, ploughing and insecticide applications were assumed to give mortalities of 50%, 75% and 90% respectively. Crop husbandries were represented by the timing and frequency of these three operations. Other operations considered were grazing and mowing, each assumed to give a 50% mortality (G. Thomas personal communication, in Topping & Sunderland 1994a) in the month in which they were applied.

The model has been used to simulate an area of Lincolnshire, UK from grid reference TF 16. This area was chosen because of its simple topography consisting of a flat landscape, with very large fields and >90% of the area covered with arable farmland (Topping & Sunderland 1995, Topping 1997). Data from the Ministry of Agriculture Fisheries and Food's 1993 June census were used to determine the proportions of land covered by the major crop types and common husbandry methods were obtained from local farmers and ADAS advisors.

Simulation of a largely pastoral system is problematic since there are no solid data to estimate the rates of mortality. However, an indication of the expected effect was obtained by simulating a typical pastoral and mixed farming landscape, with smaller field sizes and only 20% arable farmland. Grazing was assumed to be by sheep or cattle, and mortalities applied at 40% per month for cattle and 50% per month for sheep, based on the relative differences in spider fauna found under these two types of grazing (Topping 1991). Initial population densities were started in January with 2 adult spiders m^{-2} in the same way as was adopted for previous arable simulations.

Results & Discussion

The simulation of the pastoral system predicted densities in agricultural fields which were uniformly lower than those predicted by an identical simulation of a predominantly arable area. Mean maximum populations over a ten year simulation were 1.20 (SE 0.0057) spiders m^{-2} in the pastoral system compared with 1.63 spider per m^{-2} (SE 0.0061) in the arable system. This result concurs with the authors own experience of sampling for spiders in largely pastoral systems where field densities in grazed pasture are normally very low. Some very small fragments of habitat (road verges, field margins etc.) can contain high densities of spiders, but the density of *L. tenuis* can be low even in these situations, especially when surrounded by grazed fields (Topping 1991).

These simulations demonstrate that the method of working from detailed population dynamics data and extrapolating this onto a landscape scale has great potential to answer commonly posed questions concerning the effect of land management on the populations contained within the landscape. The *L. tenuis* simulations referred to here concluded that there may be potential for "set-aside" to augment populations of this species in the arable system, especially if used in small areas, perhaps to round off corners of fields or as strips along the field margin. Simulations also indicated the importance of timing of farming operations before the main dispersal period of the spiders in autumn to allow spider populations to be present in the crop at the start of the following season and highlighted the relative importance of dispersal and reproduction in determining the population size at any particular location.

Confidence in the model was increased by its ability to duplicate the expected densities in a pastoral landscape despite the original model construction being based on data from a predominantly arable landscape. This may seem to be a surprising achievement but there are two factors which significantly add to the ability of the model to predict changes in *L. tenuis* populations. The first is the overriding importance of dispersal in relation to reproduction and management. Although reproduction is responsible for a major increase in spider densities during the summer, these spiders are so dispersive that this increase is largely seen as an increase across the whole landscape and is little affected by local events. A similar conclusion was drawn by Den Boer (1970, 1981) who hypothesized that carabid populations were stabilised by dispersal over distances far larger than the heterogeneity obvious at the field scale. In *L. tenuis*, once dispersal is finished in autumn there is no further significant dispersal until late spring; there is also little or no further reproduction. This pattern effectively means that variance in spider densities during dispersal periods will be low. Outside of the dispersal period, the density is determined by local mortality as a result of management. Hence it is either the local management or the average effect of management which determines spider densities in a particular location depending upon the time of year, and it is always the overall average effect of management which determines the population density at the landscape scale.

The second factor which aided the construction of this model was that it could be constructed without the need to explicitly model the dynamics of other species. As a generalist predator, food may rarely be in short supply and taking into account that spiders are thought to be able to tolerate conditions of food stress (Anderson 1970, 1974, Nakamura 1987), they are probably less susceptible to minor fluctuations in food supply than many other species. In addition, this species does not have any significant specialist predators, and egg-sac parasitoids don't occur in numbers large enough to merit the explicit modelling of this mortality (van Baarlen et al. 1994). This lack of significant association between *L. tenuis* and other species results in a very simple model of the spatial dynamics and therefore a greater chance of the simulation correctly matching the real-life dynamic processes, i.e. there is less to get wrong.

Further refinements to this model should include some measure of the effect of topography on the dispersal distances achieved by the spiders. The movement of air at low levels is significantly affected by vegetation type and the positioning of obstacles, which would need to be taken into account if a much more accurate prediction of local densities was required. However, this type of detail may be of greater significance for less dispersive species where the distribution patterns may be generated to a much larger extent by the effective dispersal abilities of the species as well as the influence of their environment (e.g. Addicot 1978, Toft & Schoener 1983, Hanski 1986, Bengtsson 1991). These details may also be superfluous if the aim of a model is to determine the gross effect of management on total population numbers, in which case the approach taken towards modelling dispersal in this study may be sufficient.

This study demonstrates the potential usefulness of autecology based spatial models and also underlines the need for detailed autecology data for the species to be modelled. However, the use of these models will usually be restricted to the species for which they are designed, and given the need for high quality data it would be impossible to generate models of this type for a wide range of species. An alternative approach would be the investigation of theoretical frameworks for combining models of different species into the same simulation. This might be a fruitful area for further study, perhaps using autecology based simulations of some key species for validation. Such models would be invaluable in the investigation of the effects of land-use change on biodiversity and could be used by a range of agencies involved in planning land-use changes on both a local and regional scale.

Acknowledgements

The author acknowledges the support of the Natural Environment Research Council (Joint Agriculture and the Environment Grant), the Leverhulme Trust and the Scottish Office for Agriculture and Fisheries.

References

Addicot, J.F. 1978. The populations dynamics of aphids on fireweeds: a comparsion of local populations and metapopulations. *Can. J. Zool.* 56: 2554-64.

Anderson, J.F. 1970. Metabolic rates of spiders. *Comp. Biochem. Physiol.* 33: 51-72.

Anderson, J.F. 1974. Responses to starvation in the spider *Lycosa lenta* Hentz and *Filistata hibernalis* (Hentz). *Ecology* 55:576-85.

Bengtsson, J. 1991. Interspecific competition in metapopulations. *Biol. J. Linnean Soc.* 42: 219-37.

van Baarlen, P., Sunderland, K.D. & Topping, C.J. 1994. Eggsac parasitism of money spiders (Araneae: Linyphiidae) in cereals, with a simple method for estimating percentage parasitism of *Erigone* spp. eggsacs by Hymenoptera. *J. Appl. Entomol.* 118: 217-223.

Den Boer, P.J. 1970. On the significance of dispersal power for populations of carabid beetles (Coleoptera: Carabidae) *Oecologia (Berl.)* 4: 1-28.

Den Boer, P.J. 1977. Dispersal power and survival: carabids in a cultivated countryside. *Miscellaneous paper Landbouwhogeschool, Wageningen.* No. 14.

Den Boer, P.J. 1981. On the survival of populations in a heterogenous and variable environment *Oecologia (Berl.)* 50: 39-53.

Ebenhard, T. 1991. Colonisation in metapopulations: a review of theory and observations. *Biol. J. Linnean Soc.* 42: 105-21

Everts, J.W., Aukema, B., Hengeveld, R., & Koeman, J.H. 1989. Side-effects of pesticides on ground-dwelling predatory arthropods in arable ecosystems. *Environ. Pollution* 59: 203-25.

Fraser, A.M. 1982. The role of spiders in determining cereal aphid numbers. PhD thesis, University of East Anglia, England.

Greenstone, M.H. 1990. Meteorological determinents of spider ballooning: the role of thermals versus the vertical wind speed gradient in becoming airborne. *Oecologia (Berl.)* 85: 164-68.

Hanski, I. 1986. Population dynamics of shrews on small islands accord with the equilibrium model. *Biol. J. Linnean Soc.* 28: 23-36.

Humphrey, J.A.C. 1987. Fluid mechanic constraints on spider ballooning. *Oecologia (Berl.)* 73: 469-77.

Nakamura, K. 1987. Hunger and starvation. In: Nentwig, W. (ed.) *Ecophysiology of spiders.* Springer, Berlin, pp 287-95.

Nyffeler, M. & Benz, G. 1988. Prey and predatory importance of micryphantide spiders in winter wheat fields and hay meadows. *J. Appl. Entomol.* 105: 190-97.

Pullen, A.J., Jepson, P.C. & Sotherton, N.W. 1992. Terrestrial non-target invertebrates and the Autumn application of synthetic pyrethroids: experimental methodology and the trade-off between replication and plot size. *Archives of Environmental Contamination and Toxicology* 23: 246-58.

Richter, C.J.J. 1970. Aerial dispersal in relation to habitat in eight wolf-spider species. *Oecologia (Berl.)* 5: 200-14.

Riechert, S.E. & Lockley, T. 1984. Spiders as biological control agents. *Ann. Rev. Entomol.* 29: 299-320.

Southwood, T.R.E. 1962. Migration of terrestrial arthropods in relation to habitat. *Biol. Rev.* 37: 171-214.

Sparks, T.H., Hinsley, S.A., Mountford, J.O., Veitch, N., Bellamy, P.E. & de Nooijer, D.S. 1994. Landscape design: preliminary estimates of the effects of landscape permutations on wildlife. In: Dover, J.W. (ed.) *Fragmentation in agricultural landscapes,* Proceedings of the 3rd IALE(UK) conference, Myerscough College, Preston. pp.87-94.

Sunderland, K.D. 1991. The ecology of spiders in cereals. *Proceedings 6th International Symposium on Pests and Diseases of small grain cereals and maize. Halle/Salle, Germany* 1: 269-80.

Sunderland, K.D. 1996. Studies on the population ecology of the spider *Lepthyphantes tenuis* (Araneae: Linyphiidae) in cereals. *Bull. IOBC/WPRS* 19(3): 53-69.

Sunderland, K.D., Fraser, A.M. & Dixon, A.F.G. 1986. Field and laboratory studies on money spiders (Linyphiidae) as predators of cereal aphids. *J. Appl. Ecol.* 23: 433-47.

Sunderland, K.D. & Topping, C.J. 1992. The spatial dynamics of spiders in farmland. XII International Congress of Arachnology. *Memoirs of the Queensland Museum,* 33: 639-44.

Sunderland, K.D., Topping, C. J., Ellis, S., Long, S., van de Laak, S. & Else, M. 1996. Reproduction and survival of linyphiid spiders in UK cereals. In: Booij, K. & den Nijs, L. (eds.) *Arthropod natural enemies in arable land. II. Acta Jutlandica* 71 (2): 81-95.

Thomas, C.F.G. 1993. The spatial dynamics of spiders in farmland. PhD thesis. University of Southampton.

Thomas, C.F.G. 1996. Modelling aerial dispersal of linyphiid spiders. *Asp. Appl. Biol.,* 46: 217-22.

Thomas, C.F.G. 1997. Modelling dispersive spider populations in farmland. In: Powell, W. (ed) *Arthropod natural enemies in arable land. III. Acta Jutlandica* 72 (2): 79-85 (this volume).

Toft, C.A. & Schoener, T.W. 1983. Abundance and diversity of orb spiders on 106 Bahamian islands: biogeography at an intermediate trophic level. *Oikos* 41: 411-26.

Topping, C.J. 1991. Pitfall trap sampling and community analysis of grassland spiders. PhD thesis, University of Newcastle upon Tyne.

Topping, C.J. 1997. Predicting the effect of landscape heterogeneity on the distribution of spiders in agroecosystems using a population dynamics driven landscape-scale simulation model. *Biol. Agric. Hortic.* (in press).

Topping, C.J. & Sunderland, K.D. 1994a. A spatial population dynamics model for *Lepthyphantes tenuis* (Araneae: Linyphiidae) with some simulations of the spatial and temporal effects of farming operations and land-use. *Agric. Ecosystems Environ.* 48: 203-17.

Topping C.J. & Sunderland, K.D. 1994b. The potential influence of set-aside on populations of *Lepthyphantes tenuis* (Araneae: Linyphiidae) in the agroecosystem. *Asp. Appl. Biol.* 40: 225-28.

Topping, C.J. & Sunderland, K.D. 1995. Methods for monitoring aerial dispersal by spiders. In: Toft, S. & Riedel, W. (eds.) *Arthropod natural enemies in arable land. I. Acta Jutlandica* 70 (2): 245-56.

Topping, C.J. & Sunderland, K.D. 1996 Estimating the mortality rate of eggs and early instar *Lepthyphantes tenuis* (Araneae: Linyphiidae) from measurements of reproduction and development. In: Booij, K. & den Nijs, L. (eds.) *Arthropod natural enemies in arable land. II. Acta Jutlandica* 71 (2): 57-68.

van Wingerden, W.K.R.E. & Vugts, H.F. 1974. Factors influencing the aeronautic behaviour of spiders. *Bull. Br. arachnol. Soc.*, 3: 6-10.

Modelling Dispersive Spider Populations in Fragmented Farmland

C.F.G. Thomas

IACR-Long Ashton Research Station, Department of Agricultural Sciences,
University of Bristol, Long Ashton, Bristol BS18 9AF, UK

Abstract

The structure and parameterisation of a spatial population dynamic model for linyphiid spider dispersal in a one-dimensional landscape is summarised. The principal features of the population dynamics of spiders within fields, the physical and behavioural influences on dispersal between fields and the representation of landscape structure are described. Some constraints and limitations are addressed in terms of unknown or intractable elements in model construction, including components of behaviour and parameter values.

Key words: population dynamics, spider, dispersal, landscape, simulation model.

Introduction

Several species of linyphiid spider are well adapted to the agricultural ecosystem where they may have an important role in suppressing populations of aphid pests, particularly in cereal fields (Sunderland et al. 1986). Their highly dispersive behaviour enables them to spread risks among the shifting pattern of source and sink habitats that constitute the rotational patchwork of fields within the agricultural landscape (Thomas & Jepson in press). However, they are highly susceptible to insecticide sprays, especially synthetic pyrethroids, which can deplete populations by more than 90 %. In order to understand the long-term, large-scale effects that changing patterns of land use and agricultural practice may have on spider populations, computer simulations of this dynamic system are needed. These simulations may then be used to make predictions about the likely impact of such changes on the size and persistence of populations in a given landscape structure. Significant progress has been made in this direction with the recent development of two models (Halley et al. 1996, Topping & Sunderland 1994). The present paper describes the relevant components of the system and the development and structure of the model developed by Halley et al. (1996) and considers possible limitations to

Arthropod natural enemies in arable land · III *The individual, the population and the community*
W. Powell (ed.). *Acta Jutlandica* vol. 72:2 1997, pp. 79-85
© Aarhus University Press, Denmark, ISBN 87 7288 673 0

the realistic representation of some potentially important elements of the system.

Description of the Model

There are three basic elements to the system which need to be modelled: population dynamics within individual fields, dispersal of spiders between landscape elements and the physical representation of the landscape structure. Spatial and temporal aspects of the model also need to be incorporated at relevant scales.

Population dynamics within fields
Birth and development rates were estimated from published data (De Keer & Maelfait 1987a, 1987b, 1988a, 1988b, Van Wingerden 1978) for two of the commonest species gathered from the field and from laboratory studies at different temperatures and levels of food availability. A crude estimate of seasonal food availability in different crop types was made to correspond with the low, medium or high feeding levels used for the evaluation of egg production. These were adjusted for mean seasonal temperature to give a final value of egg production per spider per week during the reproductive period, as a function of food and temperature.

Mortality rates are very difficult to evaluate in the field. Natural mortality was therefore defined as a proportion of the population dying per week as a reciprocal function of the known life span. As with birth rates, death rates were modified seasonally according to mean ambient temperatures. Mortality resulting from agricultural operations was measured directly by sampling field populations before and after insecticide sprays, cutting of grass for silage, harvest and ploughing, etc. (Thomas & Jepson in press). The mortalities resulting from these events have been added to the natural mortality rate at appropriate times throughout the year for specific crop types.

Maximum carrying capacity was defined as 100 adult females m^{-2} estimated from the number of web sites it was possible to fit in that ground area. Numbers of eggs and juveniles can exceed this value at times.

Although these estimates are necessarily crude, they produce model output in fairly close agreement with field observations of population growth and decline of linyphiid spiders when considered as a taxon (Halley et al. 1996) and are easily changed by the user as more precise data from a wider variety of crop types become available. As such, the model represents the population dynamics of a generalised linyphiid spider rather than those of a specific species.

Dispersal between landscape elements

Most linyphiid spiders, especially females, are sedentary sit-and-wait predators within their webs. Dispersal by walking has been modelled as a diffusion process and shown to operate over relatively small spatial scales, even in an actively hunting species (Thomas et al. 1990). Movement at this scale may be important for relocating web sites within fields, finding mates and oviposition sites but the process is too slow, compared with the frequency of aerial dispersal, to be of significance for population exchange between fields, especially when the presence of field boundaries limits this movement still further (Thomas 1992). This process was therefore omitted from the model.

Aerial dispersal by linyphiid spiders ballooning on gossamer has been well studied (Vugts & Van Wingerden 1976, Thomas 1992), although several aspects remain obscure. There are three principal parameters of relevance to this system: i) the frequency and duration of suitable meteorological conditions for ballooning, ii) the proportion of a population that engages in dispersive behaviour when suitable conditions prevail and, iii) the distance that can be travelled during any particular dispersive episode and the consequent pattern of redistribution of spiders over the landscape.

Aerial dispersal is dependent on meteorological conditions. Wind speed below 3 ms^{-1} (measured at a height of 2m) is the principal prerequisite (Vugts & van Wingerden 1976, Thomas 1992). Analysis of continuous anemometer records have shown that such conditions occur during most weeks of the year for at least an hour. This has been verified by catches of linyphiid spiders in water traps erected in arable fields during each week of the year between May and October (Thomas 1992) and over longer periods in other years (Weyman et al. 1995) and other observations (Duffey 1956).

Laboratory studies of dispersive behaviour have shown that 40% to 60% of spiders attempt to disperse when given appropriate stimulation in wind chambers and that motivation is slightly higher in starved individuals and in those kept under crowded conditions (Legel & van Wingerden 1980, Weyman et al. 1995). Variation of this parameter produced peak population sizes in model output when set at between 40% and 60%. For most simulations this value was therefore set at 50%.

Dispersal distance is dependent on the duration of suitable weather conditions, the effects of atmospheric turbulence in determining the altitude spiders can passively attain, wind speed and the number of flights made while suitable meteorological conditions prevail. Field studies have been undertaken to measure the vertical density distribution of ballooning spiders in known wind-speed gradients and the duration of ballooning on individual days, in order to correlate aerial density of spiders with meteorological parameters and to determine the probability distribution of time intervals between successive flights (Thomas 1992). These data, together with the results of a fluid mechanic model describing the terminal velocity of a spider and silk-filament falling through the air (Humphrey 1987) have been used to calculate the probability distribution of downwind dispersal distances for given time periods. A mathematical model was then developed (Halley et al. 1996) to simulate this distribution by expressing dispersal in terms of rate equations describing the rate at which spiders on the ground become airborne and enter a horizontal airstream, and the rate at which airborne spiders are redeposited on the ground. The whole process of aerial dispersal is thus described by four parameters of landing rate, take-off rate, wind speed and hours of available dispersal time. Each of these parameters may be ascribed different values to simulate dispersal under different conditions. Variation of take-off rate may be used to simulate behavioural differences between populations in different crop types at different times of the year. Landing rate, wind-speed and dispersal time may be used to simulate variation in physical factors governing dispersal distance and redistribution.

Landscape structure
Although linyphiid spiders actively adopt specific behavioural patterns when they are motivated to disperse by ballooning, once airborne the dispersal process becomes a passive one in which, as far as is known, they are entirely dependent on the strength and direction of the wind. Certain aspects of landscape structure can be simplified as a one-dimensional process because spiders can only travel in the direction of the wind. In the majority of cases, dispersal occurs during anticyclonic weather. However, anemometer records have shown that mean wind directions are relatively stable within a dispersal event and between successive episodes when analyzed on a weekly time scale. Under some conditions, for instance in the centres of large anticyclonic systems, winds are light and variable, but in these cases horizontal displacement distances are low with the majority of movement being in the vertical direction if thermal activity is high. More frequently, wind speeds and directions at the level of an observer on the ground may appear to be light and variable due to the periodic suppression of the horizontal component of the wind during the development of plumes and thermals (Stull 1988). However at higher altitudes in the surface boundary layer, the wind-speed and direction is less affected by local forcings and follows the more general pattern of atmospheric movement. These effects can be easily observed by releasing soap bubbles in a variety of wind conditions.

During dispersal, spiders may follow curved trajectories as they move with the general air-mass. The landscape thus traversed by a spider can be effectively represented by a curvilinear line drawn across a map. At the scale at which dispersal occurs, the sequence and pattern of field types traversed by such a line is assumed to be independent of direction. The field types through which such a line is drawn can therefore be translated to a linear strip in a computer simulation. This greatly reduces the processing time required to run the model, yet still allows the model to address questions concerning the effects of the proportion of field types of which a landscape is composed, and the scale of fragmentation. The effects of positional relationships of one field type with another can still be studied but detailed questions are limited. A two-dimensional representation of a landscape, while apparently able to address such questions, is of use only if the two-dimensional nature of spider dispersal is properly understood and represented within the model. This requires detailed analysis of meteorological data which is currently being undertaken in conjunction with the analytical model of spider aerial dispersal distances.

The model landscape has therefore been structured as a linear strip, divided into a number (maximum 80) of square cells or fields, each characterised according to field type. Spiders dispersing from one end of the strip re-enter at the other end such that an infinitely repeating loop is formed. Each cell within the strip can be characterised by one of any number of different field types, although, to date, only grass, cereal and non-habitat types have been used. These types were chosen because field data are available (Thomas & Jepson in press) and they represent the dominant land cover in mixed arable landscapes. Field data of spider population dynamics in other crop types are limited but will be included as they become available.

Spatial and temporal scales
Cells within the landscape strip can be scaled to any length. Thus, a cell may represent an individual field, in the order of 100 m, or a cell size of several kilometres may be chosen as that most appropriate for the representation of extensive tracts of monoculture in a landscape, or large expanses of non-habitat area which may act as significant barriers to dispersal on regional, national or international scales.

Temporal scale is important because the relationship between the frequency of dispersal, the frequency of habitat disruption and the rate of spider growth and development can be critical. A weekly time step is used as the basic unit, since it is at this scale that significant changes occur in population structure through birth and development, and both dispersal and agricultural operations can occur in any week of the year. The time available for dispersal and the synchronisation of dispersive and disruptive events can have important effects on the exposure of a population to risks. This may be especially important in this system, since spider populations and dispersal peak in the summer and autumn at the times when cereals are most frequently sprayed with insecticides. Moreover, dispersal and crop spraying both occur during periods of low wind-speed.

A second time scale operates annually, when field types are changed to a successive crop type in a rotation sequence. This component in the model prevents the unrealistic build up of populations in static, high quality crop types or long-term local extinction from high risk habitats. Non-habitat land types are normally kept static within the landscape.

Limitations and directions for future development of the model
A lack of empirical data has led us to omit the effects of certain parameters on the age structure of the population. Although the development of eggs through to adults is structured in juvenile stages, mortality rates have been applied uniformly to the whole spider population. This may require modification in future developments of this model. Similarly, recently analyzed field data (Thomas & Jepson in press) have indicated that there are variations in the timing and intensity of dispersal according to age, sex and habitat type, with increased and earlier dispersal by females and immature spiders from senescing cereal fields, compared with grass fields.

Mortality in the model currently only operates within fields. The extent of mortality to airborne spiders during dispersal is unknown but they are preyed upon by birds (Owen & Le Gros 1954) and possibly die from desiccation and exposure to extreme cold at altitude. The model has been shown to be very sensitive to mortality rates operating in non-habitat cells in the landscape (Halley et al. 1996). However, this parameter is extremely difficult to measure in the field and may be a serious limitation to the predictive power of this and similar models.

Dispersal time is currently defined as the number of hours per week when the wind speed is below 3 ms^{-1} between 0700 and 1800 GMT. However, there is some evidence that temperature thresholds affect dispersal behaviour, such that dispersal may be significantly reduced during the winter (Thomas 1992). More field observations are required to quantify this.

These aspects, especially age and habitat-specific mortality rates, require detailed clarification and quantification to reduce errors before the model is scaled up to operate in a landscape constructed in two spatial dimensions. The advanced model would involve the incorporation of dispersal and population dynamic elements of the current model characterised for individual species and structured in representations of two-dimensional landscapes constructed with a geographical information system.

Acknowledgments

IACR-Long Ashton Research Station receives grant aided support from the Biotechnology and Biological Sciences Research Council of the United Kingdom

References

De Keer, R. & Maelfait, J.-P. 1987a. Laboratory observations on the development and reproduction of *Oedothorax fuscus* (Blackwall, 1834) (Araneida, linyphiidae) under different conditions of temperature and food supply. *Rev. Ecol. Biol. Sol (Fr)* 24: 63-73.

De Keer, R. & Maelfait, J.-P. 1987b. Life history pattern of *Oedothorax fuscus* (Blackwall 1834) (Araneida, linyphiidae) in a heavily grazed pasture. *Rev. Ecol. Biol. Sol. (Fr)* 24: 171-85.

De Keer, R. & Maelfait, J.P. 1988a. Laboratory observations on the development and reproduction of *Erigone atra* Blackwall, 1833 (Araneae, linyphiidae). *Bull. Br. arachnol. Soc.* 7: 237-42.

De Keer, R. & Maelfait, J.-P. 1988b. Observations on the life cycle of *Erigone atra* (Araneae, erigoninae) in a heavily grazed pasture. *Pedobiologia* 32: 201-12.

Duffey, E. 1956. Aerial dispersal in a known spider population. *J. Anim. Ecol.* 25: 85-111.

Halley, J.M., Thomas, C.F.G. & Jepson, P.C. 1996. A model for the spatial dynamics of linyphiid spiders in farmland. *J. Appl. Ecol.* 33: 471-92.

Humphrey, J.A.C. 1987. Fluid mechanic constraints on spider ballooning. *Oecologia (Berl.)* 73: 469-77.

Legel, G.J. & Van Wingerden, W.K.R.E. 1980. Experiments on the influence of food and crowding on the aeronautic dispersal of *Erigone arctica* (White 1852) (Araneae, linyphiidae). *Proceedings, 8th International Arachnological Congress* pp.1-6.

Owen, D.F. & Le Gros, A.E. 1954. Spiders caught by swifts. *Entomologists Gazette* 5: 117-20.

Stull, R.B. 1988. *An introduction to boundary layer meteorology.* Kluwer Academic Publishers, Dordrecht, The Netherlands.

Sunderland, K.D., Fraser, A.M. & Dixon, A.F.G. 1986. Field and laboratory studies on money spiders (linyphiidae) as predators of cereal aphids. *J. Appl. Ecol.* 23: 433-47.

Thomas, C.F.G. 1992. The spatial dynamics of spiders in farmland. PhD thesis, University of Southampton.

Thomas, C.F.G. & Jepson, P.C. (in press) Field-scale effects of farming practices on linyphiid spider populations in grass and cereals. *Entomol. Exp. Appl.*

Thomas, C.F.G., Hol, E.H.A. & Everts, J.W. 1990. Modelling the diffusion component of dispersal during recovery of a population of linyphiid spiders from exposure to an insecticide. *Functional Ecology* 4: 357-68.

Topping, C.J. & Sunderland, K.D. 1994. A spatial population dynamics model for *Lepthyphantes tenuis* (Araneae: Linyphiidae) with some simulations of the spatial and temporal effects of farming operations and land-use. *Agric. Ecosystems Environ.* 48: 203-17.

Van Wingerden, W.K.R.E. 1978. Population dynamics of *Erigone arctica* (White) (Araneae, Linyphiidae) II. *Symposium Zoological Society, London* 42: 195-202.

Vugts, H.F. & Van Wingerden, W.K.R.E. 1976. Meteorological aspects of aeronautic behaviour of spiders. *Oikos* 27: 433-44.

Weyman, G.S., Jepson, P.C. & Sunderland, K.D. 1995. Do seasonal changes in numbers of aerially dispersing spiders reflect population density on the ground or variation in ballooning motivation? *Oecologia (Berl.)* 101: 487-93.

Density and activity density fluctuations of *Oedothorax apicatus* (Blackwall) in winter wheat in northern Germany

Axel Dinter

Institute for Plant Diseases and Plant Protection
University of Hannover, Herrenhäuser Str. 2, D-30419 Hannover, Germany

Abstract

From 1989 until 1991 the population dynamics of spiders were monitored in winter wheat fields in northern Germany using an intensive D-vac sampling method (actual densities) and pitfall trapping (activity densities). Beside the dominant species *Erigone atra*, *Oedothorax apicatus* was the second most numerous species in pitfall traps. In June, July and August high numbers of *O. apicatus* adults were collected in pitfall traps. At the same time actual densities of *O. apicatus* adults increased up to ~ 2 individuals m^{-2} in 1989 and 1990 and up to ~ 5 individuals m^{-2} in 1991. During winter *O. apicatus* adults were rarely found in the winter wheat fields. Aerial activity of *O. apicatus*, as measured by water traps, was low in spring and summer 1989.

Keywords: Spiders, Linyphiidae, *Oedothorax apicatus*, D-vac sampling, pitfall trapping, actual densities, population dynamics.

Introduction

Spiders occur in a wide range of natural and agricultural habitats and have a stabilizing influence on insect populations. In a field experiment, Clarke & Grant (1968) demonstrated the important impact of spiders on the Collembola population in a beech-maple forest. Manipulation experiments by Riechert & Bishop (1990) provided experimental evidence for the impact of spiders on pest insects in mixed-vegetable garden systems. Effects of Linyphiidae and Lycosidae on cereal aphids (*Rhopalosiphum padi* L.) in grassland were recorded by De Barro (1992).

In European cereal fields, the importance of spiders as predators of cereal aphids has been documented using ELISA techniques (Sunderland et al.1987,

Arthropod natural enemies in arable land · III The individual, the population and the community
W. Powell (ed.). *Acta Jutlandica* vol. 72:2 1997, pp. 87-99
© Aarhus University Press, Denmark, ISBN 87 7288 673 0

Chiverton 1987, Zeiner 1988) and by direct analysis of the prey of spiders (Sunderland et al. 1986, Nyffeler & Benz 1988a, 1988b, Heidger & Nentwig 1989). Otherwise, knowledge of the population dynamcis of spiders in agricultural habitats is rather incomplete. There are only a few studies that continuously investigated density fluctuations or life cycles of dominant spider species. Often, only pitfall trapping was carried out, which makes an interpretation of the underlying population processes and importance of spiders as predators of pest species more difficult than with data on actual density fluctuations. De Keer & Maelfait (1988) studied the life cycle of *Erigone atra* (Blackwall) in a heavily grazed pasture in Belgium, and Dinter (1996) studied the same species in winter wheat in northern Germany. The life history of *Oedothorax fuscus* (Blackwall) was investigated by De Keer & Maelfait (1987). Topping & Sunderland (1994) developed a spatial population dynamics model for *Lepthyphantes tenuis* (Blackwall) in English farmland. This paper summarizes data on *O. apicatus* collected by pitfall trapping and D-vac sampling from 1989 to 1991 in conventional winter wheat in Germany.

Materials and Methods

From April 1989 to August 1991 the investigations were performed in a 20 ha (1989/90) and a 10 ha (1990/1991) field of winter wheat near Göttingen, Lower Saxony. Spider density dynamics were studied on three plots (48 m × 100 m) in 1989 and on two plots (72 m × 100 m and 84 m × 100 m respectively) in 1990 and 1991 which were not sprayed with insecticides. Active epigeic spiders were monitored by means of 10 pitfall traps per plot from April to August. Otherwise (in early spring, autumn and winter) 10 (1989) to 20 (1990/91) pitfall traps were used. Pitfall traps consisted of plastic cups with a diameter of 8.3 cm, and containing ethylene glycol as preservative and 'Pricol' as detergent. Actual spider densities were estimated by an intensive D-vac sampling technique described by Dinter & Poehling (1990), which is an appropriate method for obtaining an exact estimation of the real densities of *O. apicatus* and other spider species (Dinter 1995a). Until November 1989, eight D-vac samples (2 m²) were taken per date. After November 1989, an area of 3 m² was checked per date by D-vac. From April to August 1989, the ballooning activity of spiders was monitored at intervals of two to five days using 9 water traps. The water traps had a size of 46.0 cm × 28.5 cm × 5.0 cm and were filled with water and 'Pricol' as detergent. They were located at the upper vegetation level in the wheat field. Additionally, on 29 June (noon) in 1990, ballooning spiders leaving the top of the wheat tillers were sampled by sweep netting (30 samples consisting of 50 strokes each).

Results

Linyphiids were the most frequent spider family occurring in the wheat fields studied. From April to August in each year, the adult spider fauna of the investigated winter wheat plots was mainly dominated by *E. atra* (24 % to 49 % of all adult spiders in the D-vac samples compared to 41 % to 64 % of the individuals caught in pitfall traps). The percentage of *O. apicatus* adults varied between <1 % to 10 % in D-vac samples and between 7 % to 38 % in pitfall trap catches.

Although D-vac sampling was carried out during all seasons, significant numbers of *O. apicatus* adults were only detected in June, July and August. As the numbers of adults were generally low, the data for males and females are presented in a combined line in Fig. 1 (adults of both sexes were caught in comparable percentages, Dinter & Poehling (1996)). In June, the densities of *O. apicatus* adults increased and in July reached peak values of ~ 2 individuals m^{-2} (1989 and 1990) or ~ 5 individuals m^{-2} (1991). Late in July or August, the densities of *O. apicatus* adults decreased. During autumn and winter almost no adults of *O. apicatus* were found in winter wheat plots by D-vac sampling.

The pattern of pitfall trap catches of *O. apicatus* adults was roughly similar to that for D-vac sampling (Fig. 1). As in the D-vac samples, males and females were found in comparable percentages in the pitfall trap catches. From December to early March, almost no adults were collected in the pitfall traps. Thereafter, low numbers of *O. apicatus* were caught continuously. In mid-June of 1989 and 1990 or early in July 1991, numbers caught in pitfall traps increased and reached peaks in mid-July of ~ 20 (1989 and 1990) or ~ 60 adults per 10 pitfall traps per day in all three years. Then, numbers of *O. apicatus* found in pitfall traps declined. During the course of autumn, continuously small numbers of adults were collected by pitfall trapping until December when no more *O. apicatus* adults were caught.

Ballooning activity by *O. apicatus* adults remained extremely low between the end of April and the end of June in 1989. Unlike the case for the adults of *L. tenuis* and *E. atra,* which were regulary found in the water traps, only one adult *O. apicatus* male was recorded on 24th June. Mass ballooning activities of spiders climbing up the wheat tillers and leaving the plants by ballooning were

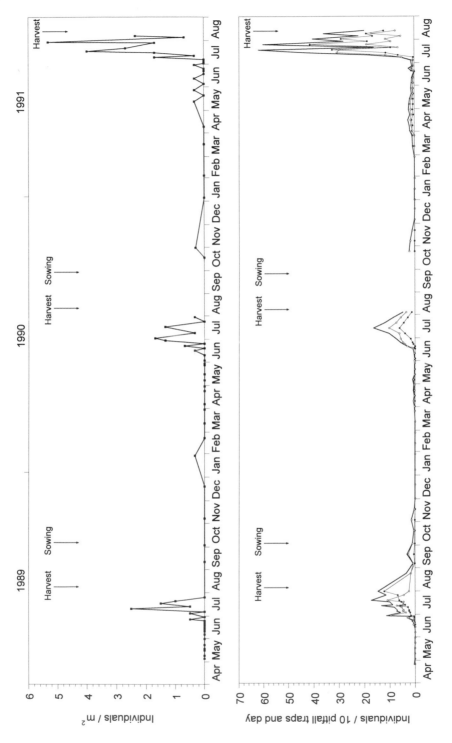

Fig. 1. Population dynamics of *O. apicatus* adults in winter wheat as determined by intensive D-vac sampling and pitfall trap catches (thick line with black squares or thick line alone = total adults; thin line with open circles = adult males; thin line with filled circles = adult females).

observed for many linyphiids (adults of *E. atra*, *L. tenuis* and *Bathyphantes gracilis* (Blackwall) and juvenile linyphiids) on June 26th and 29th in 1990 and on July 21st in 1991. In contrast, mass take-off by *O. apicatus* adults was not seen. Additionally, only two males and one female *O. apicatus* were collected by sweep netting on the 29th June 1990 when ballooning activity of spiders was observed.

Discussion

O. apicatus has a palearctic distribution (Platnick 1989). It occurs in several habitats in Central Europe, including all types of agricultural habitats where it is a common species and found in high frequency (Hänggi et al. 1995). The prey spectrum mainly consists of Collembola but aphids are also consumed (Simon 1964, Sunderland et al. 1987, Alderweireldt 1994a). Quantification of the importance of spiders as predators of cereal aphids revealed that males of *O. apicatus* and especially females of the genus *Oedothorax* are amongst the most valuable aphid predators (Sunderland et al. 1985, 1987, Janssens & De Clercq 1990).

     The occurence of *O. apicatus* as a dominant spider species, as shown in this study, coincides with the results of several other investigations in agricultural crops in Europe (Table 1). In more than half of the cited references, *O. apicatus* was the dominant spider species caught in pitfall traps. Investigations by the same method revealed that *O. apicatus* adults were more active during the night than in the day (Thornhill 1983, Alderweireldt & Desender 1990). In contrast to pitfall trap catches, actual density data for spiders are rather sparse because density estimation is more labour-intensive. But the combination of pitfall trapping and intensive D-vac sampling illustrates the considerable overestimation of *O. apicatus* adults by pitfall trapping compared to real densities (max. ~ 6 adults m^{-2}). The active searching behaviour of *O. apicatus* may be responsible for this difference. Thornhill (1983) found that this species did not build webs and was mainly active by night in sugar beet. Non-web building behaviour was also recorded by Alderweireldt (1994a) during field observations and laboratory rearing. On the other hand, my experience with rearing *O. apicatus* in the laboratory (and some visual observations in winter wheat fields) revealed that adults of this species were also active by day and often built sheet webs over holes in the soil and used them for prey capture in the laboratory (Dinter,

Table 1. Dominance (percentage and rank) of *Oedothorax apicatus* adults in agricultural habitats according to pitfall trap catches in different European countries.

Habitat	Dominance	Rank	Country	Reference
winter wheat	26 - 64 %	1	Germany	Volkmar et al. 1994
winter wheat	46 %	1	Austria	Kromp & Steinberger 1992
winter wheat	42 %	1	France	Cocquempot & Chambon 1989
winter wheat	42 %	1	Austria	Thaler et al. 1977
cereals	33 - 43 %	1	Germany	Glück & Ingrisch 1990
winter wheat	30 - 48 %	1	Belgium	Cottenie & De Clercq 1977
winter wheat	29 - 33 %	1 - 2	Belgium	Janssens & De Clercq 1986
winter wheat	22 %	2	Switzerland	Nyffeler 1982
winter wheat	18 %	1	Germany	Wehling & Heimbach 1991
winter wheat	12 %	3	Germany	Krause et al. 1993
winter wheat	12 %	3	Belgium	Pietraszko & De Clercq 1982
winter wheat		4	Ireland	Feeney & Kennedy 1988
cereal edge		10	United Kingdom	White & Hassall 1994
winter wheat	< 1 %	16	United Kingdom	Topping & Sunderland 1992
spring barley	2 - 67 %	1 - 5	Norway	Andersen 1990
sugar beet	> 60 %	1	Germany	Groh et al. 1981
sugar beet	54 - 69 %	1	Belgium	Janssens & De Clercq 1986
sugar beet	55 %	1	Germany	Kleinhenz & Büchs 1993
sugar beet		2	United Kingdom	Thornhill 1983
oilseed rape	51 %	1	Netherlands	Everts 1990
maize		6	Belgium	Alderweireldt & Desender 1990

unpubl. data). Exact measurements of real walking distances of adults of *O. apicatus* and other species would be helpful for a better understanding of the methodological differences in dominance ranking. Otherwise a high trapability of *O. apicatus* in pitfall traps could contribute to more frequent pitfall trap catches in the field compared to other species. The variability of behavioural responses of arthropods to pitfall traps was shown for linyphiids by Topping (1993) and for other arthropods by Braune (1974). Ouologeum (1996) found that *O. apicatus* adults were caught within a much shorter period of time in a laboratory arena of plain soil comprising several pitfall traps than were *E. atra* adults.

Oedothorax apicatus seems to occur in higher densites in sugar beet fields than in winter wheat, although the data result from different methods (Ulber et al. 1990, Kleinhenz & Büchs 1993, Table 2). Alderweireldt (1994b) recorded 11 adults m⁻² with fenced pitfall traps in a maize field between the end of June and mid July. The highest densities of *O. apicatus* were described by Heydemann (1962), who did ground searches in the agricultural landscape near the coast of the North Sea in Schleswig-Holstein, Germany. He detected mean densities of 250 individuals m⁻² and maximum densities of 400 individuals m⁻².

Table 2. Densities (adults m⁻²) of *Oedothorax apicatus* adults in agricultural habitats in different European countries. (Q/D: Quadrat or D-vac sampling, FP: fenced pitfalls, PE: photoeclectors, H: hand search)

Cultivation	Method	Adults/m²	Country	Reference
winter wheat	Q/D	1 - 7	United Kingdom	Sunderland 1990 (Oedothorax spp.)
sugar beet	FP	1 - 83	Germany	Ulber et al. 1990
sugar beet	PE	0 - 40	Germany	Kleinhenz & Büchs 1993
maize	FP	2 - 11	Belgium	Alderweireldt 1994 b
new polder	H	0 - 6	Netherlands	Meijer 1977
div. agr. habitats	H	250 (-400)	Germany	Heydemann 1962

Schaefer (1976) classified *O. apicatus* as a spider species which has the potential to have more than one generation per year. Under laboratory conditions (23°C), the period of time from eggsac production to spiderling emergence took ~ 9 days (Dinter & Poehling 1995). The mean total duration of juvenile development was 23 days and 27 days for males and females respectively when they were given 5 to 10 live, vestigal-winged fruit flies (*Drosophila melanogaster* L.) and Collembola of the species *Lepidocyrtus lanuginosus* (Gmelin) *ad libitum*. At 20°C and a continuous excess of prey (Collembola or fruit flies), *O. apicatus* reached adulthood in 22 days (males) and 26 days (females) (Alderweireldt & De Keer 1988). The short period required for the life cycle under laboratory conditions leads to the expectation that more than one generation of *O. apicatus* may also develop within a year under field conditions (Alderweireldt & Lissens 1988). The pattern of density changes and the course of the pitfall trap catches found in this study give indications of only one mass reproduction of *O. apicatus* per year in wheat fields. The investigated fields were colonized by *O. apicatus* adults in March. The density increase of adults in June/July in all three years seems to be caused by newly moulted adults of a new generation of adults originating from offspring of the adults that colonized the fields in spring. This is supported by the fact that almost no ballooning activity of *O. apicatus* adults was detected with water traps in the spring and summer of 1989. Furthermore, the pitfall trap catches gave no indication of an immigration of ground-active *O. apicatus* adults from outside the plots.

Shortly after reaching adulthood the newly moulted adults of *O. apicatus* probably left winter wheat fields before the disturbance of the habitat by harvest (decrease in density). Natural mortality due to old age is unlikely as in the laboratory only 3% of the females and 40 % of the males of *O. apicatus* died within 10 weeks (Dinter 1995b). Water trap catches and visual observations indicated no mass emigration of *O. apicatus* adults by ballooning. However, suction trap catches in England throughout the year caught peak numbers of *O. apicatus* males and *Oedothorax* spp. females in July (Sunderland 1990) which supports the emigration hypothesis. In the Netherlands, the highest ballooning activities of *O. apicatus,* as measured by window traps, were also observed in July by Meijer (1977). Probably the low actual density of *O. apicatus* adults compared to *E. atra* and *L. tenuis*, and the small trapping area (~ 1 m²) covered by the water traps, as well as the few opportunities to observe spiders that took off from wheat tillers, impaired the detection of ballooning activity by *O. apicatus* adults in July as the densities decreased.

The reasons for the obvious emigration from wheat fields by large parts of the population of *O. apicatus* adults in July probably are the same as assumed for the mass exodus of newly matured *E. atra* adults (natural decline in prey abundance and deteriorating microclimate as a consequence of the ripening of the wheat) (Dinter 1996). This assumed partial emigration could be adaptive since ploughing or other cropping operations reduce the probability of further successful reproduction and numerous offspring. On the other hand, ballooning carries the risk of mortality and failure to find a copulation partner and suitable habitats for reproduction and overwintering. Therefore the mixed strategy (staying and ballooning) could be an adaptation in order to maximise fitness and minimize local extinction.

While *Erigone* eggsacs were collected in high numbers by intensive D-vac sampling (Dinter 1996) no *O. apicatus* eggsacs were found in the D-vac samples (Dinter, unpubl. data) that could give information on time, intensity and duration of eggsac production or data on eggsac parasitoids. The reason for this could be that *O. apicatus* eggsacs are fixed more securely to the ground than *E. atra* eggsacs, as was found in the laboratory (Dinter, unpubl. data). Van Baarlen et al. (1994) found no eggsac parasitoids of *O. apicatus* in winter wheat and in laboratory studies *O. apicatus* females protected their eggsacs from attack by Hymenoptera. This avoidance of eggsac parasitism could be an advantage that allows this species to stay in the same habitat for several generations while species like *E. atra* and *Erigone arctica* (White) have to suffer detrimental reproductive losses due to parasitoids (van Wingerden 1973, Van Baarlen et al. 1994, Dinter 1996).

In all three years, overwintering by *O. apicatus* adults was only rarely detected within the winter wheat fields. However, the surrounding grassy field margins were a good source for collecting *O. apicatus* adults during autumn, winter and early spring and using them as a start population for mass rearing of an offspring generation in the laboratory (Dinter & Poehling 1995). The absence of adults in D-vac samples, and the decrease of *O. apicatus* adults in pitfall trap catches during autumn, probably reflects the movement of the spiders to their overwintering sites within field margins. This emphasizes the importance of field margins as overwintering sites for this spider species. Wiedemeier & Duelli (1993) identified field margins as the main overwintering sites for *O. apicatus*, which are located next to their summer habitat (fields). In winter they found up to 92 individuals m^{-2} of *O. apicatus* in field margins in Switzerland. Also in sown weed strips within winter wheat fields, *O. apicatus* adults overwinter in high densities up to 40 adults m^{-2} (mainly females) (A. Lemke, personal communication). During autumn, some *O. apicatus* adults may also emigrate by ballooning from wheat fields to find suitable habitats, which is supported by suction trap data in England (Sunderland 1990).

Acknowledgements

The author thanks Dr. K.D. Sunderland (Horticulture Research International, Wellesbourne, U.K.), A. Lemke (Institute for Plant Diseases and Plant Protection, University of Hannover) and two anonymous referees for their critical comments on earlier drafts of the manuscript. The study was funded by the German Research Council (Deutsche Forschungsgemeinschaft, DFG).

References

Alderweireldt, M. 1994a. Prey selection and prey capture strategies of linyphiid spiders in a high-input agricultural field. *Bull. Br. arachnol. Soc.* 9: 300-8.

Alderweireldt, M. 1994b. Spatial distribution and seasonal fluctuations in abundance of spiders (Araneae) occurring on arable land at Melle (Belgium). *Biol. Jb. Dodonaea* 61: 193-208.

Alderweireldt, M. & De Keer, R. 1988. Comparison of the life cycle history of three *Oedothorax*-species (Araneae, Linyphiidae) in relation to laboratory observations. In: Haupt, J. (ed.) *XI. Europäisches Arachnologisches Colloquium. (TUB-Dokumentation Kongresse und Tagungen, Berlin 1988, Heft 38.*): 169-77.

Alderweireldt, M. & Lissens, A. 1988. Laboratoriumwaarnemingen van de ontwikkeling en reproductie bij *Oedothorax apicatus* (Blackwall, 1850) en *Oedothorax retusus* (Westring, 1851). *Nwsbr. Belg. Arachnol. Ver.* 9: 19-26.

Alderweireldt, M. & Desender, K. 1990. Microhabitat preference of spiders (Araneae) and carabid beetles (Coleoptera, Carabidae) in maize fields. *Med. Fac. Landbouww. Rijksuniv. Gent* 55: 501-10.

Andersen, A. 1990. Spiders in norwegian spring barley fields and the effects of two insecticides. *Norwegian J. Agric. Sci.* 4: 261-71.

Braune, F. 1974. Kritische Untersuchungen zur Methodik der Bodenfalle. PhD thesis, University of Kiel.

Chiverton, P.A. 1987. Predation of *Rhopalosiphum padi* (Homoptera: Aphididae) by polyphagous predatory arthropods during the aphids' pre-peak period in spring barley. *Ann. Appl. Biol.* 111: 257-69.

Clarke, R.D. & Grant, P.R. 1968. An experimental study of the role of spiders as predators in a forest litter community. Part 1. *Ecology* 49: 1152-54.

Cocquempot, C. & Chambon, J.-P. 1989 Importance relative des araignées. *La défense des végétaux* 257: 13-16.

Cottenie, P. & De Clercq, R. 1977. Studie van de Arachnofauna in wintertarwevelden. *Parasitica* 33: 138-46.

De Barro, P.J. 1992. The impact of spiders and high temperatures on cereal aphid (*Rhopalosiphum padi*) numbers in an irrigated perennial grass pasture in South Australia. *Ann. Appl. Biol.* 121: 19-26.

De Keer, R. & Maelfait, J.-P. 1987. Life history of *Oedothorax fuscus* (Blackwall, 1834) (Araneae, Linyphiidae) in a heavily grazed pasture. *Rev. Écol. Biol. Sol* 24: 171-85.

De Keer, R. & Maelfait, J.-P. 1988. Observations on the life cycle of *Erigone atra* (Araneae, Erigoninae) in a heavily grazed pasture. *Pedobiologia* 32: 201-12.

Dinter, A. 1995a Estimation of epigeic spider population densities using an intensive D-vac sampling technique and comparison with pitfall trap catches in winter wheat. In: Toft, S. & Riedel, W. (eds.) *Arthropod natural enemies in arable land. I. Acta Jutlandica* 70 (2): 23-33.

Dinter, A. 1995b. *Untersuchungen zur Populationsdynamik von Spinnen* (Arachnida: Araneae) in Winterweizen und deren Beeinflussung durch insektizide Wirkstoffe. Thesis. University of Hannover. Cuvillier Verlag, Göttingen.

Dinter, A. 1996. Population dynamics and eggsac parasitism of *Erigone atra* (Blackwall) in winter wheat. *Revue suisse de Zoologie, vol. hors série*: 153-64.

Dinter, A. & Poehling, A. 1990. Nebenwirkungen von Insektiziden auf die epigäischen Spinnen im Winterweizen. Bericht zum 6. *Internationalen Symposium 'Schaderreger des Getreides'* II: 397-98.

Dinter, A. & Poehling, H.-M. 1995. Side-effects of insecticides on two erigonid spider species. *Entomol. Exp. Appl.* 74: 151-63.

Dinter, A. & Poehling, H.-M. 1996. Analyse der Populationsstruktur und -dynamik von Spinnen (Araneae) in Winterweizen – Vergleich von Intensiv-D-vac-Methode und Bodenfallentechnik. *Mitt. Dtsch. Ges. Allg. Angew. Ent.*: 10, 533-36.

Everts, J.W. 1990. Sensitive indicators of side-effects of pesticides on the epigeal fauna of arable land. Thesis, University of Wageningen.

Feeney, A.M. & Kennedy, T. 1988. The occurrence of spiders in cereal fields. In: Cavalloro, R. & Sunderland, K.D. (eds.) *Integrated Crop Protection in Cereals.* pp. 99-102. A. A. Balkema, Rotterdam.

Glück, E. & Ingrisch, S. 1990. The effect of bio-dynamic and conventional agriculture management on Erigoninae and Lycosidae spiders. *J. Appl. Entomol.* 110: 136-48.

Groh, K., Assmuth, W. & Tanke, W. 1981. Einfluss von Pflanzenschutzmassnahmen auf die Arthropodenfauna in Zuckerrübenfeldern. *Z. PflKrankh. PflSchutz, Sonderh. IX* : 199-210.

Hänggi, A., Stöckli, E. & Nentwig, W. 1995. *Habitats of Central European spiders. Characterisation of the habitats of the most abundant spider species of Central Europe and associated species.* Miscellanea Faunistica Helvetiae, 4. Centre suisse de cartographie de la faune. Neuchatel.

Heidger, C. & Nentwig, W. 1989. Augmentation of beneficial arthropods by strip-management. 3. Artificial introduction of a spider species which preys on wheat pest insects. *Entomophaga* 34: 511-22.

Heydemann, B. 1962. Untersuchungen über die Aktivitäts- und Besiedlungsdichte bei epigäischen Spinnen. *Verh. Dtsch. Zool. Ges.* 55: 538-56.

Janssens, J. & De Clercq, R. 1986. Distribution and occurrence of Araneae in arable land in Belgium. *Med. Fac. Landbouww. Rijksuniv. Gent* 51: 973-80.

Janssens, J. & De Clercq, R. 1990. Observations on Carabidae, Staphylinidae and Araneae as predators of cereal aphids in winter wheat. *Med. Fac. Landbouww. Rijksuniv. Gent* 55: 471-75.

Kleinhenz, A. & Büchs, W. 1993. Einfluß verschiedener landwirtschaftlicher Produktionsintensitäten auf die Spinnenfauna in der Kultur Zuckerrübe. *Verh. Ges. Ökol.* 22: 81-88.

Krause, U., Pfaff, K., Dinter, A. & Poehling, H.-M. 1993. *Nebenwirkungen von Insektiziden, vor allem Pyrethroiden, auf epigäische Spinnen bei der Bekämpfung von Getreideblattläusen.* Agrarökologie 9. Verlag Paul Haupt, Bern.

Kromp, B. & Steinberger, K.-H. 1992. Grassy field margins and arthropod diversity: a case study on ground beetles and spiders in eastern Austria (Coleoptera: Carabidae; Arachnida: Aranei, Opiliones). *Agric. Ecosystems Environ.* 40: 71-93.

Meijer, J. 1977. The immigration of spiders (Araneida) into a new polder. *Ecol. Entomol.* 2: 81-90.

Nyffeler, M. 1982. Field studies on the ecological role of the spiders as insect predators in agroecosystems (abandoned grassland, meadows and cereal fields). Thesis, Swiss Federal Institute of Technology. Zürich.

Nyffeler, M. & Benz, G. 1988a Prey and predatory importance of micryphantid spiders in winter wheat fields and hay meadows. *J. Appl. Entomol.* 105: 190-97.

Nyffeler, M. & Benz, G. 1988b. Feeding ecology and predatory importance of wolf spiders (*Pardosa* spp.) (Araneae, Lycosidae) in winter wheat fields. *J. Appl. Entomol.* 106: 123-34.

Ouologeum 1996. Laborversuche zu den Auswirkungen synthetischer Pyrethroide auf Spinnen (Linyphiidae) sowie zur Aussagekraft von Bodenfallenfängen bei Pflanzenschutzmittelprüfungen. Diploma study, University of Hannover,

Pietraszko, R. & De Clercq, R. 1982. Influence of organic matter on epigeic arthropods. *Med. Fac. Landbouww. Rijksuniv. Gent* 41: 721-28.

Platnick, N.I. 1989. *Advances in spider taxonomy 1981-1987.* Manchester University Press, Manchester.

Riechert, S.E. & Bishop, L. 1990. Prey control by an assemblage of generalist predators: spiders in garden test systems. *Ecology* 71: 1441-50.

Schaefer, M. 1976. Experimentelle Untersuchungen zum Jahreszyklus und zur Überwinterung von Spinnen (Araneida). *Zool. Jb. Syst.* 103: 127-289.

Simon, H.R. 1964. Zur Ernährungsbiologie collembolenfangender Arthropoden. *Biol. Zbl.* 83: 273-96.

Sunderland, K.D. 1990. The ecology of spiders in cereals. *Bericht zum 6. Internationalen Symposium 'Schaderreger des Getreides'* I: 269-80.

Sunderland, K.D., Chambers, R.J., Stacey, D.L. & Crook, N.E. 1985. Invertebrate polyphagous predators and cereal aphids. *Bull. IOBC/WPRS* 5: 105-14.

Sunderland, K.D., Fraser, A.M. & Dixon, A.F.G. 1986. Distribution of linyphiid spiders in relation to capture of prey in cereal fields. *Pedobiologia* 29: 367-75.

Sunderland, K.D., Crook, N.E., Stacey, D.L. & Fuller, B.J. 1987. A study of feeding by polyphagous predators on cereal aphids using ELISA and gut dissection. *J. Appl. Ecol.* 24: 907-33.

Thaler, K., Ausserlechner, J. & Mungenast, F. 1977. Vergleichende Fallenfänge von Spinnen und Käfern auf Acker- und Grünlandparzellen bei Innsbruck, Österreich. *Pedobiologia* 17: 389-99.

Thornhill, W.A. 1983. The distribution and probable importance of linyphiid spiders living on the soil surface of sugar-beet fields. *Bull. Br. arachnol. Soc.* 6: 127-36.

Topping, C.J. 1993. Behavioural responses of three linyphiid spiders to pitfall traps. *Entomol. Exp. Appl.* 68: 287-93.

Topping, C.J. & Sunderland, K.D. 1992. Limitations to the use of pitfall traps in ecological studies exemplified by a study of spiders in a field of winter wheat. *J. Appl. Ecol.* 29: 485-91.

Topping, C.J. & Sunderland, K.D. 1994. A spatial population dynamics model for *Lepthyphantes tenuis* (Araneae: Linyphiidae) with some simulations of the spatial and temporal effects of farming operations and land-use. *Agric. Ecosystems Environ.* 48: 203-17.

Ulber, B., Stippich, G. & Wahmhoff, W. 1990. Möglichkeiten, Grenzen und Auswirkungen des gezielten Pflanzenschutzes im Ackerbau: III. Auswirkungen unterschiedlicher Intensität des chemischen Pflanzenschutzes auf epigäische Raubarthropoden in Winterweizen, Zuckerrüben und Winterraps. *Journal of Plant Diseases and Protection* 97: 263-83.

Van Baarlen, P., Sunderland, K.D. & Topping, C.J. 1994. Eggsac parasitism of money spiders (Araneae, Linyphiidae) in cereals, with a simple method for estimating percentage parasitism of *Erigone* spp. eggsacs by Hymenoptera. *J. Appl. Entomol.* 118: 217-23.

Van Wingerden, W.K.R.E. 1973. Dynamik einer Population von Erigone arctica White (Araneae, Micryphantidae*). Prozesse der Natalität. Faun.-Ökol. Mitt.* 4: 207-22.

Volkmar, C., Bothe, S., Kreuter, T., Lübke-Al Hussein, M., Richter, L., Heimbach, U. & Wetzel, T. 1994. Epigäische Raubarthropoden in Winterweizenbeständen Mitteldeutschlands und ihre Beziehung zu Blattläusen. *Mitteilungen aus der Biologischen Bundesanstalt für Land- und Forstwirtschaft* 299: 3-134.

Wehling, A. & Heimbach, U. 1991. Untersuchungen zur Wirkung von Pflanzenschutzmitteln auf Spinnen (Araneae) am Beispiel einiger Insektizide. *Nachrichtenbl. Deut. Pflanzenschutzd. (Braunschweig)* 43: 24-30.

White, P.C.L. & Hassall, M. 1994. Effects of management on spider communities of headlands in cereal fields. *Pedobiologia* 38: 169-84.

Wiedemeier, P. & Duelli, P. 1993. Bedeutung ökologischer Ausgleichsflächen für die Überwinterung von Arthropoden im Intensivkulturland. *Verh. Ges. Ökol.* 22: 263-67.

Zeiner, C. 1988. Untersuchung zur Bedeutung von polyphagen Prädatoren als Blattlausräuber auf konventionell und ökologisch bewirtschafteten Winterweizenschlägen. Thesis, University of Kiel.

Climate Change and Aphid-Coccinellid Population Dynamics

D. Skirvin

Horticulture Research International, Wellesbourne, Warwick, CV35 9EF, U.K.

Abstract

A model describing the effect of temperature on the interaction between the cereal aphid, *Sitobion avenae* F. (Homoptera: Aphididae), and its predator, *Coccinella septempunctata* L. (Coleoptera: Coccinellidae), was constructed by modifying an existing model, which described the summer population dynamics of *S. avenae* (Carter et al. 1982). The new model incorporated a sub-model describing the population dynamics of *C. septempunctata*, and new formulations for the equations describing aphid reproduction and development. Stochasticity was introduced by sampling randomly from distributions fitted to observations of aphid and coccinellid immigration.

The model was run for three temperature regimes, based upon temperatures within current experience, and the output suggested that both coccinellid predation and increasing temperature acted to decrease aphid abundance.

The modelling process highlighted several problem areas within the model, most notably, the lack of appropriate data. This paper provides a brief description of the model and its results before discussing the problems encountered while constructing the model and the assumptions made by the model. Ways in which these problems can be addressed are examined.

Key words: cereal aphid, *Sitobion avenae*, ladybird, *Coccinella septempunctata*, climate change, simulation model, predation, population dynamics, temperature.

Introduction

It has been suggested that increased global temperatures could lead to greater pest problems in the future (Parry et al. 1989, Porter et al. 1991). However, the response of natural enemies to increased temperatures will be important in determining whether the predicted pest outbreaks will arise (Cammell & Knight 1992, Porter et al. 1991).

As a first step in examining this problem, the effect of increased temperature on the interaction between the cereal aphid, *Sitobion avenae*, and its coccinellid predator, *Coccinella septempunctata*, was investigated using a computer model

Arthropod natural enemies in arable land · III The individual, the population and the community
W. Powell (ed.). *Acta Jutlandica* vol. 72:2 1997, pp. 101-111
© Aarhus University Press, Denmark, ISBN 87 7288 673 0

based on the model of Carter et al. (1982).

A simulation modelling approach was used since it provides a useful tool in enhancing the understanding of the complex relationships between temperature and the biology of the two species being modelled. However, uncritical use of simulation models is unwise since it is impossible to include all aspects of the processes involved in population dynamics. Also, simplifying assumptions are often required to enable the modelling of complex processes. Indeed, Gilbert & Gutierrez (1973) cautioned against the uncritical use of simulation models:
"A simulation has no intrinsic value. It is useful only when it exposes our ignorance, or answers biological questions." The model described here fits into both criteria, by exposing our ignorance of coccinellid biology and attempting to answer the question of how climate change will affect pest-natural enemy interactions.

This paper will provide a brief description of the model and its main results before discussing the assumptions made and problems encountered while constructing the model. Ways of addressing these problems are then put forward.

Description of the model

The model, which is based upon an existing simulation describing the summer population dynamics of *S. avenae* (Carter et al. 1982), was amended to include new formulations of the equations describing aphid development and reproduction. A submodel describing the biology of *C. septempunctata* was also included. Temperature acts as the driving variable of the model, and the time-step is one hour.

Aphid immigration
The start date of immigration is sampled randomly from a lognormal distribution fitted to data from the Rothamsted Insect Survey (RIS) suction trap catches. The end date of immigration is then calculated from the start date using a regression equation derived from RIS data. The number of aphids entering the field on each day of the immigration period is calculated from the number of aphids caught in a RIS suction trap, by sampling randomly from a negative binomial distribution fitted to the data, and multiplying by a deposition factor (Taylor & Palmer 1972).

Coccinellid immigration
If the aphid density is greater than 10 aphids m^{-2} (Honek 1980), then coccinellid settlement can occur. There are assumed to be two waves of immigration, corresponding to the movement of overwintering coccinellids early in the season and the movement of the new generation of adults from field to field late in the season, respectively. In both waves, immigration is assumed to last for 29 days, and the

median day is sampled randomly from a normal distribution fitted to the data of Zhou et al. (1994). The number of immigrants per day in the first wave is assumed to follow a uniform distribution, while in the second wave, the numbers of immigrants per day are distributed as a normal about the number on the median day, based on the analysis of data from Zhou et al. (1994).

Aphid development
Aphid nymphs are assumed to accumulate development towards a limit. Their rate of development is proportional to temperature as described by the following equations, which were fitted to the data of Dean (1974).

$$D = 0.0 \qquad\qquad\qquad\qquad T \le 0.0$$
$$D = b_{1i}\, T \qquad\qquad\qquad 0.0 < T \le T_{1i}$$
$$D = a_i\, /\, (\, 1 + \exp(x_i + m_i\, T)) \qquad T_{1i} < T \le 25.0$$
$$D = c_i - b_{2i} \qquad\qquad\qquad 25.0 < T \le T_{2i}$$
$$D = 0.0 \qquad\qquad\qquad\qquad T_{2i} < T$$

Where D = development rate (1/time); the subscript, i, relates to the aphid instar, a, b, x, and m are parameters, shown in Table 1; and T = temperature (°C).

Adult aphids are assumed to accumulate hour degrees (H°) towards a limit, above which they die. This total is dependent upon morph (winged or wingless) and crop growth stage (Dean 1974). Fourth instar winged aphids are assumed to emigrate upon reaching the adult stage.

Table 1: Parameter estimates for the aphid development equations

Instar (i)	b_{1i}	T_{1i}	a_i	x_i	m_i	t_{2i}	c_i	b_{2i}
1	0.00101	13.45	0.272	0.160	2.155	41.0	0.0600	0.001
2	0.00112	13.00	0.0290	0.142	1.850	41.3	0.0620	0.001
3	0.00112	12.67	0.0285	0.171	2.165	39.0	0.0702	0.001
4	0.00102	12.07	0.0246	0.133	1.600	42.0	0.0517	0.001

Coccinellid development
Coccinellid development is treated in a similar way to that of the aphids. Adult coccinellids are assumed to accumulate H° towards a limit of 9975 H°, based upon their longevity after overwintering. Above this total, they die. All the other coccinellid instars are assumed to accumulate development, with development rate related to temperature in a way described by the following equations which were fitted to the data of Hodek (1973).

$$D = 0.0 \qquad\qquad\qquad\qquad T < 0.0$$
$$D = a_c / (1 + \exp(x_c - m_c T)) \qquad 0.0 \leq T \leq 35.0$$

$$D = b_c - c_c\, T \qquad\qquad\qquad 35.0 < T \leq 50.0$$
$$D = 0.0 \qquad\qquad\qquad\qquad 50.0 < T$$

Where D = development rate (1/hour); a, b, and c are parameters (Table 2) where the subscript, c, refers to the coccinellid instar; T = temperature (°C).

Coccinellid adults that have developed from eggs laid in the field are assumed to emigrate 40 hours after emerging from the pupae. (Zaslavsky & Semyanov 1986, Honek 1990).

Aphid reproduction
The number of nymphs produced per female per unit time step is calculated from the following equations fitted to the data of Dean (1974).

$$N = RT = (b_1\, T)\, T \qquad\qquad 0.0 \leq T \leq 20.0$$
$$N = RT = (a_2 - b_2\, T)T \qquad 20.0 < T \leq 30.0$$

Where N = nymphs produced per female per hour; R = Nymphs produced per female per H°; a and b are parameters (Table 3) and T = temperature (°C). The total number of nymphs produced in a time step is equivalent to N multiplied by the total number of adult aphids, since reproduction is parthenogenetic.

Table 2: Parameter estimates for the coccinellid development equations

Instar (c)	a	x	m	b	c
Egg	0.0297	5.39	0.235	0.0786	0.00157
1	0.0316	4.94	0.211	0.0970	0.00194
2	0.0532	4.76	0.191	0.154	0.00308
3	0.0400	5.30	0.229	0.125	0.00250
4	0.0196	4.93	0.204	0.0587	0.00118
Pupa	0.0201	4.61	0.181	0.0568	0.00114

Table 3: Parameter estimates for the aphid reproduction equations

Morph	b_1	a_2	b_2
Winged	0.00037	0.0220	0.00073
Wingless	0.00028	0.0170	0.00057

Coccinellid reproduction

If the aphid density is greater than 1 aphid every ten tillers (Honek 1978, 1980), a threshold which represents the density of aphids required to provide enough food to allow female coccinellids to complete ovariole maturation, then coccinellid reproduction is assumed to occur.

The egg production of female coccinellids is assumed to be related to temperature, above a threshold of $0^{\circ}C$, and consumption of aphids (Ghanim et al. 1984, A.F.G. Dixon personal communication.) according to the equations shown below:

$$E = R_c T = (0.00047C - 0.0037T)T \qquad 0.0 \le T \le 20.0, \ 10.0 \le C \le 20.94$$
$$E = R_c T = (0.0148C - 0.148 - 0.00037CT + 0.0037T)T$$
$$20.0 < T \le 40.0 \, , \ 10 \le C \le 20.94$$

Where E = number of eggs per female per hour; R_c = number of eggs per female per H°; C = daily consumption of aphids (mg) and T = temperature ($^{\circ}C$).

Predation

To calculate coccinellid predation, the number of aphids per tiller in each instar are converted to mg m^{-2} by multiplying the numbers in each instar by the weight of each aphid instar and the number of tillers m^{-2}. The proportion of the total aphid biomass contributed by each aphid instar is then calculated.

The model assumes that only a fraction of the total number of coccinellids is actively searching for prey during each hour. The percentage of active coccinellids is calculated from the following equations fitted to the data of Honek (1985), which describe the effect of temperature and satiation on coccinellid activity.

$$V = 0.0 \qquad T < 11.7 - 4.3S_n$$
$$V = 3.86T + 27.5S_n - 1.43TS_n - 45.24 \qquad 11.7 - 4.3S_n \le T \le 37.6 + 11.0S_n$$
$$V = 100 \qquad 37.6 + 11.0S_n < T$$

where V = percentage of coccinellids that are actively searching for food; S_n = satiation and T = temperature ($^{\circ}C$).

Satiation is calculated by dividing the weight of aphids eaten per day by the maximum possible biomass of aphids that can be consumed in a day (20.94 mg).

The biomass of aphids consumed by the coccinellids is calculated using a temperature- mediated functional response equation (Mack et al. 1981, Mack & Smilowitz 1982), which is shown below:

$$N = ((1/t_h) AP) / (1 / (t_h s) + A)$$

Where N = biomass of aphids consumed (mg/m^2/h); A = aphid density (mg/m^2); P =

predator density(/m^2); $1/t_h$ = handling rate (mg/predator/h) and s = coccinellid searching rate (m^2/predator/h).

The searching rate of the coccinellids is calculated from the following equations fitted to the data of McLean (1980):

$$s = 0.0 \qquad\qquad T < 0.0$$
$$s = d_c + a_c / (1 + \exp(x_c - m_c T)) \qquad\qquad 0.0 \le T \le 35.0$$
$$s = b_c + c_c T \qquad\qquad 35.0 < T \le 50.0$$
$$s = 0.0 \qquad\qquad 50.0 < T$$

Where s = searching rate (m^2/predator/hour), a, b, c, d, m, and x are parameters (Table 4) and the subscript, c, refers to the coccinellid instar. Adult coccinellids were assumed to have a similar searching rate to second instar larvae, as no data were available to describe the effect of temperature on adult coccinellids.

Table 4: Parameter estimates for equations describing coccinellid searching rate

Instar (c)	d	a	x	m	b	c
1	0.0	0.0206	8.21	0.407	0.0685	0.00137
2	0.0149	0.0602	10.1	0.438	0.249	0.00498
3	0.0422	0.108	11.3	0.503	0.500	0.0100
4	0.136	0.204	14.4	0.632	1.13	0.0226

The handling rate is calculated from the following equations which were forced through the limited data of Olszak (1988):

$$1/t_h = 0.0 \qquad\qquad T < 0.0$$
$$1/t_h = a_c / (1 + \exp(x_c - m_c T)) \qquad\qquad 0.0 \le T \le 35.0$$
$$1/t_h = b_c + c_c T \qquad\qquad 35.0 \le T \le 50.0$$
$$1/t_h = 0.0 \qquad\qquad T < 50.0$$

Where $1/t_h$ = coccinellid handling rate (mg/predator/h); a, b, c, x, and m are parameters (Table 5); T = temperature, and the subscript, c, refers to coccinellid instar.

Having calculated the biomass of prey eaten by the coccinellids, this is converted back to the number of aphids per tiller in each instar eaten by the coccinellids, and the aphid numbers are adjusted accordingly.

The model then iterates through the next hours, and updates the crop growth stage after 24 hours, unless the crop growth stage is greater than 86.3 on the Zadok's scale, at which point the model ends.

Table 5: Parameter estimates of the equations describing coccinellid handling rate

Instar (c)	a	x	m	b	c
1	0.050	9.70	0.555	0.167	0.0033
2	0.120	9.70	0.555	0.400	0.0080
3	1.60	12.2	0.694	5.33	0.107
4	1.76	12.9	0.738	5.87	0.117
Adult	0.975	13.4	0.767	3.25	0.0650

Results and Discussion

A summary of the main results of the model is shown in Table 6, and a more detailed discussion of the results and the problems of the model is given below.

The model was run for three separate temperature regimes: hot, moderate and cold, and for three temperatures within each regime: base, base plus 1°C and base plus 2°C. The regimes were obtained by classifying the years from 1965 to 1993 according to the mean monthly temperatures from April through to August. A double-Fourier curve was fitted to the daily data from the years included within each regime, and this curve was used to predict the daily maximum and minimum temperatures for each day in the model.

The model predicted that increasing the temperature within a regime leads to a decrease in aphid abundance, both in the presence and absence of coccinellids. These runs mimic the effect of an increase in temperature without a resulting change in aphid behaviour, which is unlikely to occur in reality as aphids can adapt reasonably quickly due to a plastic phenology.

The model also predicted that aphids would be most abundant in the moderate regime, in the presence of coccinellids, which seems counter-intuitive. However, it must be remembered that, in the model, aphid development peaks at 25°C, while coccinellids reach their maximal development rate at 35°C. Therefore, it is possible that in the hot regime, the warmer temperatures are above the optimal level for aphids, reducing both their fecundity and development, allowing the coccinellids to develop more rapidly and consume more aphids. This interpretation seems probable, especially since in the absence of coccinellids, the model predicts the greatest abundance of aphids to occur in the hot regime.

As the temperature both between and within regimes is increased, the predicted day on which the aphid numbers are greatest becomes earlier, and is even earlier in the presence of coccinellids. This prediction suggests that as temperature increases, aphid and coccinellid populations will be able to increase more rapidly which fits in with the fact that in both species development and reproduction are

Table 6: Mean values and standard errors (from 100 runs) of the output from all three regimes, the two values in the aphid columns represent the values with and without coccinellids respectively. Coccinellid numbers include only larvae and adults.

Regime: Temperature	Maximum number of aphids per tiller		Date of maximum aphid abundance		Total number of aphid days in a season (/tiller)		Maximum number of active coccinellid stages (/m^2)	Date of maximum coccinellid abundance	Total number of coccinellid days in a season (/m^2)
Cold: base	10.2 ±0.77	17.9 ±1.08	188.5 ±1.99	194.5 ±2.03	192.9 ±14.6	352.4 ±19.3	31.8 ±1.77	195.0 ±2.03	327.7 ±15.3
Cold: base plus 1°C	7.9 ±0.54	15.6 ±0.78	189.7 ±0.52	190.9 ±1.97	143.3 ±10.7	295.5 ±15.0	26.9 ±1.11	197.5 ±0.49	228.0 ±9.88
Cold: base plus 2°C	6.8 ±0.46	13.9 ±0.81	183.8 ±2.67	186.9 ±1.92	112.7 ±8.41	244.2 ±15.2	24.5 ±1.13	192.5 ±2.81	175.6 ±7.98
Moderate: base	20.2 ±1.96	34.2 ±2.54	184.2 ±0.48	189.9 ±0.59	331.3 ±27.6	569.5 ±29.6	33.6 ±1.31	191.3 ±0.57	384.0 ±16.6
Moderate: base plus 1°C	14.1 ±1.20	31.1 ±1.98	182.4 ±0.56	185.4 ±0.50	240.3 ±19.4	512.5 ±23.9	29.8 ±1.51	189.9 ±0.56	315.5 ±14.0
Moderate: base plus 2°C	14.7 ±1.26	24.2 ±1.18	181.4 ±0.47	184.0 ±0.45	241.9 ±19.5	463.0 ±17.9	30.3 ±1.14	190.6 ±0.52	300.6 ±13.8
Hot: base	11.7 ±0.93	36.3 ±1.56	181.4 ±0.57	184.0 ±1.86	199.8 ±15.9	594.5 ±18.6	28.9 ±1.12	183.1 ±0.59	321.7 ±13.7
Hot: base plus 1°C	8.8 ±0.85	31.4 ±1.45	174.5 ±0.57	178.2 ±0.55	144.7 ±14.4	507.3 ±17.3	25.5 ±1.27	182.0 ±0.57	292.3 ±15.3
Hot: base plus 2°C	7.35 ±0.61	24.4 ±0.92	172.0 ±1.83	173.5 ±2.53	118.5 ±10.5	400.6 ±13.6	26.1 ±1.31	179.0 ±1.88	286.6 ±15.0

temperature dependent.

As mentioned earlier, this model was a first step towards investigating the effect of climate change on aphid-natural enemy interactions. The model predictions must therefore be treated with caution, and will probably only act as indicators towards the type of responses we may see to climate change. The model is by necessity a simplification, and therefore there are several areas which could be improved. Firstly, aphid and coccinellid immigration are not represented as fully as they could be. Although there is a great deal of data on aphid immigration in terms of suction trap catches, the exact relationship between the suction trap catches and the number of aphids landing in a field is not fully understood. The suction trap catch is representative of a wide area and is therefore unable to account for field-to-field variation.

Data on coccinellid immigration is extremely scarce due to the fact that few studies have looked at field-to-field movement of coccinellids, and that *C. septempunctata* does not often appear in suction trap catches. which made the determination of coccinellid immigration patterns very difficult, and they are represented only very simply in the current model. Since climate change is likely to lead to warmer winters (Cammell & Knight 1992, Porter et al. 1991), which will almost undoubtedly affect immigration of both aphids and coccinellids, it is important that immigration for both species is represented as accurately as possible in the model.

The most obvious approach to gathering data on the timing of immigration and the number of immigrants is to introduce a long-term monitoring programme. This is already partly provided by the Rothamsted Insect Survey, although more work needs to be performed on the relationship of the suction trap catch to numbers in the field. The programme would also be able to provide information on the annual variation in immigration, as well as providing a large set of data which could be used to validate both this model and any future models.

The model discussed here is limited to a single aphid species and a single natural enemy species, when in the field there are often several species of aphid and a large complex of natural enemies. In order to discover how climate change will affect aphid outbreaks in cereals, it will be necessary to include the other main cereal aphids, *Rhopalosiphum padi* and *Metopolophium dirhodum*, since these two species often cause outbreaks at times when *S. avenae* does not reach outbreak levels (Carter et al. 1980).

Including the natural enemy complex is likely to prove difficult due to the lack of appropriate data. However, it is felt that it is an important component in determining aphid population dynamics since one single natural enemy is unlikely to be responsible for controlling cereal aphids. Indeed it has been observed that the natural enemy responsible for the majority of control of aphid populations varies with environmental conditions (Chambers et al. 1982, 1986, Thompson 1930, Vickerman

& Wratten 1979).

The equations used in both the aphid and coccinellid submodels are based upon limited data, and the curves fitted to this data have been extrapolated into areas not covered by the data. This procedure will inevitably introduce errors into the model predictions, and indeed some of the extrapolations are rather tenuous, such as the linearization of coccinellid development, handling and searching rate curves from the maximum at 35°C to zero at 50°C. However, this process currently has very little effect on the model since the temperatures used in the model do not often exceed 30°C, but it is an area which needs to be addressed before the model is used for further investigation of climate change.

Also much of the data used to develop the equations describing coccinellid biology come from studies performed in countries other than Britain, and it is possible that these studies will not accurately represent the responses of British coccinellids. This is perhaps most noticeable in the development data (Hodek 1973) which comes from studies performed in Czechoslovakia. The studies show that optimal development occurs around 35°C, a temperature which is rarely reached in Britain.

The way to address the problems listed above is to conduct an intensive study of the biology of British coccinellids, using a combination of both field and laboratory experiments, over a wide range of temperatures, especially temperatures below 10°C or above 30°C, where data is most noticeably lacking.

This model has acted to highlight the areas of coccinellid and aphid biology where further work is required, and it also draws together much of the work performed on coccinellids and aphids. It should therefore provide a firm grounding upon which to base future studies of aphids and coccinellid population dynamics.

References

Cammell, M.E. & Knight, J.D. 1992. Effects of climate change on population dynamics of crop pests *Adv. Ecol. Res.* 22: 117-55.

Carter, N., Dixon, A.F.G. & Rabbinge R. 1982. *Cereal Aphid Populations: Biology, Simulation and Predation,* Pudoc, Wageningen.

Carter, N., McLean, I.F.G., Watt, A.D. & Dixon, A.F.G. 1980. Cereal aphids: A case study and review. *Appl. Biol.* 5: 271-348.

Chambers, R.J., Sunderland, K.D., Stacey, D.L. & Wyatt, I.J. 1982. A survey of cereal aphids and their natural enemies in winter wheat in 1980. *Ann. Appl. Biol.* 101: 175-78.

Chambers, R.J., Sunderland, K.D., Stacey, D.L. & Wyatt, I.J. 1986. Control of cereal aphids in winter wheat by natural enemies: aphid specific predators, parasitoids and pathogenic fungi. *Ann. Appl. Biol.* 108: 219-31.

Dean, G.J. 1974. Effect of temperature on the cereal aphids *Metopolophium dirhodum* (Wlk.), *Rhopalosiphum padi* (L.) and *Sitobion avenae* (F.). *Bull. Ent. Res.* 63: 401-09.

Ghanim, A.E.B., Freier, B. & Wetzel, T. 1984. On the feed intake and oviposition of *Coccinella septempunctata* L. at different availability of aphids of the species *Macrosiphum avenae* (Fabr.) and *Rhopalosiphum padi* (L.). *Archiv für Phytopathologie Pflanzenschutz* 20: 117-25.

Gilbert, N. & Gutierrez, A.P. 1973. A plant-aphid-parasite relationship. *J. Anim. Ecol.* 42: 323-40.

Hodek, I. 1973. *Biology of Coccinellidae.* Dr W. Junk, The Hague.

Honek, A. 1978. Trophic regulation of postdiapause ovariole maturation in *Coccinella septempunctata* (Col.; Coccinellidae). *Entomophaga* 23: 213-16.

Honek, A. 1980. Population density of aphids at the time of settling and ovariole maturation in *Coccinella septempunctata. Entomophaga* 15: 427-30.

Honek, A. 1985. Activity and predation of *Coccinella septempunctata* adults in the field. *Z. angew. Entomol.* 100: 399-409.

Honek, A. 1990. Seasonal changes in the flight activity of *Coccinella septempunctata* L. (Coleopteran, Coccinellidae). *Acta Entomol. Bohemoslov.* 87: 336-41.

Mack, T.P., Bajus, B.A., Nolan, E.S. & Smilowitz, Z. 1981. Development of a temperature-mediated functional response equation. *Environ. Entomol.* 10: 573-79.

Mack, T.P. & Smilowitz, Z. 1982. Using temperature-mediated functional response model to predict the impact of *C.maculata* adults and third instar larvae on green peach aphids. *Environ. Entomol.* 11: 46-52.

Mclean, I.F.G. 1980. *Ecology of natural enemies of cereal aphids.* PhD thesis, University of East Anglia, Norwich.

Olszak, R.W. 1988. Voracity and development of three species of Coccinellidae, preying upon different species of aphids In: Niemzyk E., Dixon A.F.G. (eds.) *Ecology and Effectiveness of Aphidophaga*, pp. 47-53. SPB Academic Publishing, The Hague.

Parry, M.L., Carter, T.R. & Porter, J.H. 1989. The greenhouse effect and the future of U.K. agriculture. *Journal of the Royal Agricultural Society of England* 150: 120-31.

Porter, J., Parry, M.L. & Carter, T.R. 1991. The potential effects of climatic change on agricultural insect pest. *Agricultural and Forest Meteorology* 57: 221-40.

Taylor, L.R. & Palmer, J.M.P. 1972. Aerial Sampling. In: van Emden, H.F. (ed.) *Aphid Technology* pp.189-234. Academic Press, London.

Thompson, W.R. 1930. The principles of biological control. *Ann. Appl. Biol.* 17: 306-38.

Vickerman, G.P. & Wratten, S.D. 1979. The biology and pest status of cereal aphids in Europe: a review. *Bull. Ent.. Res.* 69: 1-32.

Zaslavsky, V.A. & Semyanov, V.P. 1986. Migratory behaviour of coccinellid beetles In: Hodek I. (ed.) *Ecology of Aphidophaga*, pp. 225-27. Dr. W. Junk, The Hague.

Zhou, X., Carter, N. & Powell, W. 1994. Seasonal distribution and aerial movement of adult coccinellids on farmland. *Biocontrol Science and Technology* 4: 167-75.

Simulation of population dynamics of woolly apple aphid and its natural enemies

P.J.M. Mols

Dept. Entomology, Agricultural University Wageningen, P.O.Box 8031, 6700EH
Wageningen, the Netherlands

Abstract

Woolly apple aphid (*Eriosoma lanigerum* Hausmann) is one of the most important apple pests in the Netherlands. Weather conditions and natural enemies determine whether woolly apple aphid (WAA) will reach pest status. In some years, WAA escapes control by natural enemies and has to be controlled by spraying with Pirimicarb. To prevent unnecessary spraying, and to promote biological and natural control of WAA, more knowledge has to be obtained about the role of natural enemies, weather and their interactions on the development of WAA populations. The monophagous parasitoid *Aphelinus mali* (Hald.) has been introduced into most apple growing areas to control WAA with more or less success. Parasitization levels by *Aphelinus mali* in spring are often too low to be effective in time for the farmer, although during the summer paratisitization may be high (Mueller et al. 1992). In early spring the oligophagous Pine ladybird beetle, *Exochomus quadripustulatus L.,* and during the summer the very polyphagous European earwig, *Forficula. auricularia L.,* play an important role (Mueller et al. 1988).

To be able to quantify the role of these factors on population fluctuations of WAA, a simulation model has been developed which aims to explain quantitatively the influence of these various factors on the population density of WAA. Feeding experiments have been done with earwigs and ladybird beetles to assess their potential role as control agents. Especially important for the prediction of future control are the WAA/natural enemy ratios when the natural enemies emerge or start to climb the trees.

Key Words: *Eriosoma lanigerum,* woolly apple aphid, *Aphelinus mali, Exochomus quadrpustulatus, Forficula auricularia,* ladybird, earwig, population dynamics, simulation model, predation, parasitism, biological control.

Introduction

In apple orchards aphids play an important role. The most important species are

Arthropod natural enemies in arable land · III The individual, the population and the community
W. Powell (ed.). *Acta Jutlandica* vol. 72:2 1997, pp. 113-126
© Aarhus University Press, Denmark, ISBN 87 7288 673 0

the rosy-apple aphid (*Dysaphis plantaginea*), the woolly apple aphid (*Eriosoma lanigerum*), the green apple aphid (*Aphis pomi*), the apple grass aphid (*Rhopalosiphum insertum*) and the rosy leaf curling aphid (*Dysaphis devecta*). Their numbers fluctuate strongly from year to year depending on biotic (natural enemies, host plant) and abiotic (temperature, rainfall etc.) factors. In the Netherlands, natural control of apple aphids is not dominated by a single factor, as none of these factors separately can reliably keep aphid densities below acceptable levels for the fruit growers. Therefore, the aphids are often controlled by means of spraying with the aphicide Pirimicarb. To prevent unnecessary spraying and to promote biological and natural control of pests, more knowledge has to be obtained about the role of weather and natural enemies and their interactions on the development of aphid populations. To gain more insight into the role of the various factors, research has been started to unravel the population dynamics of the woolly apple aphid (*Eriosoma lanigerum* Hausmann), one of the important aphid pests in the Netherlands.

Woolly apple aphid (WAA) originates from North America and has spread over the fruit growing areas of the world during the last centuries. The life cycle of WAA in the Netherlands is completely asexual because the winter host, *Ulmus americana,* is absent. The aphid reproduces parthenogenetically throughout the year.

The parasitoid *Aphelinus mali* (Hald.) has been introduced into most of these areas to control WAA with more or less success, probably depending on the climatological conditions. The WAA has 10-11 generations per year, while the parasitoid has only 4-5. The aphid is also attacked by various predator species. Exclusion experiments have been used to try to quantify the relative roles of the various natural enemies of WAA (Mueller et al. 1988). It seems that in springtime the Pine ladybird beetle, *Exochomus quadripustulatus* L., and during the summer the European earwig, *Forficula auricularia* L., and sometimes the seven-spot ladybird, *Coccinella septempunctata* L., play an important role. Other predators such as lacewings, hoverflies and the mite *Allotrombium,* have sometimes been observed feeding on WAA but their role is not clear. These different natural enemies may interact strongly. The earwig preys upon all stages of WAA and on other aphid species, but also on parasitized aphids and on mummies of *A. mali*. Therefore, it may also affect the role of *A. mali*. The ladybird *E. quadripustulatus* has a preference for WAA, but also eats other aphids. Mummies with *A. mali* are not eaten but larval stages of *A. mali* in the aphids are. Thus we have an interesting system comprising an aphid species whose population fluctuations are governed mainly by weather factors, tree condition, a monophagous parasitoid *A. mali*, oligophagous ladybird beetles and a very polyphagous earwig. What is the relative role of these factors in the population dynamics of WAA? Is it mainly the abiotic factors that determine the numbers of WAA or are they mainly regulated by biotic factors?

To be able to quantify the role of these factors on population fluctuations of WAA, an age-structured simulation model has been developed. Feeding experiments have been done with earwigs and ladybird beetles to assess their potential role as control agents, and to quantify the WAA/natural enemy ratios at the moment that the natural enemies emerge or become active in the trees. This is important for deciding whether or not a chemical treatment is necessary.

The Model

The model is implemented within the INSIM environment (Mols & Diederik 1996). INSIM (INsect SIMulation) is menu driven and needs the biological information of the insect species. INSIM generates age-structured models and therefore includes modules that calculate the number and development of insects by use of "boxcars", to account for the relative dispersion of development. The core of the programme is the so-called fractional boxcar train which is used to mimic the dispersion of the individuals passing through a life stage. The number of life stages determines the basic structure of the insect model. In these models the flow of individual insects from one stage to the next is taken into account by the so called distributed-delay process (de Wit & Goudriaan 1978, Goudriaan & van Roermund 1989). By means of tables, life-history data are read into the programme. For a simple phenological or population model the information required is: life cycle, developmental rate and standard deviation of each insect stage, sex ratio, life expectancy of the adult and age dependent reproduction. The stages are coupled to each other in a spreadsheet which makes the programme flexible. The weather data, the minimum and maximum daily temperature, are stored in an environment file. The temperature for each time step is calculated from a sinusoidal curve through the minimum and maximum temperature. The output of the chosen variables can be presented graphically or numerically on the screen.

For the simulation of complicated predator- or parasitoid-prey interactions specific programming was needed. This was achieved by writing the programme in Quick-Basic.

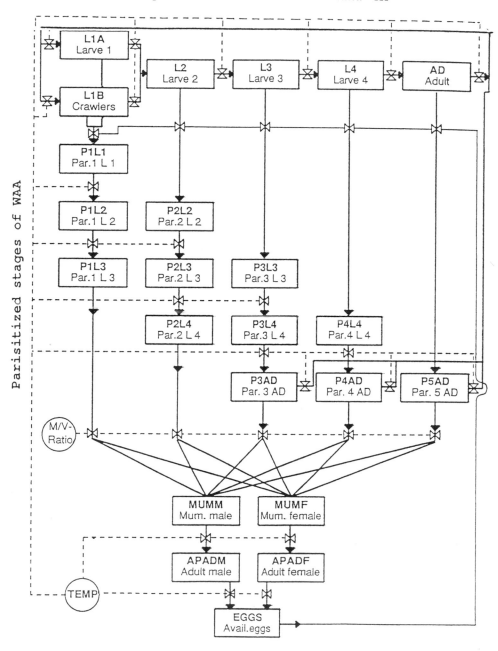

Fig. 1. Relational diagram of the development of woolly apple aphid and its parasitoid *Aphelinus mali.*

General set up of the model

To show the relationships involved, a relational diagram of the WAA-*A. mali* model is shown in Fig. 1 as an example. The aphid has four nymphal stages. The development of the stages, longevity of the adults and reproduction is governed by temperature. Natural mortality depends also on temperature, and frost especially is an important mortality factor. The first nymphal stage is divided into sessile aphids and mobile aphids (the crawlers). When temperature exceeds a threshold for crawling, aphids are produced that have a much higher mortality than sessile crawlers.

A. *mali* is able to parasitize and to develop in all the stages of WAA. The developemental rate is governed by temperature but also by the WAA stage. Therefore, all the parisitized aphid stages had to be included into the model with their respective developmental rate for the parasitoid larvae. The coupling of *A. mali* with the aphid is done by a functional response curve. The preference of the parasitoid for the different aphid stages is also included. The sex of the parasitoid is determined by the size of the host (Mueller et al. 1992).

Up till now models have been produced for WAA-*A. mali*, WAA-*F. auricularia* and WAA-*E. quadripustulatus*. A model combining WAA and the three natural enemies is under construction. Information on interactions between natural enemies and WAA is lacking and must be obtained.

In the following sections, general information is given about the life history data of each insect and the relevant relationships that are built into the models.

Woolly apple aphid

a) The developmental rates of the nymphal stages with their dispersion and adult longevity were derived from Walker (1985) and Bodenheimer (1937).

b) Age dependent reproduction was derived from Walker (1985), Evenhuis (1958), Bonnemaison (1965) and Ehrenhardt (1940).

c) Mortality, especially of the crawler stage, is related to temperature. The mortality of WAA in winter differs for the various developmental stages. The younger the stage, the higher its resistance to frost, which may partly explain the greater chance of WAA outbreaks following mild winters (Jancke 1935, Ehrenhardt 1940, Kjellander 1953).

d) Above a threshold of 15°C the crawlers leave the colony and swarm out in search of young shoots, where they try to establish a new colony (Hoyt &

Madsen 1960). From our experiments it became clear that mortality is very high during this phase. In the laboratory, 40% succeeded in founding a new colony, while in the field only 5-8% was able to do so, even when protected against natural enemies. Laboratory tests with an artificial rain maker, which simulated rain at various wind speeds, did not give any significant relationship with colony founding in the range tested (rain 20-25 ml/hour and wind of 0, 2, 7 and 18 m/sec tested for 5-30 min.). In the field, heavy rainstorms had a large effect in one year, but in another year a single weather factor could not be identified as critical but rather a combination of factors were important. Therefore, in the model an average survival of 8% is assumed for the crawler stage.

e) The influence of the apple tree is not included in the model. However, the apple tree, as host plant of WAA, may have a strong influence on the performance of WAA by:

1) Resistance: Apple varieties exist which are partially resitant to WAA. Resistant rootstocks inhibit the hibernation of WAA on the roots during winter. Up till now the relationships hold for apple cultivars Golden Delicious and James Grieve.

2) Plant condition (governed by ferlilization and moisture) may affect developmental rate, longevity and reproduction, but data are lacking.

3) WAA induced galls on the tree. In these galls development rates may be higher. A detailed relationship between aphid development rate and gall development is not included in the model.

The effects of these factors on life history data such as developmental rate and reproduction can be added later by a multiplication factor.

The parasitoid *Aphelinus mali.*

a) Induction and break of diapause. Trimble et al. (1990) investigated post diapause developmental rate (Y) in relationship to temperature (T). He found a linear relationship with the equation:

$$Y = -.00682 + 0.0073 \, T.$$

b) Adult longevity depending on temperature (Bonnemaison 1965; Evenhuis 1958).

c) Parasitization rate in relation to weather factors and aphid density (Evenhuis 1958). Data about a functional response is lacking. A simple Holling type I functional response is assumed with a level depending on maximum parasitization/day in relation to temperature and aphid availability. The maximum egg production capacity is 85 eggs per female (Evenhuis 1958). The maximal attained egg production per day depends on temperature and availability of aphids as hosts.

d) Egg and larval development up to the mummy stage (Mueller et al. 1992). These development rates in the host depend on temperature and on the age of aphid stages (Mueller et al 1992). Development time is shortest in fourth stage WAA nymphs and in adult aphids, and longest in first stage nymphs.

e) Sex ratio depending on the size of the host stage (Mueller et al. 1992). It is strongly biased to males when small stages are parasitized and to females when eggs are deposited into adults.

The earwig *Forficula auricularia.*

Earwigs are very polyphagous, feeding also on fungi, pollen and young leaves, but they have a preference for aphids as food. They have one generation per year. A female can produce one or two eggbatches per year. The first in January-March and the second May-June. The following information is incorporated into the model:

a) Developmental rates and standard deviation of the third and fourth larval stage (Buxton & Madge 1976).

b) Survival of larval and adult stages in relation to temperature (unpublished observations).

c) Phenology of the earwig stages and their survival in the tree. At the start of the third larval stage the earwigs climb up into the tree where they remain as they develop to adulthood.

d) Searching and feeding rate with respect to WAA as food (Noppert et al. 1987, and unpublished results). With simulation models which integrate digestion rates at various temperatures and maximum meal size, an estimate is made about the potential daily consumption of WAA as food for the different earwig stages (Figs. 2 & 3). No preference for the different WAA stages was found. Consumption by the earwig is according to the proportional presence of the different WAA stages in the population. Data on earwig preference for alternative food is lacking .

The ladybird *Exochomus quadripustulatus*

a) Break of reproductive diapause takes place at the end of February/beginning of March (unpublished observations).

b) Survival of the adult stages in relation to temperature (unpublished observations).

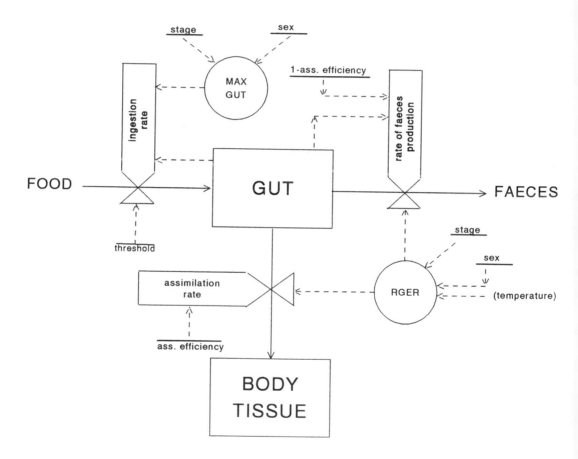

Fig. 2. Relational diagram of the feeding model of the predators. The model is used to estimate potential daily consumption of woolly apple aphids.

c) Feeding rate with respect to WAA as food. With simulation models which integrate digestion rates at various temperatures and maximum meal size, an estimate was made of the potential daily consumption of WAA as food for adult ladybirds (Figs. 2 & 3). They feed only during the daytime. Consumption by the ladybirds is assumed to be in accordance to the proportional presence of the different WAA stages in the population. Preference for different WAA stages was not included. Although they can complete larval development by feeding on aphids, they need some alternative food to reach full adult weight (otherwise they develop into small adults). Therefore, predation by the larval stages is not included in the model. Also, field numbers are in most cases relatively low at the time when WAA population growth is high.

d) Sex ratio (m/f) is 1.

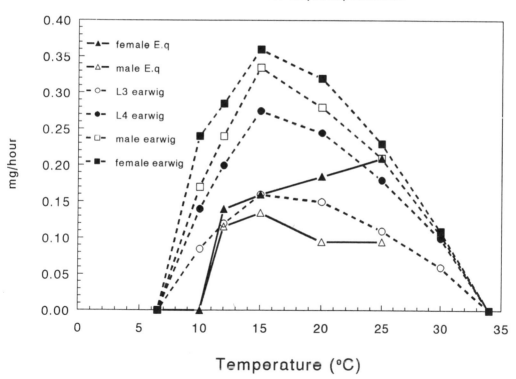

Fig. 3. Simulated predation rates expressed in mg/hour of the earwig (*Forficula auricularia* and the ladybird (*Exochomus quadripustulatus*) for different temperatures.

Results of simulations.

1) Temperature and WAA population development.

The population development of WAA in the absence of natural enemies depends mainly on the prevailing temperature. The temperature during the winter strongly determines the mortality and hence the stage composition of the population. After severe winters, first instar nymphs are the main survivors and consequently reproduction starts just after the development period needed to reach the adult stage is fulfilled. During mild winters, the adults also survive and thus reproduction starts immediately when temperatures exceed the required threshold.

Simulations have been validated with observations on growth of WAA populations under constant temperatures in a greenhouse and under field conditions in an orchard. Simulations and experimental results are in close agreement.

2) *Aphelinus mali* and WAA.

The parasitoid *A. mali* may control WAA only when it exceeds a specific ratio to WAA at the moment of emergence from diapause. Therefore, simulations were done to find the numbers of *A. mali* able to control WAA in the course of the season under field conditions (Table 1.).

Table 1. The results of simulations for the years 1986-1995 are given. The simulation started 1st January with 100 WAA in a stable stage distribution (L1=65 , L2=16, L3=8, L4=6, adult=5). Tsum is the number of day degrees above 9.4 °C.

Break of diapause			Tsum	Aphid number	*A. mali* females required to achieve control
Year	Daynumber	Date			
1986	133	13 May	91	38	1.2
1987	126	6 May	92	15	0.5
1988	128	7 May	91	696	14
1989	129	9 May	92	934	20
1990	117	27 April	96	1209	27
1991	106	16 April	124	138	4
1992	136	15 May	97	578	10
1993	118	29 April	94	446	10
1994	132	12 May	92	489	10
1995	126	6 May	97	789	18

With a sex ratio (m/f) of *A. mali* of 1, an *Eriosoma/Aphelinus* mummies ratio of less than 23 is required at the break of diapause to achieve control later in the season.

Table 1 clearly shows the influence of winter conditions on WAA growth. The winter and spring of the year 1990 were relatively warm. Only a few older stages of WAA died and reproduction started early. The moment *A. mali* became active the population of WAA had reached such numbers that many parasitoids were necessary to attain control. This agreed with field observations in an unsprayed plot of the orchard where WAA reached high numbers and caused severe damage to the trees. Only at the end of August was the parasitoid able to control the population.

3) Earwigs and WAA.
Earwigs enter the tree in the third larval stage during June, depending on the previous temperatures. The number of WAA present at that moment determines the amount of earwigs needed to achieve control.The results of simulations are shown in Table 2.

The   average WAA/earwig ratio has to be 50 to achieve control. Simulations were done for those years for which data about the phenology of earwigs were available.

Table 2. The simulation starts with the same conditions as in Table 1.

50 % of earwigs enter the tree			Aphid number	Number of earwig L3 larvae required to achieve control	Ratio
Year	Daynumber	Date			
1981	172	20 June	12023	222	54
1984	181	1 July	4186	79	53
1985	175	23 June	56	2	28
1986	169	17 June	794	15	53
1987	172	20 June	341	11	31
1988	154	1 June	14892	280	53
1989	158	5 June	7567	117	64
1990	162	10 June	13364	447	52
1991	172	20 June	3124	50	63

4) *Exochomus quadripustulatus* and WAA.
Simulations were done with ladybirds and WAA under field conditions. At the start of the simulation on 1st March, 100 WAA in a stable age distribution were present on one tree. Numbers of the ladybird were varied until control over WAA

was attained at the end of April. "Control" means that numbers of WAA at the end of April are equal or smaller than the numbers at the beginning of March when *E. quadripustulatus* becomes active. In May, mortality of the ladybirds increases strongly and WAA always escaped control of the ladybird. In this way a preliminary estimate of WAA/ladybird ratio could be estimated. On average, 0.86 ± 0.34 (mean$\pm$SD) were necessary to control 100 WAA, giving a ratio of 116.

Growth of WAA on small apple trees (1 m high) in cages, with 0,1 or 2 *E. quadripustulatus* per cage, were not significantly different from the outcomes of the simulations.

In the winters of the years 1989, 1990, 1991 and 1992, *Exochomus* was present in the trees at an average 2-4 ladybirds per tree. In 1994, six ladybirds were counted per tree while in the spring of 1995 only 0.5 per tree were present. The initial density of WAA at that time of the year is between 50-250 aphids per tree (James Grieve, 25 years old, 3 m high). The conclusion is that by their feeding early in the year the ladybird *E. quadripustulatus* may reduce the initial amount of WAA considerably.

Discussion and Conclusions

Although not all interactions between WAA and its natural enemies have been studied thoroughly, some conclusions can be drawn from the simulations. Winter conditions determine the initial amount and the stage distribution of the aphids before the natural enemies can become active. This determines the potential of WAA to reach pest status in the course of the year.

The ladybird *E. quadripustulatus* is important at the start of the growing season and determines the amount of WAA before the parasitoid *A. mali* can take over. *E. quadripustulatus* can live on other aphid species as well and has been observed feeding in summertime on the green apple aphid *Aphis pomi*. In fact it is known as a good predator of scale insects on different trees, and in Southern Europe it feeds on Olive scale. This implies that alternate food sources may influence its role in controlling WAA. Its numbers in spring are mainly determined by the number of WAA in the previous year and it thus occurs alternately with WAA. This makes it a less reliable control factor. For modelling, the role of the larval stages of this ladybird is ignored for several reasons. The consumption of the *E. quadripustulatus* larval stages is low and takes place in a period when the WAA population is rapidly growing. Also, numbers per tree are low because of the low egg production. Control of WAA has to be attained later in the season by the action of earwigs and *A. mali*.

Earwigs are important from the moment they arrive on the tree from the beginning of June to the end of August. Their numbers, relative to the number of aphids present at that moment, determines their success as control agents. In an IPM orchard, numbers of earwigs per tree differ from year to year and fluctuate between 20-100 per tree independently of the amount of aphids. The number of earwigs was assessed by mark-recapture methods. This number is in many years sufficient to control WAA, depending on weather conditions and the other natural enemies.

The parasitoid *A. mali* is not well synchronized with WAA under the Dutch climatological conditions. After a warm winter it will be completely outnumbered by WAA if the predators are not also present. The combination of predators and the parasitoid controls WAA population growth in most years. Only after warm winters, in the absence of *E. quadripustulatus,* can a pest status be reached.

Studies concerning the interactions of natural enemies have still to be done. It is not just a summation of their separate roles towards WAA that is relevant, some interactions may also be counter-productive. For example, earwigs eat *A. mali* mummies. The condition of the host plant, which may influence the development rate and fecundity of WAA, is still an unknown factor in the system, especially when an orchard has no watering or vertigation system, so that dry spells may occur that hamper the sap flow in the trees and thus the food intake of the aphids. Another important factor is the presence of alternative prey because this may hamper the efficacy of the predators towards WAA.

Acknowledgements

I thank Dick Diederik who programmed the model and many students who did an exellent job collecting data at the experimental orchard "de Schuilenburg".

References

Bodenheimer, F.S. 1937. *Studies on the physical ecology of the woolly apple aphid* (Eriosoma lanigerum) *and its parasite* Aphelinus mali *in Palestine*. Hebrew University, Jerusalem.
Bonnemaison, L. 1965. Observations ecologiques sur *Aphelinus mali* Haldeman, parasite du

puceron lanigère (*Eriosoma lanigerum* Hausm.). *Ann. Soc. Ent. Fr.* 1:143-76.

Buxton, J.H. & Madge, D.S. 1976. The evaluation of the European earwig (*Forficula auricularia*) as a predator of the damson-hop aphid (*Phorodon humuli*). I : Feeding experiments. *Entomol. Exp. Appl.* 19: 109-14.

Ehrenhardt, H. 1940. Der Einfluß von Temperatur und Feuchtigheit auf die Entwicklung und Vermehrung der Blutlaus. *Arb. physiol. angew. Ent.* 7:150-68.

Evenhuis, H.H. 1958. *Een oecologisch onderzoek over appelbloedluis,* Eriosoma lanigerum *(Hausm.), en haar parasiet* Aphelinus mali *(Hald.) in Nederland.* Proefschrift RU Groningen, H. Veenman & Zn, Wageningen.

Goudriaan, J. & Roermund, H.J.W, van 1989. Modelling of ageing, development, delays and dispersion. In: Rabbinge, R., Ward, S.A. & van Laar, H.H. (eds.) *Simulation and systems management in crop protection*, pp. 115-57. PUDOC, Wageningen.

Hoyt, S.C. & Madsen, H.F. 1960. Dispersal behaviour of the first instar nymphs of the woolly apple aphid. *Hilgardia* 30(10): 267-99.

Jancke, O. 1935. Sur Kaltempfindlichkeit der Blutlaus. *Nach. Bl. Dtsch. Sch. D.* 15: 47-47.

Kjellander, E. 1953. Investigation on the biology of the apple woolly aphis together with some experiments on the control of the aphis. *Statens Vaxskyddsant., Meddel.* 64: 1-51.

Mols, P.J.M. 1992. Forecasting an indispensable part of IPM in apple orchards. *Acta Phytopathol. Entomol. Hung.* 27: 449-60.

Mols, P.J.M. & Diederik, D. 1996. INSIM a simulation environment for pest forecasting and simulation of pest-natural enemy interaction. *Acta Horticultura* 416: 255-62.

Mueller, T.F., Blommers, L.H.M. & Mols, P.J.M. 1988. Earwig (*Forficula auricularia* L.) predation on the woolly apple aphid *Eriosoma lanigerum. Entomol. Exp. Appl.* 47: 145-52.

Mueller, T.F., Blommers, L.H.M. & Mols, P.J.M. 1992. Woolly apple aphid (*Eriosoma lanigerum* Hausm., Hom., Aphidae) parasitism by *Aphelinus mali* Hal. (Hym., Aphlinidae) in relation to host stage and host colony size, shape and location. *J. Appl. Ent.* 114: 143-54.

Noppert, F., Smits, J.D. & Mols, P.J.M. 1987. A laboratory evaluation of the European earwig (*Forficula auricularia* L.) as a predator of the woolly apple aphid (*Eriosoma lanigerum* Hausm.). *Meded. Fac. Landbouwwet. Rijksuniv. Gent* 52 (2a): 413-31.

Trimble, R.M., Blommers, L.H.M. & Helsen, H.H.M. 1990. Diapause termination and thermal requirements for postdiapause development in *Aphelinus mali* at constant and fluctuating temperatures. *Entomol. Exp. Appl.* 56: 61-69.

Walker, T.T.S. 1985. The influence of temperature and natural enemies on population development of woolly apple aphid, *Eriosoma lanigerum* (Hausman). PhD thesis Washington State University.

Wit, C.T. de & Goudriaan, J. 1978. *Simulation of ecological processes.* Pudoc, Wageningen.

Simulation of carabid beetle foraging and egg production.

P.J.M.Mols

Department of Entomology, Wageningen Agricultural University, P.O.Box 8031,
6700EH Wageningen.

Abstract

A simulation model was developed to study the foraging behaviour and egg production of the carabid beetle *Pterostichus coerulescens* L (=*Poecilus versicolor* Sturm) in relationship to prey density and prey distribution. Foraging behaviour is divided into its most dominant components: searching, acceptance of prey and feeding, which are in turn related to the most important internal and external factors. Hunger or its opposite, the relative satiation level (RSATL), is the internal factor that determines the "motivation" for a large part of the behaviour and it results from the physiological state of the beetle. RSATL is estimated by means of a simulation model (Mols 1988) and it is correlated with the behavioural components that play a role in searching and predation. The hunger level has a strong influence on locomotory activity, walking speed, duration of area-restricted search and prey acceptance. Three types of searching behaviour were distinguished: 1) straight high-speed walking when RSATL is below 5%, 2) intermediate walking when RSATL exceeds 5% and 3) Intensive, winding, area-restricted search after consumption of a prey item.

The searching model shows the advantage of the intensive winding or tortuous walk (TW) when prey is aggregated and its disadvantage in random prey distributions. When the searching model is coupled to the motivation model, the advantage of TW shows when prey is aggregated and when overall prey density is low (less than 1 prey m^{-2}).

Walking speed, time spent walking and success ratio (defined as the number of prey captured/prey discovered), which also depend on the relative satiation level, in combination with prey density and prey aggregation determine the predation rate and the egg production. The model is set up in such a way that characteristics of other predator species can be implemented easily.

Key Words: *Pterostichus coerulescens,* simulation model, foraging behaviour, predation, feeding, prey distribution, egg production, locomotory activity.

Arthropod natural enemies in arable land · III The individual, the population and the community
W. Powell (ed.). *Acta Jutlandica* vol. 72:2 1997, pp. 127-138
© Aarhus University Press, Denmark, ISBN 87 7288 673 0

Introduction

The ultimate aim of this study was to gain insight into the impact of spatial distribution and density of prey on predatory behaviour and on the resulting egg production of the carabid beetle *Pterostichus coerulescens* L. Thus, a better understanding of the survival strategy of this carabid beetle would be obtained, reflected in its ability to cope with specific prey distributions varying from random to aggregated.

The carabid beetle *Pterostichus coerulescens* L.(= *Poecilus versicolor* Sturm) is a predator of small arthropods (Hengeveld 1980) such as aphids, spiders, caterpillars, larvae of Diptera feeding in the litter and also larvae of the heather beetle, *Lochmaea suturalis* Thomson, which are usually highly aggregated in their distribution. Sandy soils with a low humus content and open, dry localities with sparse vegetation, where sunlight can easily reach the soil, are preferred. Prey density in the beetle habitat may be temporarily very low as in poor heathland and moorland or sometimes excessively high as during heather beetle outbreaks. In previously cultivated land, prey density may be moderate to high depending on the nitrogen content of the soil.

Flight has only rarely been observed in *P. coerulescens*. Therefore, they walk to perform all vital behavioural functions such as searching for food, finding a mate and escaping from predation by birds, toads and rodents (Larochelle 1974 a,b).

The density and distribution of prey affects searching, the rate of feeding, and the rate of reproduction, and so also affects population dynamics in the field.

To understand the dynamics of the processes governing searching and predatory behaviour, information is required on the motivational drives for predatory behaviour. In many species behaviour is governed by the relative satiation level of the gut (Holling 1966, Nakamura 1972, Fransz 1974, Rabbinge 1976, Sabelis 1981, Kareiva & Odell 1987). For *P. coerulescens*, the relative satiation level of the gut (RSATL) is defined as the actual gut content divided by the apparent gut capacity. The latter is the weight of a meal needed to satiate a starving beetle (Mols 1988). The apparant gut capacity cannot exceed the maximal gut size, and is contrained by the size of other organs and tissues needed for egg formation and storage of reserves. The actual gut content changes by ingestion, excretion and resorption. The rates of changes are predominantly affected by ambient temperature and day-length, the latter determining the onset of vitellogenesis.

The internal factors which determine the motivation were integrated in a simulation model (Mols 1988). The output of that model was compared with the results of experiments and showed that it was possible to estimate continuously the motivation of the beetle, which is the most important state variable that dictates walking and predatory behaviour.

Therefore, the relationships between the relative satiation level and locomotory activity, walking behaviour and successful encounters with prey (success ratio) had to be quantified experimentally and integrated into a model. With this model it is possible to simulate searching and predation behaviour and egg production under different sets of environmental conditions, prey densities and prey distributions.

Sensitivity analysis can be used to estimate the value of specific behavioural components in the total predation process, thus giving insight into which specific behaviour is advantagous for the predator. Functional and numerical response curves can be calculated depending on the spatial distribution of the prey, which is an important feature determining the population dynamics of this carabid.

Simulation

To simulate predation rate and the resulting egg production at different densities and distributions of the prey, a model was constructed that integrates the most important components of predation and egg production by the beetle.

In general, the predation rate is determined both by the encounter rate (Er) with prey, and by the proportion killed by the predator, called the succes ratio (Sr) (Fransz 1974, Sabelis 1981), and these are determined by the relative satiation level (RSATL).

A general description of the components of the simulation model is given below.

a)   Walking velocity in relation to RSATL.
b)   Turning rate with the standard deviation and kurtosis of the frequency distribution of the turning rate and their relationship with walking velocity.
c)   Duration of intensive searching in relation to RSATL
d)   Locomotory activity in relation to RSATL and diurnal rhythmicity.
e)   Relative satiation level from ingestion and digestion of food.
f)   Ovary growth and egg production
g)   Success ratio in relation to RSATL

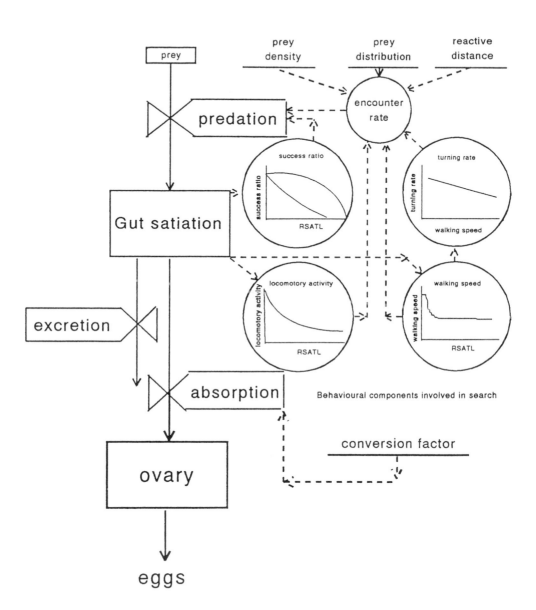

Fig. 1. The relational diagram of the searching and egg production model of a carabid. In the circles the most important behavioural relationships determining predation are given.

h)   Reaction distance to prey.
i)   Prey cluster scanning and individual prey scanning.
j)   Prey density and prey distribution.

The variables from (a) to (f) are properties of the beetle, (j) concerns prey properties and (g) and (h) concern properties of both. A relational diagram of the model is given in Fig. 1.

Velocity.
The walking speed of the beetle is expressed in cm/sec. The majority of prey items taken by *P. coerulescens* show a very low walking speed compared with the beetle. In those cases the speed and the turning rate of the prey can be ignored. However, the walking speed of mobile prey such as spiders and ants should be included in the resulting velocity, if these are important prey items.

The average speed of the beetle is introduced as a forcing variable at the beginning of the simulation. Beetles do not walk with a constant speed but show variations in velocity. Therefore, for each time step a value is taken at random from a uniform distribution around this average value (Mols 1993). During these experiments it was observed that the beetle showed three distinct types of walking: a) walking at high speed (5 cm/sec) when hungry (RSATL, 5%); b) walking at an intermediate speed of 2.5 cm/sec when some food was present in the stomach (RSATL>5%); c) a slow, very winding, walk after eating a prey item. The velocity increased from zero to the average velocity depending on RSATL.

Turning rate
The turning rate is expressed in degrees/unit time. It determines the tortuosity of the walking pattern. When the pattern is very winding, the effectiveness of searching per unit of walking distance decreases because the same spots will be visited more often.

The direction of walking is calculated by randomly drawing a direction from the experimentally established distribution of the turning rate. This is done again at each timestep, and the next angle is added to the former direction which then gives the new direction.

As the speed (V) of the beetle is known (for each time step this is the distance between two sets of coordinates XP_1, YP_1 and XP_2, YP_2 it is possible to calculate the new coordinates of the beetle with the cosine rule.

XP =XP + COS(DIR)*V
YP =YP + SIN(DIR)*V

The walking direction (DIR) of the beetle depends both on the former direction and on the turning rate (A) drawn randomly from the Tukey distribution.

DIR=DIR + A
A= AV + SIGMA * (P^{KURT} - (1-P)^{KURT})/KURT

This formula calculates the turning rate/2 sec as a function of the average turning rate (AV), the standard deviation (SIGMA) and kurtosis (KURT) of the turning rate distribution.

In the frequency distribution of the turning rates the average turning rate is zero because of the equal number of left and right turns taken (the negative and positive signs of the angles). From analysis of walking patterns the relationship of SIGMA and KURT with the walking speed is:

SIGMA= exp(-0.173V+0.208)
KURT = -0.0661V+0.2

Thus, when the walking speed (V) increases, the distribution of the turning rate changes from wide to narrow. Except for the intensive searching period the speed depends on RSATL. Therefore, the width of the turning rate distribution also partly depends on RSATL.

The model simulates the walking of beetles in an arena of an arbitrarily chosen size. The moment the beetle reaches the border of the arena it rebounds into the field.

Duration of intensive searching.
Just after consumption of a prey item, the beetle resumes searching at a very low speed and the pattern is very tortuous. This period may last 11 minutes when the beetle is still very hungry. When no new prey are encountered, the speed increases gradually to a level that corresponds with a specific level of satiation. This is 5 cm/sec when RSATL< 5% and 2.5 cm/sec above this value. With increase of the speed the walking pattern becomes less winding. When RSATL is about 50%, tortuous walking lasts about 4 minutes and then the speed returns more quickly to the level of 2.5 cm/sec. When RSATL exceeds 80%, intensive searching no longer occurs.

Locomotory activity.

The locomotory activity is the proportion of time that the predator is mobile. The time a beetle spends walking during a specific timestep can be calculated from the experimental relationship between the relative activity coefficient (AC) and RSATL (AC=(actual activity)/(maximum activity of hungry beetles per time period)) (Mols 1993). Per timestep the activity period can be calculated from:

$$TWALK = AC*FREQ*MAXACT$$

```
TWALK = The total time a beetle is locomotory active during a
        timestep.
AC = The relative activity coefficient related to the relative
     satiation level.
FREQ = The fraction of the total daily locomotory activity
       realized during that timestep.
MAXACT = The duration of locomotory activity of a hungry
         beetle expressed in hours per day at a specific
         temperature
```

The relationship of AC with RSATL follows a hyperbole with the equation:

$$AC= 1/(5.7RSATL+1)$$

The frequency distribution of locomotory activity follows a circadian rhythmic pattern with the highest activity at noon and no activity during the night. The total daily locomotory activity of a hungry beetle (MAXACT) at 20 °C is 5.3 hours.

In the motivational model the timestep of integration (DELT) is a quarter of an hour, which is sufficient to calculate the decrease of RSATL by digestion. The walking part of the programme is running during that period for TWALK seconds.

Relative satiation level

In this simplified model a fixed maximum gut capacity of 1 is used and RSATL is only determined by ingestion and by gut-emptying. Prey size is given as a fraction of the maximum gut capacity. Thus each time a prey is consumed, RSATL is increased with that fraction. Gut emptying is simulated with the relative gut emptying rate at 20 °C (Mols 1988).

Ovary growth and egg production

When food is ingested, a fraction of it is used for egg production. This conversion factor is obtained experimentally by measuring egg production following specific amounts of food ingested. In the model, when the ovaries have matured, the number of eggs produced is calculated from the surplus of food ingested multiplied by the conversion factor and divided by the individual egg weight.

Success ratio and RSATL

The relationship between the success ratio and RSATL determines the chance that a discovered prey will be captured. The relationship is different for different types of potential prey. The relationship found for maggots is used in the model (Fig 1.) (Mols 1993).

Reaction distance

The reaction distance is the distance at which a predator detects a prey. It determines the searching path width of the predator.

In the model, the beetle is represented as a circular object with a specific radius. The prey is considered in the same way. Together with the distance of prey discovery they constitute the reaction distance (radius predator + radius prey + discovery distance) which varies mainly between 1 and 2 cm. In the model the reaction distance is set at the start of the simulation and remains constant.

Prey scanning and prey cluster scanning

To increase the speed of prey scanning, this process is divided into two parts. Firstly, whether or not a cluster is within reaction distance is determined and secondly, when this is true, which individual prey item of this cluster is within reaction distance.

Cluster scanning: to determine whether or not a cluster is within the reaction distance of the beetle the coordinates of the centrally located prey of a cluster are used. If the distance measured between the centre of the beetle and the centre of the cluster is smaller than the sum of the reaction distance of the beetle and the cluster radius, only then the distance to each individual prey in that nearest cluster will be measured. If this condition is not fulfilled, the next walking step will be taken and cluster scanning will start over again.

Preyscanning: when a beetle moves from one position to the next during one time step, an area is covered consisting of a stripwidth of 2 times the reaction distance and of a length equal to speed/sec times timestep. All the prey in this strip is considered to be encountered. This strip is rectangular and, because of the tortuosity of the track, parts at the beginning and at the end of this strip will overlap. Prey in these overlaps will thus be counted twice. To avoid double counting a specific procedure is followed (Mols 1993).

Prey density and prey distribution

In the case of prey aggregation, prey density will differ from place to place. To avoid confusion in terminology the following terms will be used:

"Overall density" is the total number of prey items per m^2.

"Within cluster density" is the number of prey items in a cluster divided by the surface of a prey cluster.

In the programme, the distribution of the prey can be arranged such that it may vary from random to very aggregated. This is done by putting the prey in discrete prey clusters. The numbers of prey in each cluster, the number of clusters, and the size of the clusters, can be varied. The prey is considered to be immobile. In the clusters they are randomly distributed, and the clusters themselves are also randomly distributed in the arena. The arena is a square which can be varied in size. The clusters are circular.

For the establishment of functional response curves, prey density in the model has to remain constant. This is done by putting "consumed" prey randomly back into the cluster. If the density does not have to remain constant, which is the real situation after prey consumption, prey is removed from the cluster.

Simulation of searching, predation and egg production

The model can be run for one or more days depending on the aim of the simulations. It can also be run for a whole season which lasts for about 35 days at 20°C to assess egg production. Because prey size, cluster size and within prey density can be varied, functional and numerical responses can be obtained for different prey distributions over a range of prey densities for different sets of predator characteristics.

Discussion

A brief summary of the results will be discussed here, but for a more extensive discussion see Mols (1993).

The advantage of specific walking behaviour.

After consuming a prey item, a hungry beetle slows down its speed to start an intensive search by using a tortuous walking pattern. It was observed that walking speed increased during this period, and that the duration of intensive search decreased with an increase of the satiation level. What is the advantage of this behaviour? In general, it appears that intensive search is only advantagous when the prey are aggregated in their distribution, otherwise it is a waste of time.

Considering the rate of discovery that results from the walking behaviour without the feedback of the motivational part, it can be stated that the intensive searching pattern itself is not very efficient because much time is spent recrossing previously visited areas. The efficiency of the walking pattern is defined by the area searched divided by the distance walked. If intensive searching was done at the same low speed, this was only advantagous when consumed prey were replaced in the cluster. But in reality the prey is depleted locally by prey consumption. After having consumed a few prey items, it is better to move on and this is done by increasing the speed and by shortening the intensive search period. Thus a balance must exist between speed and searching efficiency. The most efficient search appeared to occur at velocities above 3 cm/sec because of the low number of recrossings. The constraint is that at increasing speed the residence time in a prey cluster becomes shorter. From simulations it appears that intensive searching is of most use within clusters smaller than 30 cm diameter.

High-speed walking is very useful when large distances have to be covered. It only occurs when the beetle is hungry. It is profitable when prey is aggregated or when large prey is randomly distributed at a low density. In these situations high-speed walking between the clusters results in a substantial increase in predation because the distance between them is covered in the shortest time. The time between encounters with clusters is more than halved, because at higher speed the beetle walks straighter. Intermediate speed is advantageous in areas where prey can be encountered regularly, because it results in a longer stay in these profitable areas.

The effect of locomotory activity.
Factors which significantly influence the locomotory activity include temperature (Thiele 1977), hunger level and the diurnal activity of the beetle (Luff 1978, Mols 1993). As the locomotory activity of the beetle decreases like a hyperbole with an increase of the satiation level, thus decreasing strongest at low satiation levels, this points to an adaptation to clustered prey. By decreasing its locomotory activity after consuming a prey item (even a small one), the beetle stays longer in a location profitable for feeding. While resting, it digests the prey and utilises it to produce eggs. In combination with the slow and tortuous walk, this is highly profitable. In contrast, long periods of locomotory activity, together with high speed walking during periods of hunger, increase the chance of finding areas where food is more abundant.

Effect of motivation.
Relative satiation level determines walking behaviour, success ratio and

locomotory activity of the beetle and appears to be the most important motivation in the searching and predation process.

By an increase of the satiation level, the duration of the intensive searching period decreases and at a RSATL above 80% it disappeares completely. This occurs above prey densities of 4 prey m^{-2}. In this way, the walking pattern becomes more efficient and also the periods that have to be walked to attain the same effect become shorter .

When walking behaviour is coupled to motivation, it may be concluded that the searching behaviour of *P.coerulescens* is adapted to various types of prey distribution, and that it shows its greatest profitability at overall low prey densities with aggregated prey. Especially when the prey is small, i.e. 5-10% of the maximum gut size, tortuous walking is very profitable.

Motivation has also a strong influence on the success ratio. With increasing satiation level the success ratio decreases. The decrease is especially strong above 60% of RSATL. The success ratio also depends on the prey type but the satiation level also plays an important role here.

Because of decreasing motivation for locomotory activity and prey acceptance the functional response levels off.

Field situation

How does this model relate to the field situation where factors such as variable temperatures, rain, soil and vegetation structure will be important?

Walking velocity is positively related to temperature (Mossakowski & Stier 1983). If the relationship between velocity and turning rate remains the same at different temperatures, the model takes care of the effect of fluctuating temperatures. But if the relationship changes this has to be carefully studied before reliable statements about the the field situation can be given.

The influence of soil surface structure and vegetation is only incorporated into the experiments to a small extent. Dense vegetation and a rough surface hamper walking speed and increase the turning rate so that these may deviate a lot from the relationship established experimentally between velocity and turning rate. On the other hand, the agreement between the dispersal found in the field (Baars 1979) and that found by simulation indicates that this effect could be less important than that of temperature.

The conclusion of the simulations based on this model is that it appears that the beetle can respond to a range of prey distributions by changing its behaviour such that even at low average prey densities a reasonable egg production is attained.

Education

The model is a valuable educational tool as it shows the walking behaviour and the reactions of the predator to its prey on the screen. Using the model, the effects of the most important behavioural characteristics can be tested seperately and their contribution to the functional and numerical responses can be established. Functional and numerical responses can be calculated for different prey distributions and prey densities. The effect of the underlying behavioural traits on these responses becomes more clear.

References

Baars, M.A. 1979. Patterns of movement of radioactive carabid beetles. *Oecologia (Berl.)* 44: 125-40.

Fransz, H.G. 1974. *The functional response to prey density in an acarine system*. Pudoc, Wageningen.

Hengeveld, R., 1980. Polyphagi, oligophagi and food specialisation in ground beetles (Coleoptera, Carabidae). *Neth. J. Zool.*, 30 (4): 564-84.

Holling, C.S.,1966. The functional response of invertebrate predators to prey density. *Mem. Entomol. Soc. Can.* No. 48.

Kareiva, P. & Odell, G. 1987. Swarms of predators exhibit 'preytaxis' if individual predators use area restricted search. *Amer. Nat.* 130: 233-70.

Larochelle, A.. 1974a. Carabid beetles (Coleoptera, Carabidae) as prey of North American frogs. *Gt. Lakes Entomologist* 7: 147-48.

Larochelle, A.. 1974b. The American toad as champion carabid beetle collector. *Pan-Pacif. Entomol.* 50: 203-0.

Luff, M.L. 1978. Diel activity patterns of some field Carabidae. *Ecol. Entomol.* 3: 53-62.

Mols, P.J.M. 1988. Simulation of hunger feeding and egg production in the carabid beetle *Pterostichus coerulescens* L. (= *Poecilus versicolor* Sturm).*Wageningen Agricultural University Papers* vol. 88-3,1- 99 .

Mols, P.J.M. 1993. Foraging behaviour of the carabid beetle Pterostichus coerulescens L. (=*Poecilus versicolor* Sturm) at different densities and distributions of the prey. *Agricultural University Wageningen Papers* 93(5): 105-201.

Mossakowski, D. & Stier, J. 1983. Vergleichende Untersuchungen zur laufgescwindigkeit der Carabiden. *Report 4th Symp. Carab.* 1981: 19-33

Nakamura, K. 1972. The ingestion in Wolf spiders II. The expression of degree of hunger and amount of ingestion in relation to spiders hunger. *Res. Popul. Ecol.* 14: 82-96.

Rabbinge, R. 1976. *Biological control of the fruit tree spider mite*. Pudoc, Wageningen.

Sabelis, M.W. 1981. Biological control of two-spotted spider mites using phytoseiid predators. Part 1. *Agricultural Research Report*, 910, Pudoc, Wageningen.

Thiele, H.U. 1977. *Carabid beetles in their environments*. Springer Verlag, Heidelberg.

Discrimination of Belgian *Sitobion avenae* (F) populations by means of RAPD-PCR

D. Stilmant, Th. Hance & Ch. Noel-Lastelle

Unité d'Écologie et de Biogéographie, 4-5 Place Croix-du-Sud,
1348 Louvain-la-Neuve, Belgium

Abstract

RAPD-PCR technique is widely used for the discrimination of taxa or strains and the study of paternity. Our aim was to use this technique to compare genetical distances between populations of the grain aphid *Sitobion avenae* (F) originating from different winter wheat fields, at two geographical levels: intra - and inter - regional. This was done in order to define what a population is, for this species, from the genetical point of view. Three aphid populations were sampled in each of two Belgian regions, Louvain-la-Neuve and Ciney, separated by 50 km. These three populations were sampled in a one square kilometre site. RAPD-PCR (Random Amplified Polymorphic DNA - Polymerase Chain Reaction) was performed using 10 primers on nine virginoparous aphids of each population. Similarities between individuals were calculated with the Jaccard coefficient, and between the populations with the Steinhaus. Eighty bands were obtained. The largest part of the genetic variation (86.5%) was observed at the population level, only 1.5% additional variation was obtained by integrating the different populations of a same region. The variation residual (12%) was related to the inter-regional scale. The clustering performed on the genetic similarities between the individuals showed that 81.5% of the Ciney individuals and 70.4% of the Louvain-la-Neuve individuals stayed in their regional group. However, no clear population clusters were observed within these regions. Each region, can thus be regarded as a metapopulation.

Key Words: *Sitobion avenae*, cereals, population, RAPD-PCR, genetics.

Introduction

During the last decade, several molecular genetic methods, such as RFLP, microsatellite repetition, intergenic spacer length in the rRNA cistron, defined PCR and RAPD-PCR DNA fingerprint (Black IV et al. 1989, Black IV et al. 1992, Carvalho et al. 1991, De Barro et al. 1994, 1995a, 1995b, Loxdale et al. 1996, Powers et al. 1989, Shufran et al. 1991, Williams et al. 1990), have been

Arthropod natural enemies in arable land · III The individual, the population and the community
W. Powell (ed.). *Acta Jutlandica* vol. 72:2 1997, pp. 139-148
© Aarhus University Press, Denmark, ISBN 87 7288 673 0

developed in order to study nuclear and extra-nuclear DNA polymorphism. These techniques also allow the analysis of the population dynamics of species with low isozyme polymorphism, such as aphid species (May & Holbrook 1978).

Black IV et al. (1992) have shown the usefulness of the RAPD-PCR technique to detect polymorphism in aphids. They reported significant variation between aphid biotypes in *Schizaphis graminum* Rondani or between aphid clones in *Acyrthosiphum pisum* Harris. More recently, De Barro et al. (1995b) used the multilocus (GATA)4 probe to study the population dynamic of *Sitobion avenae* (F.) and *Metopolophium dirhodum* Walker at two geographical scales (< 1km and > 1 km) and between two kinds of host plants (wheat and road-side grasses). They found that, at the beginning of the season, *S. avenae* clones were more homogeneous among wheat fields than among the corresponding patches of road-side grasses. This structure was not shown for *M. dirhodum*. However no relationship was revealed between genetic similarities and geographical distances separating aphid populations. Using the RAPD-PCR technique on *S. avenae* sampled on wheat and corresponding bordering cocksfoot in two sites separated by 60 km, De Barro et al. (1995a) had shown that the scheme found in March changed during the season, at the end of which the genetic segregation agreed more with the geographic origin than with the host origin. However only one primer was used in this study and this with a low annealing temperature that may give a great number of bands with a low specificity.

Our aim was to determine the genetic distances, obtained on the base of ten primers, between *Sitobion avenae* populations, at the intra - (scale < 1 km) and inter - regional (scale > 50 km) levels.

Materials and Methods

Louvain-la-Neuve and Ciney, two Belgium regions separated by 50 km, were sampled in June 1994. In each region, three populations of *S. avenae*, separated by a mean distance of 0.7 km, were collected and labelled L1, L2, L3 and C1, C2, C3 respectively. Each population corresponded to one winter wheat field. Nine virginoparous aphids were individually analyzed per population. DNA extraction and amplification followed the procedures of Landry et al. (1993). Ten primers (Table 1) were used for each individual. The amplification products were separated in a standard electrophoresis agarose gel and bands were detected on an ultra-violet light transilluminator. A band presence/absence matrix was then constructed, and similarities between individuals calculated with the Jaccard coefficient (eq.1) (Legendre & Legendre 1979, Legendre & Vaudor 1991). At the population level, the Steinhaus coefficient (eq. 2) (Stevens & Wall 1995) which integrates band frequencies, was used.

Table 1. Sequences of the ten Primers and number of bands considered to calculate the genetical distances.

Name	Sequence (5' -> 3')	Number of bands
A07	GAAACGGGTG	11
A11	CAATCGCCGT	9
A16	AGCCAGCGAA	8
BAM	ATGGATCCGC	7
CO1	TTCGAGCCAG	7
CO4	CCGCATCTAC	5
CO6	GAACGGACTC	9
C18	TGAGTGGGTG	6
ECO	ATGAATTCGC	6
E14	TGCGGCTGAG	11

$$Sj(A\&B) = \frac{\text{nbr. bands } (A \& B)}{\text{bands } (A \& B) + \text{bands } (A, \text{not } B) + \text{bands } (B, \text{not } A)} \qquad (eq.1)$$

$$Sj(A\&B) = \frac{\bullet \text{ min. nbr. of bands (popA \& popB)}}{\bullet \text{ bands for pop } A + \bullet \text{ bands for pop } B} \qquad (eq.2)$$

Moreover, the comparisons of mean similarity coefficients (s) observed at the different geographical levels were performed using normalised Student test (m) (Dagnelie 1970). A Lance and Williams mean association cluster analysis (see Legendre & Legendre 1979, Legendre & Vaudor 1991) was also performed at the individual and the population level.

The genetic and geographic distances between aphid populations were compared with a correlation analysis, at the local level. At the inter-regional level, population mean comparisons were necessary.

Results

From the 80 bands obtained using the ten primers, the mean similarity coefficient observed between two individuals was 0.678. When we compared this coefficient integrated over different geographical levels we found a significantly

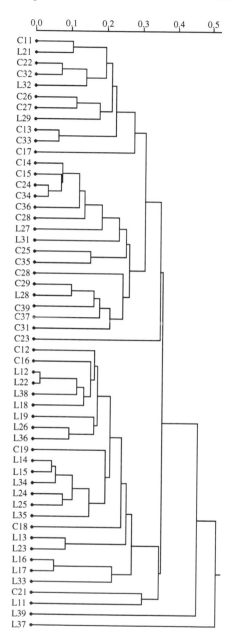

Fig. 1. Cluster (Lance & Williams - Mean association) of the 54 *S. avenae* individuals sampled. C: Ciney, L: Louvain-la-Neuve, first number: population number in the given site, second number: individual number. Scale: Jaccard coefficient of similarity.

greater similarity (m=8.23; p<0.001) between two individuals from the same intra-regional level (s=0.699) than between individuals from the two different regions (s=0.658). However, this similarity was not significantly greater (m=0.877; p>0.10), when two individuals came from the same population (s=0.704) than when they came from two different populations of the same region (s=0.697).

The largest part of the genetic variation (86.5%) was observed at the population level, only 1.5% additional variation was obtained by integrating the different populations of the same region (metapopulation). The inter-regional scale explained the residual variability (12%).

The cluster analysis of the individual similarity matrix revealed that 81.5% of the Ciney and 70.4% of the Louvain-la-Neuve individuals were correctly classified according to their region of origin (Fig. 1). Even so, within these regional units, no clear population unit, corresponding to our fields, could be observed.

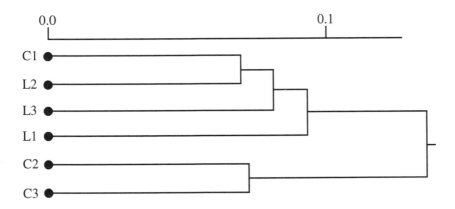

Fig. 2. Cluster (Lance & Williams - Mean association) of the 6 *S. avenae* populations studied. C: Ciney, L: Louvain-la-Neuve, number: population number in the given site. Scale: Steinhaus coefficient of similarity.

Five of the six Ciney individuals that were incorrectly classified came from the first field of this site, and four of the six incorrectly classified Louvain-la-Neuve individuals came from the second field of this site. This explained the clustering obtained with the coefficient of Steinhaus, where the second and the third population of Ciney were grouped and well separated from the four other ones (Fig. 2). The first population of Ciney was closer to any population from Louvain-la-Neuve than to the other Ciney populations.

The lack of relationship between the geographical distances and the genetical distances at the regional scale (Spearman's rank correlation coefficient = 0.51; NS) suggest that the three populations, or sub-populations, of each region were part of the same metapopulation (Fig. 3). However, we observed a significantly greater genetical distance between two populations from two different regions than from the same region (m_{obs}=7.43; p<0.001), suggesting that each region supports a distinct *S. avenae* metapopulation.

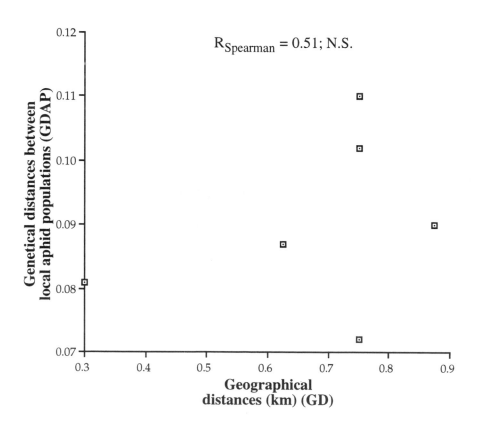

Fig. 3. Relationship between the genetical distances separating *S. avenae* populations and the corresponding geographical distances at a regional level.

Discussion

Since its development (Williams et al. 1990), the RAPD-PCR technique has been used for many purposes, like the discrimination of taxa (Demeke et al. 1992; Hance & Neuberg submitted) and strains (Edwards 1994, Edwards & Hoy 1995), or study of paternity (Fonrdk et al. 1993, Hadrys et al. 1993). In the work of De Barro et al. (1995a) as in the present one, this technique has been used to investigate the population structure of a parthenogenetic species: *S. avenae*.

In spite of the low number of individuals sampled in each field (counterbalanced by a better definition of each of them using ten primers) our results lead to conclusions close to those of similar studies: the majority of the genetic diversity was observed at a local scale.

Shufran et al. (1991) in a study on the length polymorphism of the intergenic spacer in the rRNA cistron of another aphid, *Schizaphis graminum*, observed that the majority of the genetic diversity (82.3%) was found on a single leave of sorghum or wheat. The other sampling levels (plants, fields and, in their case, counties) explained very little additional variability, respectively 0.9%, 0.6 to 3.6% and 1.2%. However, the temporal variation was higher with 11%. This factor was not tested in the present study because individuals were collected, from both sites, only in June 1994. The major difference to our results was the low additional variability observed at the regional scale (1.2% contrary to the 12% observed in our study). Thus the populations of *S. graminum* seem to be more homogeneous through a greater geographical scale than those of *S. avenae*. This may be due to the more extensive and homogeneous type of agricultural practices in Kansas than in Belgium.

Black IV et al. (1989), using the same technique (length polymorphism of the intergenic spacer in the rRNA cistron), also observed the majority of genetic variation of *Aedes albopictus* (Skuse), a sexually reproducing insect, at the population scale (65.2%). But, variation between populations in a same region (5.3%) was higher than the variation between countries (4.2%), suggesting that the mosquito populations were more isolated than aphid populations. This may be due to the dependency of mosquitos on water. Yet such an inference has to be considered with caution, since mosquito populations were sampled at different times and had been reared in the laboratory before the analysis. The rest of the variability came from a higher geographical scale, i.e. the comparison between the different countries (25.3%).

Our results also agree with those of De Barro et al. (1995a). Indeed, they used the RAPD-PCR technique on *S. avenae* sampled on wheat in two sites separated by 60 km. They showed that, at the end of the season (mid-July), that corresponds approximately to our sampling period (end of June), the genetic segregation agreed well with the geographic origin. However, they sampled only one site in each locality.

The low degree of differentiation observed between the three sites of the same localities may also be explained by the results of De Barro et al. (1995a, 1995b). They found a great similarity between the *S. avenae* colonizing wheat in different sites separated by less than 1 km to more than 80 km (De Barro et al. 1995b), but this similarity decreased during the season (De Barro et al. 1995a). Two reasons may explain this: (1) a secondary colonization from roadside grasses supporting a greater *S. avenae* genetic polymorphism (De Barro et al. 1995a) or (2) better performances of some *S. avenae* clones under local (scale < 1 km) selection pressures.

In the present work, we conclude that the three populations studied at each site, at this time of the season, were part of the same metapopulation (Blondel 1986) without any real barrier between them. The proximity of the first Ciney population to those of Louvain-la-Neuve is difficult to explain. One hypothesis is that the selection of similar clones in the two regions results from feeding on the same winter-wheat variety (Service & Lenski 1982, Hance & Noel-Lastelle unpublished data). Nevertheless, in our study, no direct relationship was shown between population genotype and winter wheat variety.

Acknowledgements

This study was supported by the National Fund for Scientific Research (FNRS, Belgium). We thank Dr. G. Boivin and two anonymous referees for their very helpful comments on previous drafts of this manuscript.

References

Black IV, W., McLain, D. & Rai, K. 1989. Patterns of variation in the rDNA cistron within and among world populations of a mosquito, *Aedes albopictus* (Skuse). *Genetics* 121: 539-50.

Black IV, W., DuTeau, N., Puterka, G., Nechols, J. & Pettorini, J. 1992. Use of the random amplified polymorphic DNA polymerase chain reaction (RAPD-PCR) to detect DNA polymorphisms in aphids (Homoptera: Aphididae). *Bull. Ent. Res.* 82: 151-59.

Blondel, J. 1986. *Biogéographie évolutive*. Masson, Paris.

Carvalho, G., Maclean, N., Wratten, S., Carter, R. & Thurston, J. 1991. Differentiation of aphid clones using DNA fingerprints from individual aphids. *Proc. R. Soc. Lond. B.* 243: 109-14.

Dagnelie, P. 1970. *Théorie et méthodes statistiques. Applications agronomiques. Volume II: les méthodes de l'inférence statistique.* J. Duculot, Gembloux.

De Barro, P., Sherratt, T., Carvalho, G., Maclean, N., Nicol, D. & Iyengar, A. 1994. An analysis of secondary spread by putative clones of *Sitobion avenae* within a Hampshire wheat field using the multilocus (GATA)4 probe. *Insect Molecular Biology* 3: 253-60.

De Barro, P.J., Sherratt, T.N., Broookes, C.P., David, O. and Maclean, N. 1995a. Spatial and temporal genetic variation in British field populations of the grain aphid, *Sitobion avenae* (F.) (Hemiptera: Aphididae) studied using RAPD-PCR. *Proc. R. Soc. Lond. B.* 262: 321-27.

De Barro, P., Sherratt, T., Carvalho, G., Maclean, N., Nicol, D. & Iyengar, A. 1995b. The use of the multilocus (GATA)4 probe to investigate geographic and microgeographic genetic differenciation in two aphid species over Southern England. *Molecular Ecology* 4: 375-82.

Demeke, T., Adams, R. & Chibbar, R. 1992. Potential taxonomic use of random amplified polymorphic DNA (RAPD): a case study in *Brassica. Theor. Appl. Genetics* 84: 990-94.

Edwards, O. 1994. Discriminating between biotypes of the walnut aphid parasite, *Trioxys pallidus* (Hymenoptera: Aphidiidae), using random amplified polymorphic DNA (RAPD) markers. PhD thesis, University of California, Berkeley.

Edwards, O. & Hoy, M. 1995. Random amplified polymorphic DNA markers to monitor laboratory selected, pesticide-resistant *Trioxys pallidus* (Hymenoptera: Aphidiidae) after release into three California walnut orchards. *Environ. Entomol.* 24: 487-96.

Fondrk, M., Page, R. & Hunt, G. 1993. Paternity analysis of worker honeybees using random amplified polymorphic DNA. *Naturwissenschaften* 80: 226-31.

Hadrys, H., Schierwater, B., Dellaporta, S., Desalle, R. & Buss, L. 1993. Determination of paternity in dragonflies by random amplified polymorphic DNA fingerprinting. *Molecular Ecology* 2: 79-87.

Landry, B., Dextraze, L. & Boivin, G. 1993. Random amplified polymorphic DNA markers for DNA fingerprinting and genetic variability assessment of minute parasitic wasp species (Hymenoptera: Mymaridae and Trichogrammatidae) used in biological programs of phytophagous insects. *Genome* 36: 580-87.

Legendre, L. & Legendre P. 1979. *Ecologie numérique. Tome 2: La structure des données écologiques.* Masson, Paris: Les Presses de l'Université du Québec.

Legendre, P. & Vaudor, A. 1991. *Le progiciel R: Analyse multidimensionnelle, analyse spatiale.* Département de sciences biologiques, Université de Montréal.

Loxdale, H., Brookes, C. & De Barro, P. 1996. Application of novel molecular markers (DNA) in agricultural entomology. In: Symondson, W. & Liddell, J. (eds.) *The Ecology of Agricultural Pests : Biochemical Approaches.* Systematics Association, U.K.: Chapman & Hall, (in press).

May, B. & Holbrook, F. 1978. Absence of genetic variability in the green peach aphid, *Myzus persicae* (Hemiptera: Aphididae). *Ann. Entomol. Soc. Amer.* 71: 809-12.

Powers, T., Jensen, S., Kindler, S., Stryker, C. & Sandall, L. 1989. Mitochondrial DNA divergence among greenbug (Homoptera: Aphididae) biotypes. *Ann. Entomol. Soc. Amer.* 82: 289-302.

Service, P. & Lenski, R. 1982. Aphid genotypes, plant genotypes, and genetic diversity: a demographic analysis of experimental data. *Evolution* 36: 1276-82.

Shufran, K., Black IV, W. & Margolies, D. 1991. DNA fingerprinting to study spatial and temporal distributions of an aphid, *Schizaphis graminum* (Homoptera: Aphididae). *Bull. Ent. Res.* 81: 303-13.

Stevens, J. & Wall, R. 1995. The use of random amplified polymorphic DNA (RAPD) analysis for studies of genetic variation in populations of the blowfly *Lucilia sericata* (Diptera: Calliphoridae) in southern England. *Bull. Ent. Res.* 85: 549-55.

Williams, J., Kubelik, A., Livak, K., Rafalski, J. & Tingey, S. 1990. DNA poymorphisms amplified by arbitrary primers are useful as genetic markers. *Nucleic Acids Research* 18: 6531-35.

Spatial association for counts of two species

Joe N. Perry

Department of Entomology & Nematology, IACR Rothamsted Experimental Station,
Harpenden, Herts. AL5 2JQ, England

Abstract

The current status of the SADIE (Spatial Analysis by Distance IndicEs) system for measuring and testing spatial pattern in data from a single species is described. Examples are given of the use of SADIE for analysis and for modelling. For ecological data in the form of counts of two species, sampled at identical specified locations, it may be possible to detect spatial association or dissociation between the two species. The problem of measuring and testing for association is discussed and a partial solution offered. Fortran software is freely available from the author.

Key words: Spatial pattern, regularity, crowding, aggregation, distance indices, scale, association, segregation, heterogeneity.

Introduction

Data for invertebrates, weeds and diseases in agriculture and ecology are gathered at a variety of levels of spatial information. At the highest level are maps, where the location of each individual is known. These are less common than counts of individuals of a certain species at a particular location. At the lowest level are summary statistics such as the sample mean and variance of a frequency distribution. For counts, previous approaches considered only the relationship between variance and mean. However, although the set of counts of cyst-nematodes in six soil cores: 0, 1, 4, 56, 484, 4095, may be highly-skewed and obviously non-Poisson, their spatial arrangement may be completely random. Conversely, a set of counts of carabid beetles in pitfall traps: 0, 0, 1, 1, 2, 2, 2, 2, 3, 3, 5, may conform closely to a Poisson distribution, but if sampled in that order along a line transect show an obvious linear trend departing strongly from randomness. To overcome such problems Perry (1995a) introduced a new method to detect and measure spatial pattern, termed Spatial Analysis by Distance IndicEs (SADIE).

Arthropod natural enemies in arable land · III The individual, the population and the community
W. Powell (ed.). *Acta Jutlandica* vol. 72:2 1997, pp. 149-169
© Aarhus University Press, Denmark, ISBN 87 7288 673 0

Consider the four different arrangements of 36 individuals in a 3x3 grid (Table 1). The observed arrangement in (b) is clearly clustered towards the top-left of the grid. By contrast, the randomly permuted counts of the observed arrangement, in (c), show, as expected, no evidence of spatial pattern. Of course, since the counts in (b) and (c) are identical, so are their sample means ($m=4$) and sample variances ($s^2=16$), so here there is no relationship at all between variance and spatial pattern. Variance does have some relationship with aggregation and with regularity, but variance imparts useful spatial information only at their extremes. For example, arrangement (a) has maximal variance, and is as crowded as is possible; arrangement (d) has minimal variance and is as regular as is possible. The SADIE approach is to use these biologically intuitive extreme arrangements as baselines against which to measure aggregation. Specifically, the degree of aggregation may be measured by calculating the minimum amount of effort, equated to distance moved, that the sampled individuals would need to expend in order to achieve an extreme pattern, such as complete crowding (a), when all individuals occur in a single sample unit, or complete regularity (d), when each unit has the same number of individuals. This enables discrimination between arrangements such as (b) and (c), that traditional approaches could not separate. The concept has also been extended to data in the form of maps (Perry 1995b). In summary, the two advantages of SADIE for counts are this improved intuitive basis, compared with traditional more abstract, mathematical approaches, and its use of all the spatial information in the sample. The purpose of this paper is to summarise progress to date in the development of methodology and software for analysis and modelling for single populations, and to extend the concept to the problem of spatial association between two populations.

Table 1. Three arrangements of 36 individuals in a 3x3 grid (artificial data).

a : Crowded	b : Observed	c : Permuted	d : Regular
36 0 0	13 7 3	3 4 1	4 4 4
0 0 0	5 4 1	1 13 5	4 4 4
0 0 0	1 1 1	1 7 1	4 4 4

Methods

Analysis of spatial pattern for maps with single species data
For mapped data, the SADIE technique requires two-dimensional coordinates of each individual to be specified, within a given rectangular area. The movement of these observed locations to a final regular arrangement is effected by means of an iterative algorithm (Perry 1995b) that relocates each point towards an

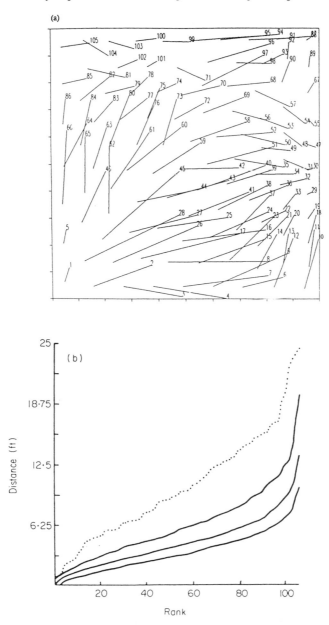

Fig. 1. Data for 105 beetle larvae (Perry 1995b), originally described by Fleming & Baker (1936). (a) an IAF plot. (b) an EDF plot; the dotted line relates to the observed data; the solid lines relate to the randomizations and are, from top, the upper 97.5th centile, the mean, and the lower 2.5th centile.

arrangement in which points occupy a triangular lattice. The algorithm operates by simple rules based on the construction of Voronoi polygons. The distance between the corresponding positions of each individual in the observed and final arrangement is noted and the sum, over all the n observed individuals gives the distance to regularity, D. A test of randomness and an index of non-randomness may be found as follows.

Firstly, n random points are generated independently of one another within the sample area. The algorithm is run as above for this set of random points, the distance to regularity computed, and this value, D_{rand}, is stored. This is repeated, usually about 100 times. The resulting set of values of D_{rand} is then ordered and a one-sided randomization test of complete spatial randomness is effected by comparison of D with D_{rand} (e.g. Perry & Hewitt 1991) against the alternative that the observed arrangement is either aggregated or regular. As we would expect intuitively, relatively large values of D result from relatively aggregated patterns and relatively small values of D from regular patterns. If the average value of D_{rand} over the randomizations is denoted as E_p, then an index of aggregation, I_p, is calculated from D / E_p. Values of $I_p > 1$ indicate aggregation, $I_p < 1$ indicates regularity and $I_p \approx 1$ suggests randomness.

It is visually useful to plot the observed location (as a numeral), linked to the final position with a straight line, for each location, in an 'initial and final' (IAF) plot because this aids the identification of clusters and of areas of relatively low density (Perry 1995b). As an example, Perry (1995b) considered a subset of Fleming & Baker's (1936) Japanese beetle larvae, for which $n = 105$, $I_p = 2.32$ and $P_p = 0.0063$ (Fig. 1a).   Another useful diagnostic plot is the empirical distribution function (EDF) plot, analogous to that of Diggle (1983), in which the individual distances shown in the IAF plot are ranked and cumulated. The observed data (dotted line) greatly exceeds the upper 97.5th centile (top bold line) of the randomizations (Fig. 1b).

Analysis of spatial pattern for counts with single species data
For count data, the SADIE technique requires only the two-dimensional coordinates of each sample unit and its associated count to be specified, but places no restriction on the arrangement of the sample units themselves. The distance to regularity, D, is found by means of the transportation algorithm from the operational research literature (Perry 1996b). The solution found by the algorithm, that finds the minimum total distance that individuals would need to move to achieve regularity, usually involves the movement of a fractional number of individuals, because complete regularity implies that each unit has a number equal to the sample mean, which is not usually itself an integer; however, this presents no problems of interpretation. As an example, for the observed and

0	0	0	1	0	0	1	1	1	0	1	0	2	0	10
0	0	0	0	1	0	2	1	7	2	0	0	1	0	0
0	0	0	3	11	3	1	0	28	16	0	0	1	31	22
0	0	0	2	9	39	24	9	24	11	1	0	0	0	16
0	0	0	0	15	3	15	12	46	7	0	1	0	0	0
0	0	0	0	7	19	18	3	2	2	0	0	0	0	5
0	0	0	0	4	1	0	1	0	0	6	0	2	3	2
0	0	0	0	13	3	3	0	2	0	0	0	0	0	0
0	0	1	0	1	0	0	0	0	0	0	0	0	0	0
0	0	0	0	0	0	1	0	0	0	0	1	0	0	0
7	0	1	0	3	0	15	4	4	3	0	0	0	0	0
0	2	0	2	8	16	23	18	5	0	5	0	6	1	0
0	2	0	2	12	0	13	6	0	1	0	32	12	42	0
0	0	0	1	5	21	27	4	25	2	0	12	9	1	0
0	1	0	1	7	22	19	15	13	0	4	0	0	0	18

Fig. 2. Counts of cereal cyst-nematodes, *Heterodera avenae*, at 7.14m spacing, collected by B. Boag (Perry 1996b).

highly-aggregated data in Table 1(b), $D = 26.26$, a relatively large value; for the randomly permuted arrangement in Table 1(c), the value of D was only 15.49. By contrast with mapped data, the randomizations for counts are made by permuting the observed counts among the sample units, and not by randomly relocating the individuals. Hence, (c) is an example of a permutation of (b) that would be used as a single randomization. This conditioning on the counts allows inferences to be made about the spatial pattern of the observed arrangement, because it allows automatically for the heterogeneity of the counts' frequency distribution. This purely statistical heterogeneity displays non-randomness of a non-spatial form; it implies the existence of spatial pattern at a smaller scale than that to which the sample unit count relates, and must therefore be ignored. As was the case for maps, D is compared with the average value from the random permutations, here denoted E_a, to yield an index of aggregation, I_a, formed from D / E_a, and a randomization test of randomness based on the probability, P_a.

The distance to crowding, C, is found by direct search over all the sample units; the permuted randomizations yield an index, J_a and probability, Q_a, in similar fashion (Perry 1996b). For example, for the arrangement in Table 1(b), $C = 32.96$ (counts move to the top-left unit, with count 13), while for that in Table 1(c), $C = 25.49$ (counts move to the central unit, with count 13).

Consider, as a more realistic example, counts of *Heterodera avenae*, the cereal cyst-nematode, estimated from soil cores by B. Boag (Perry 1996b), on a 15x15 grid (Fig. 2). It is possible to construct (Fig. 3) an equivalent plot, for counts, of the IAF plot for mapped data because the output from the transportation algorithm gives the optimal number of individuals required to move from each of the cells with initially more individuals than the sample mean, to cells with initially fewer, to achieve regularity. Such flows may be plotted, possibly, as here, excluding relatively trivial flows, to create an IAF plot (Fig. 3). There are clearly two large, equally-sized central clusters, one at the top and one at the foot of the sample area, and the plot gives a good visual indication of two further smaller clusters, one each to the right of the larger ones. The original data, for which $I_a = 1.46$ and $P_a = 0.005$, confirms this visual impression readily. Notice that units with particularly large counts are identified by the emanation of several lines. Also, since the maximum flow between any two units must be equal to the sample mean, here $m = 4.15$, minus the minimum count, and since the minimum count here, as in most animal datasets, was zero, the maximum flow was also $m = 4.15$. In Fig. 3, for clarity, only flows larger than four are shown, but there is a flow identified with each sample unit. This concept will be returned to later, in connection with the problem of association. For the time being, note that for each sample unit there is either an outflow or an inflow; that this flow has an overall value that is either positive (outflow) or negative (inflow), denoted here as f; that the individual flows, either into or out of a unit, may be summed as vectors; and that the overall vector flow for a unit has a strength, denoted here as F, and a direction, denoted here as q, where q is measured clockwise from the vertical direction (y increasing). Of course, F is always less than or equal to f, and only equal to it if all the individual flows are in exactly the same direction. As an example, consider the count of 18 in the

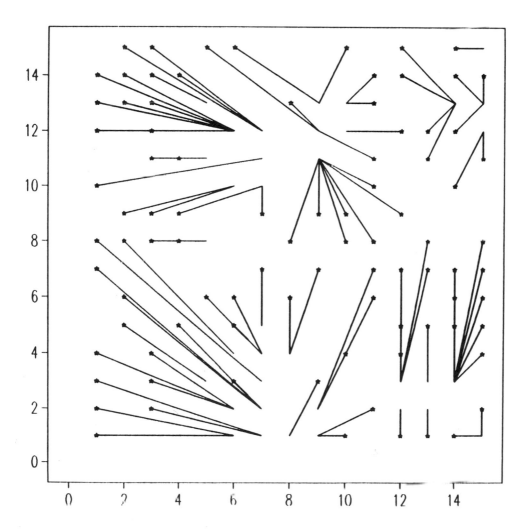

Fig. 3. An IAF plot for the nematodes of Fig. 2 (Perry 1996b), showing only those flows greater than four. Destination units are indicated by *.

extreme lower-right corner ($x=15$, $y=1$) of the grid (Fig. 2). There were four flows (only two of which are shown in Fig. 3) with values: 2.44 to ($x=14$, $y=2$); 3.12 to ($x=15$, $y=3$); 4.15 to ($x=14$, $y=1$) and 4.15 to ($x=15$, $y=2$). The total of these four flows, $f(15,1)$, is 13.65, and, of course, the remainder, after deduction of these four flows from 18, is, as required, 4.15, the sample mean. Here, the vector strength of flow is easily calculated from its four components as $F(15,1) = 10.74$, and the vector direction of the flow to be from ($x=15$, $y=1$) towards ($x=14.35$, $y=2$), i.e. in the direction of $\theta = -33.15°$.

Consider now the performance of J_a, the index based on the distance to crowding, C, for the nematode data. J_a cannot detect non-randomness with any power if the data contains more than a single cluster, because C itself is not a good measure of non-randomness for multi-clustered data. Indeed, here, $J_a = 1.03$ and $Q_a = 0.265$. When sample units form large contiguous rectangular grids, Perry (1996b) showed how changes in the values of both I_a and J_a with sub-grid size may be considered simultaneously, to yield useful information regarding cluster-size and inter-cluster distance. For the nematode data, sub-grids of size $r×r$ units yielded median values of J_a that achieved a maximum of about 1.3 between $r=5$ and $r=7$, whilst I_a stabilized around 1.5 for $r>9$. These results implied that the main contribution to the spatial pattern came from clusters with an approximate diameter of five to six units, separated by a distance of about four units. The IAF plot confirmed this conclusion. Further work by Ewen Bell, University of Leicester is currently addressing problems of interpretation of distance to crowding when there are two or more clusters, using a completely different approach outlined briefly in Perry et al. (1996).

Modelling of spatial pattern for counts with single species data
There are now many software packages that allow the simulation of discrete statistical frequency distributions, such as the Poisson, negative binomial or beta-binomial, that are used commonly to describe counts in agriculture and ecology. However, the lack of a methodology to study the spatial pattern of counts, now provided by the SADIE system, has prevented, hitherto, the development of techniques to simulate spatial arrangements of these counts with given levels of aggregation or regularity. The automatic generation of such arrangements will prove useful in simulation models and in evaluations to compare the efficiencies of different sampling plans for pest monitoring. Perry (1996a) described an algorithm to generate an arrangement of a given set of counts of a single population over predefined locations, that yields the closest possible degree of aggregation to any desired specified level. In particular, this may be used to permute a set of observed counts between sample units, to form a different arrangement, but one with a very similar degree of aggregation (defined through

the distance to regularity) to the original. Furthermore, because of the multiplicity of possible permutations, for most sets of data it is possible to find hundreds of such different arrangements. The algorithm works by selecting pairs of counts at random and exchanging them if certain criteria are met, starting from a random permutation of the counts.

In addition to aggregation, as measured by distance to regularity, an important feature of any observed arrangement concerns the degree to which the counts occupy units towards the 'edge' of the sampled area. This may be formally measured by δ, the distance from the centroid of the counts to the centroid of the sample units. Because the distance to regularity is inflated by relatively large values of δ, it is important to select the new arrangement to have a very similar value of δ to that of the observed; fortunately this restriction is relatively easy to impose. As an example, consider counts of *Ceutorhynchus assimilis,* a weevil, collected in 19 yellow-bowl water traps on 26 June 1992 in Furzefield, a field of oil seed rape at Rothamsted Experimental Station by A.K. Murchie (Perry 1996a), (Fig. 4a) for which $D = 716.3$ and $\delta = 0.45$. One realization of the algorithm to produce a permutation with a similar degree of aggregation is in Fig. 4b ($D = 716.8$; $\delta = 0.34$); note the lack of correlation between the observed and permuted sets. Such an algorithm will prove useful because, while sampling is expensive and time-consuming, it allows the generation of many sets of simulated data from each observed set, each simulated set having features very similar to the observed. Perry (1996a) gave an example in which data were generated to test competing sampling schemes to detect cyst-nematodes.

Modelling of spatial association for counts with data from two species
Another common source of data occurs when two populations are studied, with a count from both populations being available at each sample unit. The two populations may be spatially dissociated, as for an insect host in refuge from its parasitoid attacker, where the counts are negatively correlated; or they may be positively associated, as for the incidence of diseased plants and the plant's pathogen; or they may occur at random with respect to one another. Perry (1996a) described another algorithm to generate an arrangement of a given set of counts over predefined locations, at which there are known counts of a second set, with any desired level of association between the two. Such an algorithm will prove useful in simulation models that study species interactions or in the generation of data with a known structure to test competing indices and tests of association. Briefly, let K_{1i}, $i = 1, \ldots, n$, represent those counts of set one to be assigned; K_{2j}, $j = 1, \ldots, n$, those (ranked) counts from set two at known locations; and k (>0) be an association parameter with values larger than unity implying dissociation and vice-versa, and $k=1$ indicating random placement of

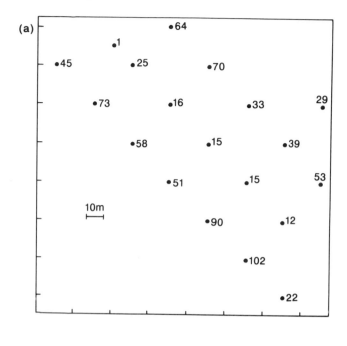

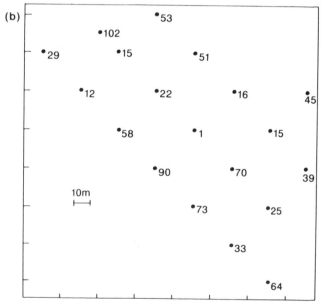

Fig. 4. *C. assimilis* counts collected by A.K. Murchie (Perry 1996a): (a) observed, (b) permutation with similar *D* value.

one set with respect to the other. The algorithm associates labels j with i, stage by stage. Suppose that $p_{r(2)}$ denotes the probability that the rth ranked count, of those labels of set two as yet unassigned, is associated with the highest ranked unassigned label of set one, and suppose that $p_{s(2)}$ denotes the probability that the sth ranked count, of those labels of set two as yet unassigned, is associated with the highest ranked unassigned label of set one. Then, at any stage, the algorithm ensures that the probability $p_{r(2)}$ is k^{r-s} times the probability $p_{s(2)}$.

The algorithm, in brief, follows. Step 1: rank K_{1i} and K_{2j}; step 2: put t equal to the number of counts yet to be assigned; step 3: draw a uniform random number on (0,1); step 4: select the minimum integer w for which $u < \sum_{v=1}^{w} k^{v-1}(k-1)/(k^t-1)$ for $k \neq 1$, or for which $u < w/t$ if $k=1$, and associate the wth as yet unassigned count of set two with the highest ranked yet unassigned count of set one. Decrease t by unity and, unless t is zero, return to step 2. For example, we may wish to simulate rearrangements of A.K. Murchie's $n=19$ counts of the parasitoid *Trichomalus perfectus* (set one) (Perry 1996a): $\{0, 1^2, 2^3, 3, 4, 7, 8^2, 10^2, 12, 14, 22, 24, 32, 35\}$, caught in the same traps as its weevil host *C. assimilis* (set two, Fig. 4a), to demonstrate different degrees of association. The algorithm generated rearrangements with the desired degrees of association for the three values of k chosen: 0.2 (strong association), 1.0 (random pairing), and 2.0 (moderate dissociation) (Table 2). The algorithm is non-parametric, and non-spatial in the sense that it does not utilise the spatial information concerning the location of the counts of set two.

Table 2. Three simulated arrangements of nineteen *T. perfectus* counts showing varying degrees of association with observed counts of *C. assimilis,* in the same units (Perry, 1996a).

X-coordinate of trap	77	75	69	75	73	71	81	73	67	79	77	81	71	79	73	75	77	79	70
Y-coordinate of trap	44	46	52	54	56	50	48	48	54	50	52	52	54	42	52	50	48	46	55
C. assimilis	102	90	73	70	64	58	53	51	45	39	33	29	25	22	16	15	15	12	1
T.perfectus (k=0.2)	32	35	24	12	22	14	10	8	4	10	8	7	3	2	1	2	2	1	0
T.perfectus (k=1.0)	3	2	14	2	7	1	12	8	0	24	32	35	22	10	8	2	4	10	1
T.perfectus (k=2.0)	1	2	2	0	1	3	2	8	10	8	4	10	7	22	32	14	12	24	35

Analysis of spatial association for counts with data from two species
One of several extensions planned for the SADIE system is to develop a test and index for spatial association between two sets of counts that share the same locations. The definition of association is not easy, because it may not be possible to consider it in isolation from the spatial pattern of its two component species. For example, the presence of a parasitoid population may clearly affect the spatial pattern of its host, which will attempt to relocate in refuges. This

escape response by the host may then cause an alteration in the spatial pattern of the parasitoid, which strives to seek out the new locations of the host aggregations. This process is dynamic, and may result in a ceaseless shifting of positions of both species. Then again, there may be absolutely no direct effect of one species on another, yet both may demonstrate marked pattern in response to some third species, or environmental component that causes a degree of spatial association. Without further information these two cases, of direct and indirect association, are indistinguishable in the data as defined. Fortunately, the fact that those dynamic interactions that are the root cause of association or dissociation cannot be disentangled does not affect the process of testing, for the null hypothesis is one of lack of any association or dissociation, i.e. that the location of either one species are random with respect to those of the other. However, if this null hypothesis is correct, and there are neither direct nor indirect effects and no association of any kind between the species, both species may still exhibit strong spatial pattern, possibly in response to a set of other species or environmental effects that are themselves entirely independent and differ between the two species. These individual patterns may in any event be described and tested in each species in isolation from the other, and this may be done before the problem of spatial association is addressed. It seems sensible therefore to eliminate their effects, which may otherwise confuse the issue, by assessing association conditional upon the observed spatial pattern, with a similar rationale to that of conditioning on the observed counts which was described for the randomization tests above.

The following example lends support to this idea. Consider the two sets of artificial counts of two species in a 3x3 grid (Table 3), where the numbers in bold represent counts of species 1 and those in italic counts of species 2. At first sight, the spatial association appears very strongly negative. However, suppose we now condition on the spatial pattern exhibited by both species. This, for both species,

Table 3. Sets of aggregated counts of species 1 (bold) and species 2 (italics) in a 3x3 grid (artificial data).

	Column 1		Column 2		Column 3	
Row 1	**9**	*1*	**8**	*3*	**5**	*4*
Row 2	**7**	*2*	**6**	*6*	**2**	*7*
Row 3	**4**	*5*	**3**	*8*	**1**	*9*

is as aggregated as is possible. Indeed, for species 1, the only other arrangements with aggregation (measured through the value of *D*) comparable to that observed would be the three simple rotations of the observed arrangement that would relocate the high-density counts in the top-right, bottom-right and bottom-left corners, respectively. By a similar argument, there are only four possible arrangements with such extreme aggregation for species 2, including that observed. Hence, if we condition on the observed aggregation in the arrangements of both species 1 and species 2, there are only a total of (4 x 4 =) 16 combinations possible, in exactly four of which the degree of association is maximal, and in exactly four of which it is minimal (as in the observed arrangement). With this constrained randomization, the probability of obtaining this degree of dissociation by chance, on the null hypothesis that there was really none acting between the species, is four in sixteen (P = 0.25), and does not now appear at all extraordinary. Hence we would conclude that, given the observed aggregation of each of the two species in isolation, the observed arrangement of them together shows some evidence of dissociation, but not significant evidence (P = 0.25). This concept, of performing the randomization tests after conditioning on the observed arrangements was suggested also by Harkness & Isham (1983). Although they had no means to find such constrained arrangements; they too considered some form of rotation as the basis for randomization. They sought some ideal function that would measure the association between the nest locations of two ant species: "Ideally this function for the observed nest processes would be compared with the same function calculated for two independent processes having the same distributions as the *Cataglyphis* and *Messor* nests. Since we do not know what these marginal distributions are ... An alternative possibility would be to perform toroidal translations of the *Cataglyphis* data ... with the *Messor* data".

For maps, the above concept of a rotation provides an easy way to generate a randomization that preserves the degree of aggregation of a spatial pattern identically and preserves also other important properties such as that of d, the distance of the centroid from the edge, described above. Consider, for example, some artificial data in which species 1, represented by filled circles, and species 2, represented by open circles, are observed in a circular area (Fig. 5a). To effect a randomization we could first hold all the individuals of species 2 in their observed positions, then rotate the observed positions of species 1 clockwise about the centre of the circular area, by a random angle, α, where $0° < \alpha < 360°$. Fig. 5b shows the randomized pattern resulting for a value of α of approximately 35°. A further source of valid randomizations, that preserves the qualities given above, and that may be used in concert with rotations, is a reflection of the points in a randomly chosen diameter. Following the generation of a set of such randomizations of species 1, the positions of species 1 could be held instead, and species 2 rotated by random degrees. This yields two sets of randomizations and it is important to realise that the inferences pertaining to each may be distinct. For example, it is perfectly possible for every individual of species 1 to have an individual of species 2 as its nearest neighbour, but for most nearest neighbours of species 2 to be individuals of their own species.

For counts, such arguments are more problematic, as Harkness & Isham (1983) found. However, although it is not possible in general to use the method of rotation and reflection to provide randomizations, this is fortunately unnecessary, because the first algorithm described above can easily provide random rearrangements of the counts of a species with the desired degree of aggregation and value of d to match the observed set.

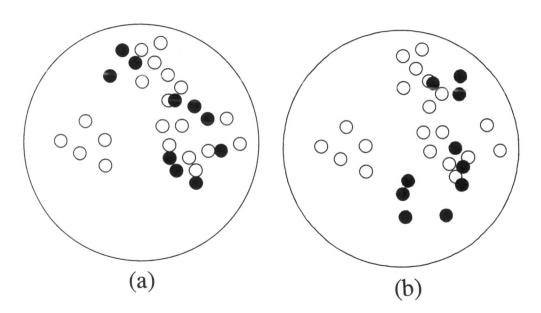

Fig. 5. (a) Circular area with mapped, artificial observed data, for species 1 (filled circles) and species 2 (open circles). (b) Species 2 in observed positions; species 1 randomized to new positions using a random rotation about the centre of the circle.

After appropriate randomizations have been generated that preserve D and δ, some index of association is required, to compare the observed association with those randomizations. Previously, the correlation coefficient (usually calculated after transformation of both sets of counts to $\log_{10}(n+c)$ scales) has been used often in this role. However, the use of $c = 1$ is arbitrary, and the fact

Table 4. Two sets of counts, (a) and (b), of species 1 (bold) and species 2 (italics) in a 5x5 grid (artificial data), demonstrating different degrees of association. Only non-zero counts are shown.

	(a)					(b)				
Row 1			**3**	*2*	*4*				*2*	*4*
Row 2				**4**	**2**					
Row 3			**1** *1*		*3*	**3**		**1** *1*		*3*
Row 4							**4**			
Row 5							**2**			
	Col.1	Col.2	Col.3	Col.4	Col.5	Col.1	Col.2	Col.3	Col.4	Col.5

that the correlation coefficient uses only the information from two parallel lists of counts, and cannot make use of information concerning their locations, is a serious drawback. As an example, consider the two sets of artificial counts of two species in a 5x5 grid (Table 4), where a blank entry denotes a zero count (Perry et al. 1996). Once again, species 1 is denoted by bold counts and species 2 by counts in italics. In both sets, the same counts are used for species 1 and for species 2, and for species 1 the counts are in identical positions. In both sets, there is a single coincidence (row 3, column 3) of a non-zero entry for species 1 with a non-zero entry for species 2; hence the correlation coefficient (-0.0826, $\log_{10}(n+1)$ scale) is identical for both sets, and non-significant. Furthermore, the degree of aggregation for species 2 is identical for both sets, because their arrangements are identical save for a rotation about the central cell and a reflection in the diagonal. Sets (a) and (b) may therefore be fairly compared and have many aspects of spatial pattern in common. However, visually, the species counts in set (a) clearly appear associated, and those in set (b) highly dissociated. Although the parallel lists of species counts are identical in the two sets, the proximity of counts of different species that are not coincident gives the strong impression of non-random association. Hence, methods are required that allow for this spatial information, from nearby but not necessarily coincident units, to be utilised fully.

One possible method is now described: Consider the two sets of artificial counts of two species in a 3x4 grid (Table 5), where the counts of each species comprise the integers 1 to 12 occurring once only. Set (a) displays strong positive, and set (b) strong negative association. Sets (c) and (d) display the totals for each sample unit over the two species. (Had the two species not had identical sample means they would have required scaling to achieve this). The positively associated counts of set (a) reinforce one another in the totals for set(c), which range widely, from 3 to 23, and any aggregation in the individual species' patterns is enhanced. By contrast, the negatively associated counts of set compensate for each other to produce a very even set of totals for set (d),

Table 5. Two sets of counts, (a) and (b), of species 1 (bold) and species 2 (italics) in a 3x4 grid (artificial data). (c) shows the total count (medium type) over the two species for each unit in set (a); (d) shows the totals for set (b).

[a]				[b]			
Row 1	**6** *6*	**12** *11*	**11** *12*	**9** *10*	**6** *5*	**12** *1*	**11** *3*	**9** *7*
Row 2	**7** *5*	**8** *9*	**2** *3*	**5** *8*	**7** *4*	**8** *6*	**2** *10*	**5** *8*
Row 3	**1** *2*	**3** *1*	**4** *4*	**10** *7*	**1** *12*	**3** *11*	**4** *9*	**10** *2*
	Col.1	Col.2	Col.3	Col.4	Col.1	Col.2	Col.3	Col.4

[c]				[d]			
Row 1	12	23	23	19	11	13	14	16
Row 2	12	17	5	13	11	14	12	13
Row 3	3	4	8	17	13	14	13	12
	Col.1	Col.2	Col.3	Col.4	Col.1	Col.2	Col.3	Col.4

covering a much narrower range, from 11 to 16, and disguising any aggregation present in the pattern of either species. These features may be isolated quantitatively by treating the totals as a single species count and analysing their distance to regularity, as follows. The distance to regularity of the observed totals, say T, is computed and stored. Permutations of the counts for species 2 are found with very similar values of D and δ to those observed, and, for each permutation, k, the total is formed of this set and of the original counts of species 1, as above, say $T_{k(1)2}$. A randomization test is performed, as described above for the single species case, from the observed T and the frequency distribution of the values of $T_{k(1)2}$. Furthermore, an index of association, $I_{t(1)2}$, may be formed from $T / E_{t(1)2}$, where $E_{t(1)2}$ represents the average value of $T_{k(1)2}$ over the randomizations. Values of $I_{t(1)2} > 1$ indicate positive association, values < 1 negative association (dissociation), and values close to unity indicate random placement of one species with regard to the other. The procedure is then repeated, with sets one and two reversed, for the observed total T and the randomized values $T_{k(2)1}$. In most cases both the test and index give very similar values for the two sets of randomizations (1)2 and (2)1, and they may then be combined to give an average probability, P_t, and index, I_t. For example, for the 25 counts of Table 4, the value of D for the totals of set (a) was 36.6 while for (b) it was 20.8. For set (a) the value of I_t was 1.28 ($P_t = 0.145$); for set (b) $I_t = 0.74$ ($P_t = 0.847$). In each case the value of the index is correctly indicating the type of association, but the test is not powerful enough to yield a significant result. A similar randomization test, based on the observed correlation coefficient, performed much worse, as expected, giving probabilities of 0.463 and 0.426 for sets (a) and (b) respectively.

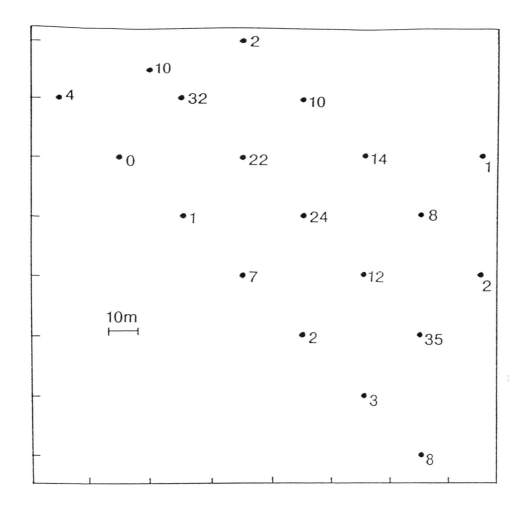

Fig. 6. *T. perfectus* counts collected by A.K. Murchie (Perry 1996a), simultaneously with, and in the same traps as the *C. assimilis* counts in Fig. 4(a).

As a further example, but now using real data, the next section considers the association between A.K. Murchie's counts of the host weevil *C. assimilis* given above (Fig. 4a), and counts of its parasitoid *Trichomalus perfectus*, collected simultaneously in the same traps (Fig. 6).

Results & Discussion

Whereas *C. assimilis* was relatively sparse in the six traps along the centre of the field and dense in the two diagonal rows of edge traps, the reverse was true of *T. perfectus* (Figs 4a and 6). This dissociation was shown by the highly significant value of both the simple correlation coefficient ($r = -0.57$) and Spearman's rank correlation coefficient ($S = -0.71$). Since the mean density for *C. assimilis* was about four times that of *T. perfectus*, the counts were first scaled to have the same totals. The strong negative association was detected at a significant level by the above method of totals ($I_t = 0.707$, $P_t > 0.995$).

The method of totals given above appears for some sets of data to have relatively low power to detect borderline cases of association. An alternative test and index are being developed, that use information concerning the relative size of the counts from the two sets in coincident units, but that attempt additionally to utilise the spatial information in those coincident units. Consider, for example, how the IAF plots would appear for the data analyzed above, in Fig. 4a and in Fig. 6. For *C. assimilis*, the vector direction of the flows would be from the edge rows towards the centre row, whereas for *T. perfectus* the reverse would be true. Considering each unit individually, in the majority of the units the vector direction of the flows for the two sets would be diametrically opposed. The same is true for the other negatively associated sets in Tables 3 and 5b. By contrast, for the positively associated counts in Fig. 5a, the flows from individual units would be very similar in strength *f*, vector strength *F*, and in vector direction θ, for the two species. These ideas may be

Table 6. Flow statistics for three sample units from the sets of counts of *C. assimilis* and *T. perfectus*, shown in Figs 4a and 6. The vector flow directions, θ, are shown in Fig. 7.

X-coordinate of trap	67	70	73
Y-coordinate of trap	54	55	56
C. assimilis: count	45	1	64
T. perfectus: count	4	10	2
C. assimilis: scaled *f*	8274	-156418	79391
T. perfectus: scaled *f*	-98373	-5691	-129267
C. assimilis: scaled *F*	8274	-99342	77296
T. perfectus: scaled *F*	-98373	-5691	-129267
C. assimilis: $\theta°$	71.6	6.0	-123.3
T. perfectus: $\theta°$	-90.0	-45.0	45.0
θ	106647	150727	208658
ψ	161.6	51.0	168.3

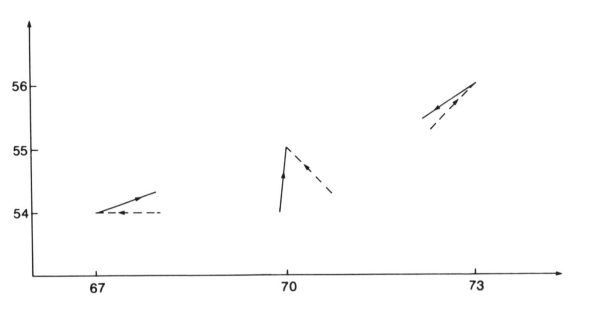

Fig. 7. The vector directions of flow computed from the full IAF plots for a small subset of three sample units: (67,54), (70,55) and (73,56) for *C. assimilis* (solid lines, observed data shown in Fig. 4a) and *T. perfectus* (dashed lines, observed data in Fig. 6).

quantified, for each sample unit, into several statistics, two of which are denoted ϕ and ψ. The first, ϕ, is the sum of the values of f, for each sample unit, of the two species multiplied together, after suitable scaling. The second, ψ, is defined through the sum for each sample unit of $\omega = |\theta_{C.\ assimilis} - \theta_{T.\ perfectus}|$; $\psi = \omega$ if $\omega < 180°$, but $\psi = |360 - \omega|$ if $\omega > 180°$ (note that $0° < \psi < 180°$), again after suitable scaling for each unit, this time by the product of F for each species. Relatively small values of both ϕ and ψ are expected for spatially associated data, and relatively large values of both for dissociated data such as these *C. assimilis* and *T. perfectus* counts.

As an example, the vector directions, θ, are plotted (Fig. 7) for a small subset of the full IAF plots for *C. assimilis* and *T. perfectus* comprising the topmost three sample units, for which the statistics: f, F, θ, and unscaled values of ϕ and ψ are also computed (Table 6). The statistics, in general, are large and behave as expected, the only discrepancy being caused by the slightly small count of 10 for *T. perfectus*.

Further work is required to define precisely how these statistics may best be combined, firstly over the sample units and then with each other, into a single index and test. One possibility being investigated currently is an index derived from suitably scaled plots of $\sum_{j=1}^{n} \phi_j$ versus $\sum_{j=1}^{n} \psi_j$.

Some further work is also required in the current SADIE software for counts, to allow for edge effects on the index I_a. Since it is relatively easy to generate randomizations with specified values of δ, it is planned to generate a set of such randomizations constrained to have the same value of δ as that observed, and from them to construct an index K_a in exactly the same way as the unconstrained index I_a is constructed currently. Inferences based on K_a would relate to counts that are identical in size and have a similar overall position with regard to the edge of the sample area to those observed, but which occupy positions that are otherwise spatially random. A substantial difference between I_a and K_a would imply the presence of significant edge effects.

Thus far, versions of the SADIE programmes have been distributed as Fortran code, developed on mainframe computers. Over the coming months it is hoped to develop versions for distribution as executable files, suitable for running on PCs.

Acknowledgements

I am very grateful to B. Boag and A. Murchie for permission to use their data and for the latter's patience during my overlong development of ideas to measure association. IACR Rothamsted receives grant-aided support from BBSRC. I also thank Niels Holst for his idea to use directional flow statistics from IAF plots as a basis for an index of association.

References

Diggle, P.J. 1983. *Statistical analysis of spatial point patterns*. Academic Press, London.
Fleming, W.E. & Baker, F.E. 1936. A method for estimating populations of larvae of the Japanese beetle in the field. *J. Agric. Res.* 53: 319-31.
Harkness, R.D. & Isham, V. 1983. A bivariate spatial point pattern of ants' nests. *Appl. Statistics* 32: 293-303.

Perry, J.N. 1995a. Spatial aspects of animal and plant distribution in patchy farmland habitats. In: D.M. Glen, M.P. Greaves & H.M. Anderson (eds.) *Ecology and Integrated Farming Systems*, pp. 221-42. Wiley, Chichester.

Perry, J.N. 1995b. Spatial analysis by distance indices. *J. Anim. Ecol.* 64: 303-14.

Perry, J.N. 1996a. Simulating spatial patterns of counts in agriculture and ecology. *Computers and Electronics in Agriculture* 15:93-109.

Perry, J.N. 1996b. Measuring the spatial pattern of animal counts with indices of crowding and regularity. *Ecol. Monographs* (submitted).

Perry, J.N, Bell, E.D., Smith, R.H. & Woiwod, I.P. 1996. SADIE: software to measure and model spatial pattern. *Asp. Appl. Biol.* 46: 95-102.

Perry, J.N. & Hewitt, M. 1991. A new index of aggregation for animal counts. *Biometrics* 47: 1505-18.

The use of single-species spatially explicit models

C. J. Topping[1], K. Booij[2], R. A. Daamen[2], A. Dinter[3], U. Heimbach[4],
P. J. Kennedy[5], V. Langer[6], J. N. Perry[5], W. Powell[5], D. Skirvin[7], D.
Stilmant[8], C. F. G. Thomas[9] L. Winstone[9]

[1] Department of Landscape Ecology, NERI, Kalø, Grenåvej 14, DK-8410, Rønde,
Denmark.
[2] Research Institute for Plant Protection, Binnenhaven 12, P.O. Box 9060, 6700 GW
Wageningen, The Netherlands.
[3] Institut für Pflanzenkrankheiten und Pflanzenschutz
der Universität Hannover, Herrenhäuserstr. 2, 30419 Hannover, Germany
[4] Biolog. Bundesantalt für Land- und Forstwirtschaft, Messeweg 11/12, D-38104
Braunschweig, Germany.
[5] IACR-Rothamsted, Dept. Entomology & Nematology, Harpenden, Herts. AL5 2JQ,
UK.
[6] Royal Veterinary & Agricultural University, Department of Agricultural Sciences,
Agrovej 10, DK-2630 Taastrup, Denmark.
[7] Horticulture Research International, Wellesbourne, Warwick CV35 9EF, UK.
[8] Unité d'Ecologie et de Biogéographie, Université Catholigue de Louvain 4-5, Place
Croix du Sud, B-1348 Louvain-la-Neuve, Belgium.
[9] IACR-Long Ashton Research Station, Long Ashton, Bristol, BS18 9AF, UK.

Abstract
A review of modelling approaches involving spatial components, as applied to single
species populations, is presented. Reasons for modelling populations and
considerations pertinent to the choice of model type are discussed. Further sections
deal with the advantages and disadvantages of individual as opposed to population
based approaches, species choice and data collection requirements,
parameterisation and model validation and analysis. Finally, areas for future
research are outlined.

Key words: spatially explicit models, population dynamics, landscape structure,
spatial scale, dispersal, simulation model.

Arthropod natural enemies in arable land · III The individual, the population and the community
W. Powell (ed.). *Acta Jutlandica* vol. 72:2 1997, pp. 171-191
© Aarhus University Press, Denmark, ISBN 87 7288 673 0

Introduction

This paper presents the results from a discussion session of the 3rd EU-Workshop on beneficial predators and parasitoids. The discussion was restricted to spatial modelling of, primarily, single species (i.e. excluding community and predator/prey models), to try to reduce the scope of the material covered. However, the potential for discussion of the remaining areas is vast and so this paper can only present an overview of some of the main approaches, pitfalls and considerations when using spatial modelling. This paper deliberately ranges far outside the boundaries of models of beneficial arthropods in order to present as full a description of available approaches as possible. Many of the techniques mentioned may have rarely, if ever, been used for beneficials, but this does not mean that they are not relevant, and so equal weight is given to all types of spatial modelling approaches.

Why model populations?

There are many reasons for producing models as a framework for scientific research. Primarily models are constructed to conceptualise hypotheses concerning the relationships between organisms and their environment. Such models have been used to analyse patterns with the aim of determining mathematical analogues for the underlying processes (e.g. Taylor's power law (Taylor 1961, 1965, 1971)) or to simplify systems to make general statements concerning their behaviour under various conditions (e.g. Levin 1970). For instance, Jones (1977) and Jones et al. (1980) used a spatially explicit simulation model of *Pieris rapae* as a tool to predict and explain distributions of this pest species. Other models may simply be used to describe and simplify complex systems (e.g. Crowley et al. 1987) or can be used to explore systems which are either too large, too complicated or too precious to be manipulated experimentally (e.g. Linden-Mayer & Lacy 1995). Models may also be employed to study long-term impacts of factors such as the use of pesticides (e.g. Jepson & Sherratt 1991, Sherratt & Jepson 1993), the effects of other human management factors (Constanza et al. 1990), climate (Woiwod & Harrington 1994), or changes in habitat management (Ellis & Seal 1995). Models may be constructed for education (e.g. Lewis 1995) or extension purposes (Liu et al. 1995, Royle et al. 1988). Recently, there has been an upsurge in interest in the response of organisms to spatial heterogeneity in their landscapes. Many types of ecological questions, such as reserve design, landscape planning, habitat fragmentation, and understanding biodiversity, all ultimately require an appreciation of the mechanisms behind spatial distribution patterns of organisms (Wiens et al. 1993). The need to understand the causes of spatial distributions of beneficial organisms is particularly acute since by understanding what controls the fluctuations in population density, it may be possible to use the information to manage those populations. In order to understand how to bring about such changes it may be necessary to develop an analogy to the real world (a model), study its properties, and then use the results to direct management decisions.

Models can be made which consider ecosystems, communities, populations or individuals. Models of the population dynamics of more than one species are generally very complex and may only produce robust predictions when the environment of the organisms is strictly controlled, as in glasshouse conditions (e.g. Nachman 1991). Even less detailed modelling of the interactions between two or more closely linked species in a spatially and temporally heterogeneous and complex environment is a very difficult undertaking, though modelling a single species in such an environment can be a productive exercise. In pest control situations, single species modelling may be considered as unrealistic, or not useful, due to the perceived need to have the dynamics of the pest closely linked to the natural enemy and vice versa. However, there are a number of good reasons for modelling the dynamics of single species populations without closely addressing their impact on the dynamics of other species. Firstly, the species may have some significance unrelated to its effect on another species, for example it may represent the ecological characteristics of functional or taxonomic groups, or because it is deemed worthy of conservation (e.g. Thomas & Harrison 1992, Thomas & Jones 1993, Lavers & Haines-Young 1996). Alternatively, it may illustrate a process, such as the influence of predation on the control of a pest population (Petersen & Holst 1997, Winder et al. 1997). A third reason for single species modelling is that the population dynamics of other species may not be closely associated with the model organism. This is probably the case for the majority of polyphagous predators for which it is likely that the overall level of food availability is important rather than the fluctuations in abundance of any one particular prey item.

The use of spatial models for studying population fluctuations has increased recently, largely following a revival of the metapopulation ideas of Levins (1969, 1970) (see Gilpin & Hanski 1991) and an increase in the awareness of scale as a factor in ecology (Wiens 1989). Published papers which explicitly address questions of scale have steadily increased since 1985

(Schneider 1994). A number of theoretical ideas have emerged, including metapopulation theory. However, the extent to which these new paradigms accurately describe the relationships between organisms may be questioned. For instance, Taylor (1991) investigated claims in the literature that predator-prey systems were persisting as a result of metapopulation dynamics, but found little evidence to support this hypothesis. Care must therefore be exercised in the choice of model.

Types of model

Choice of the type of model to use could be dependent upon the level of simplification achievable whilst maintaining enough realism to answer the questions which are posed to the model system. There are many advantages in keeping models as simple and as easy to understand as possible. Simpler models are easier to explain, parameterise and test than complicated ones. However, a model that is simplified to the point where it no longer conforms to the key features of the dynamics of the system it represents is useless. No one type of model could be recommended for all cases, therefore this section does not aim to guide the choice of model, but to describe the range of modelling possibilities which should be considered, from the simple to the complex.

One question which can be posed is whether the chosen model needs to consider space explicitly. Occupancy models (Levins 1970, Caswell 1978, Crowley 1979, Hanski 1983) make the simplifying assumption that there is no spatial correlation in the state (occupied or not) of habitat patches (Hanski 1991). A test of this assumption can therefore be used to determine whether models of this type might be suitable. If spatially correlated extinctions are likely to occur, for example, as a result of mortality caused by simultaneous harvest of crops in an arable system, then spatial aspects will need to be modelled explicitly. Similarly, if colonisation rates are spatially correlated, for example, as a result of differential landscape permeability, then occupancy models may not be suitable. Occupancy models also make a second simple assumption; that all patches are either occupied or not and are all of the same quality. In more advanced, structured metapopulation models (e.g. Hastings and Wolin 1989, Hanski and Gyllenberg 1993), variation in sub-population structure and size is taken into account. In some of these (Perry & Gonzalez-Andujar 1993, Gonzalez-Andujar & Perry 1995), not only are the subpopulations expressed quantitatively within explicit space, but a stochastic compartmentalisation allows for local extinctions at the end of each generation. However, variation in population size is also the result of the internal dynamics of the population and may not be determined solely by the environmental structure, and so this is also probably unrealistic for the majority of cases. If this factor is considered important, then it may be incorporated by the inclusion of sub-models for the internal dynamics of patches (e.g. Drechsler & Wissel 1994, 1995), but there is a risk of losing analytical tractability.

Generally, there is great variation between patches in size and quality, often exemplified by a mainland population and smaller satellite populations. These cases are often more closely described using models derived from the equilibrium theory of island biogeography (MacArthur & Wilson 1967) (e.g. Gilpin & Diamond 1976, 1981, Fritz 1979, Thomas 1984) or by a source-sink model (see Pulliam 1988). Ruxton (1995) also points out that models which neglect spatial heterogeneity may be unrealistic in their basic assumptions concerning species' behaviour. If populations are thought to be spatially heterogeneous then space may have to be considered explicitly in the model, but it is acknowledged that where this is the case the model is no-longer easy to study analytically (Hanski & Gilpin 1991).

If the model must incorporate space explicitly, then the next decision to be taken is the scale on which the model will operate (i.e. the resolution at which the model will operate). This will be determined by both the aim of the model and the spatial dynamics of the organism. Where the aim of the model is to look at large scale processes (e.g. the incidence of gypsy moth outbreaks in the USA, (Zhou & Liebhold 1995)) then the scale must be large in relation to the home-range of the organism. Models requiring answers to smaller scale questions will require a higher degree of resolution (such as two arable fields in rotation (C.J.H. Booij personal communication), or a single ecotone (Vermeulen & Opsteeg 1993ab)). A further consideration in determining the scale of the model will be the inherent properties of the patterns of abundance which are being modelled. For instance, marine birds may be associated with prey distribution at spatial scales of five kilometers or more, but not with prey distribution at smaller scales (Piatt 1990).

Models operating on scales much larger than the spatial dynamics of the organisms can employ a number of approaches. The cellular automata approach to spatial modelling (Langton 1990, Phipps 1992) allows species with low dispersal compared to the resolution of the model to be represented using relatively simple calculations. The limitation of these models is that the influence of spatial arrangement is only exerted on immediate neighbours in a raster representation of landscape; although this limitation can be relaxed somewhat as demonstrated by Zhou & Liebhold (1995) in their modelling of Gypsy Moth outbreaks. A different approach is needed when the dispersal distances of the organism to be modelled are much greater than the resolution of the model. In such cases, immigration from outside the system should be linked to movement within the system being modelled (e.g. Topping & Sunderland 1994). When using these approaches, it is important to know enough about how the organisms move and how this relates to the landscape structure. For instance, Gross et al. (1995) found that even with animals that can move over large distances the actual pattern of movement may be dependent on factors which operate at very local spatial scales. The movement of syrphids, which migrate over large distances, but respond to the local presence of nectar and larval food sources, is an example.

In evaluating modelling approaches in landscape ecology, Sklar & Constanza (1990) considered that the current landscape models could be divided into two, almost mutually exclusive groups. These are the stochastic models and the process-based models. Stochastic models look for ways to combine spatial structure, patterns and information with probabilistic distributions, whilst the process-based models aim to simulate the system using as much realism as the current computers allow. The stochastic models rely on transition probabilities (e.g. Turner 1987, 1988). There are, however, problems with the use of such models in the simulation of landscape patterns as the landscape changes are not strictly Markovian[1], because they are not constant through time, and they may be determined by economic rather than natural factors (Burnham 1973, Alig 1986). These models could be linked to changes in factors such as climate and to rules concerning the effect of vegetation dynamics which would provide a much more realistic transition matrix, but in doing so these models would tend to progressively overlap with the process-based approach (Sklar & Constanza 1990). Hence, probabilistic models are largely used in a descriptive way and need further development if they are to become useful predictive tools.

Process-based approaches describe the processes within compartmentalised landscape units and the processes that spatially link these compartments. These models tend to have high demands on computer resources but have advantages over other types of models in that they are flexible, spatially explicit, realistic and dynamic (Sklar & Constanza 1990, Constanza et al. 1990).

---

[1] Markovian — in order to be Markovian a change in state of one habitat patch should be dependent only on the existing state and not how that state was arrived at.

Fahrig (1990) also argues for process-based approaches in preference to analytical models because of the limitations of analytical methods for the development of general hypotheses in a landscape context. She also disparages the highly-complex process-based simulations which cannot be simplified to produce general hypotheses because of the fear of 'oversimplification'. Fahrig (1990) concludes that an approach between these two extremes will be most profitable for the generation of underlying principles and processes, the results of which should be amenable to standard experimental statistical approaches. An important justification for adopting this approach is given by Dunning et al. (1993) who state that large-scale spatially-explicit population models are important aids in bridging the gap between ecology and other sciences, because the scale necessitates the need to incorporate economic and geographic variables not usually considered by ecologists. This may be a particularly important point when considering beneficials, which by definition must be affected by agricultural practices and the many political and economic factors linked to farming.

A population or individual approach?

A further decision to be taken in the type of model selected is what to use as the basic organism unit of the model. Metz & Diekmann (1986) distinguished between p-states and i-states in models which consider structured populations: p-state models represent states characterising the whole population, i-state models represent the states of individual organisms. There can be an alternative subdivision into i-state configuration, and i-state distribution models (Metz & Diekmann 1986). The i-state configuration model follows individuals as separate entities whilst the i-state distribution model uses continuous functions which correspond to the i-state configuration. Note that a p-state model could result from looking at either the i-state configuration or distribution approach from a population perspective. On a conceptual level it may be easier to understand models based on individuals rather than on populations or cohorts (Botsford 1992). However, individual-based models are very demanding on computer resources and it may be possible to achieve the same outcome using traditional i-state distribution models such as those used by Halley et al. (1996), Topping (this volume), Thomas (this volume). Whether this is a justification for the use of the i-state distribution model is debatable. Huston et al. (1988) made the point that such models violate two basic biological principles i.e. that there is variation between organisms in both characteristics and in position. The incorporation of this variation may be very significant in describing the overall population dynamics (e.g. DeAngelis et al. 1979, Huston & DeAngelis 1987, Crowder et al. 1992).

DeAngelis & Rose (1992) pointed out that distribution models have been used successfully for several decades to give insights into biological problems associated with age structure (e.g. Leslie Matrix Models, delay-differential equations (Nisbet & Gurney 1982)), and that configuration-based models have also been used successfully to investigate certain types of problems e.g. forest succession (Shugart 1984). DeAngelis & Rose (1992) also compared the advantages and disadvantages of each approach for a variety of situations and recommended that the *i*-state configuration model be used for small populations subject to large variation in environmental stochasticity and encounters with other individuals, and that *i*-state distribution models are best suited to large populations where all individuals are relatively similar.

Largely following Murdoch et al. (1992) and Botsford (1992), the arguments for and against an individual based approach can be summarised as:
For: 1) Individual models can greatly increase our understanding of systems because they incorporate a description of the actual mechanisms which determine the distribution of characteristics within and between; 2) They are conceptually simple to understand (although they may be highly complex in detail); 3) They may be testable both at the level of the individual mechanisms and against the results of the whole simulation; 4) Their realism may make some problems tractable that are otherwise analytically difficult.
Against: 1) The models can be so complex that we lose understanding of how a particular process or mechanism affects the overall outcome of the model; 2) The complexity of the model may make it untestable because almost any scenario might be accommodated by an adjustment of the functions or parameters; 3) The complexity and detail incorporated may mean that the data requirements of the model exceed the data supply. This will also result in a lack of testability. Wennergren et al. (1995) use this fact to criticise this modelling approach for endangered species, although when working with beneficials, a lack of individuals from which to gather data should not normally be a problem; 4) Even with the present increase in computing power, ambitious models may quickly become too large; 5) Such detailed models may result in complex simulations which are incapable of producing general conclusions which might advance ecological theory.

However, there is an intrinsic conflict between developing general theory and modelling specific cases, and it is possible that general concepts could be obtained from a collection of the latter.

Which species to use ?

The choice of species to model can be dictated by a number of criteria. These may include:

1) The study species has potential economic importance such as being harmful (pests) or beneficial (e.g. *Adalia bipunctata)* or has a high conservation value (e.g. bumblebees) in addition to an agricultural value.

2) It is common over a large geographical range. Many species common in agroecosystems have very large geographical ranges (e.g. *Lepthyphantes tenuis* (Sunderland 1996), *Pterostichus melanarius* (Thiele 1977), the skylark (*Alauda arvensis*) (Cramp 1988)), with the advantage that scientists from many countries can work on the same species.

3) The species chosen may have a life-history which is representative of a particularly important group. This is, at least in part, the reasoning behind the choice of IOBC pesticide testing species. Other examples are the use of particular species of carabid or spider to denote generalist predators.

4) The species may have specific interactions which could elucidate the spatial phenomena behind predator / prey and parasitoid / host interactions (van der Werf 1995).

5) A recent suggestion is that most advances are usually achieved by the reductionist approach of a designated study group which carefully studies a single species in a well structured and model-guided approach (e.g. Holst & Ruggle 1997).

6) A pragmatic reason for choosing a study species might be that we know a lot about it already and therefore the amount of extra data-gathering required would be minimal.

7) A second practical reason for the species choice might be that data collection is especially easy for that particular species. This is almost certainly the reason why papers based on pitfall trapping of carabids are so numerous in the ecological literature on beneficial arthropods.

Parameterisation

Having decided upon a modelling approach and subject species, it would be rare to find that all the required information for a proposed model is in existence in the literature. Even for comparatively well studied groups such as Carabidae, with a database containing over 3,000 references, there are considerable gaps in knowledge. There is also the problem that different authors consider different species or present data from different geographical areas, making the use of information from a wide range of published sources very difficult (Skirvin, Perry & Harrington 1997). Usually, therefore, data must be gathered before the model can be constructed. Ideally, the model will be considered before a programme of data collection is implemented. A close interactive process between modellers and data collectors within a single project is essential. Data collection will depend upon the type and complexity of the model, which in turn will depend on the aims of the model. However, some general points should be mentioned.

Primarily, any spatially explicit population model must include measures on movement or dispersal. This can be an extremely demanding requirement. Simply measuring movement rates can be very difficult for most species of beneficials. Methods such as radio tracking (Wallin & Ekbom 1988), mark, release and recapture (Thomas 1995, Mader et al. 1990, Lys & Nentwig 1991), inclusion trapping (Topping et al, 1992) and interception trapping (Duelli & Obrist 1993) have all been used with varying success. For some species there is no simple way to measure dispersal (e.g. Topping & Sunderland 1994). In such cases it may be possible to obtain dispersal data from genetic approaches such as gel electrophoresis (Reh & Seitz 1990) or RAPDs (Stilmant et al. 1997, A'Hara et al. in prep). Even if a movement or dispersal rate can be obtained, the way in which this might vary, dependent upon landscape structure, must be considered. For instance, Sawyer & Haynes (1985, 1986) found that a simple diffusion model could not simulate the movement patterns of *Oulema melanopus* (L.) because the beetles altered their behaviour according to spatial heterogeneity in their local environment. Linear features such as hedgerows and roads can provide either barriers to the movement of beneficials (Mader 1984, Mader et al. 1990) or possibly movement corridors (Maelfait & De Keer 1990), although the evidence for the utilisation of such corridors is weak (Dawson 1994a,b, Fry 1994). Dispersal rates across 'hostile' habitats may also need to be assessed if the aim of the study is to investigate the effects of habitat fragmentation. This has been shown to be of particular importance in a number of butterfly species (e.g. Wood & Samways 1991) and may also be relevant to dispersing predators. For instance, the movement of syrphids can be inhibited by tall hedges and roads (MacLeod 1994) and it has been suggested that the number of predatory species in orchards may be due to the vegetational diversity and structure around the orchard (Szentkiralyi & Kozar 1991).

A further complication of dispersal is that it is often difficult to identify the cause of the phenomenon. Proximal and ultimate causes of dispersal should be distinguished. Proximal causes of dispersal may be defined as the escape from adverse conditions, or the abiotic factors that trigger dispersal movement (e.g. Duffey 1963, 1979, Weyman et al. 1994, van Huizen 1979); ultimate causes of dispersal may be defined as those linked to long-term population survival in relation to the habitat (e.g. den Boer 1990). If the aim of the model is to simulate the population accurately then both proximal and ultimate causes of dispersal must be known. If the aim is a more general, less detailed model then it may not be necessary to fully attribute all proximal causes. However, the argument used by Huston et al. (1988) for an individual approach to ecological modelling because of the possibility of positive feedback in biological systems, may also be relevant when considering the level of detail to incorporate. Such considerations are especially relevant where the detailed local factors may significantly alter survival or reproductive success in space, leading to similar positive feedback situations.

The behaviour of the study organisms may also complicate the parameterisation of dispersal. Even if the theoretical dispersal capability of an animal is known this does not necessarily mean that the animal will travel that far. In studies on isolated orchards, the introduction of beneficials was required even though sources of colonists were within the expected dispersal range of the predators (Gruys 1982). This type of phenomonen could be caused by the behavioural component of active dispersal. For instance, the immigration of green lacewing adults into fields has been shown to be more dependent upon wind direction than habitat quality (Duelli 1980). Passive dispersal, as may occur in spiders (Thomas 1997, Topping 1997), is a relatively simple process to model; active dispersal (e.g. coccinellids, syrphids & butterflies) requires an understanding of the choices made by the dispersing animal and is, therefore, a much more complex and 'data hungry' process, e.g. Jones (1977) requires 17 parameters for his *Pieris rapae* model.

When modelling well studied species, such as many Carabidae, the modelling of all life-stages should be considered. In many Carabidae very little is known about the larval ecology, and sampling for larvae is an enormous logistic problem because of low larval densities per volume of soil.

Model validation and analysis

Apart from serving to structure knowledge and to describe processes, models should preferably have a predictive value for real and new conditions of the system, especially where those conditions are contained within the parameter space for which estimates based upon existing data are precise. If prediction involves extrapolation beyond the parameter space, then no model can give confident results. In order to have confidence in these predictions, testing of models or model-units against reality is necessary. This can be done either by a detailed comparison of predictions and actual data or by comparisons of patterns with large scale trends of output when changing major input parameters. With long-term studies, model improvement is often an iterative process of modifying inputs or even model structure as new knowledge becomes available.

Sensitivity analysis, where the response of model output to changes in input parameters is determined, is particularly important where the model is used analytically to look at a (sub)process and to evaluate the importance of particular parameters. It is often used to test the robustness and stability of the model in relation to our imperfect knowledge about input parameters. This is important, as in many ecological systems parameters are not fixed but may vary in response to other elements of the system, often outside the scope of the study. Determining the extent to which variation in such parameters does not significantly affect model output is, therefore, often necessary. In some cases uncertainty analysis can be used. This is a special case of sensitivity analysis which can be applied in two different ways depending upon the complexity of the model. For a relatively simple model, the contribution of known or estimated error of input parameters to the variance in model output is evaluated (Jansen et al. 1994). For more complex models, Monte Carlo techniques can be used by choosing parameter values using a stochastic selection method and running a very large number of simulations.

Model validation can mean many things to different modellers. It can be restricted to a comparison of predicted and actual results or may be considered valid if they simply successfully formalise the concepts and hypotheses behind the model. Law & Kelton (1982) provide a three step approach to validation:

1) **The 'face validity'**, i.e. does the model appear reasonable – this is established most usually by publication in scientific journals or other peer reviewed activities.

2) **Testing model assumptions** – model assumptions should always be explicitly defined and tested fairly.

3) **Evaluation of model predictions** – a fundamental principle is that the data used for validation were not used in the construction of the model.

Validation may not be done for many reasons, including a subconscious fear of the model failing. The excuse that models are mostly used for data structuring and not for predicting reality is often too easily used. Perhaps the stimulating part may have been the development of the model and so often there is little further interest in validation. Also, the validation may be meaningless because, out of necessity, the input parameters may be inaccurate or merely guessed, so that errors cannot be estimated.

Even when validation is desired and feasible, the costs of the process may be extremely high. Building models is usually much less labour intensive than collecting the data. Research budgets are under pressure and are usually offered for open competition, and funding may not often be granted for validation because it is not seen as innovative research. The result can be that competing scientists succeed in winning contracts at the expense of quality – possibly at the expense of proper validation. A compounding factor is that when model production is driven by political considerations and not by scientific ones, validation is not generally a priority.

Future research areas

One factor which is certain to affect ecological modelling in the future is the development of new computer hardware. Recent years have seen an enormous increase in processing power available to the scientist in the form of fast PCs and the development of parallel processing. Parallel processing may be of particular interest since as Palmer (1992) points out, "In nature, events are concurrent and the influence of one event upon another propagates through yet other events.". Parallel processing may therefore provide the means by which ecological modellers can grapple with the concept of concurrency and must be a step towards greater realism in ecological simulation models. There have also been coincident advances in programming techniques. Of particular interest to the ecological modeller is the use of object-orientated techniques (see Zeilger 1990) and their application to a wide range of ecological problems (e.g. Folse et al. 1989, Holst & Ruggle 1997). This approach provides the basis for great flexibility in the design and maintenance of models, a framework for concurrent models, and is intuitively easy for the non-programmer to conceptualise.

A fruitful area of work will probably be the development of methods for integrating information over different spatial scales. This may be particularly important in the development of models aimed at understanding constraints for, and functions of, biodiversity where the integration of a wide range of species with different life-history strategies will probably be required. One approach worth considering is the use of hierarchical methods (Salthe 1985, O'Neill et al.

1989, de Vasconcelos et al. 1993), which are based on the fact that landscapes and ecological systems are often scaled in space and time. This method models landscape-scale responses to events by considering their effect on lower level entities which are then integrated into higher levels using a hierarchy. Using these methods it may be possible to integrate the effects of landscape changes across a wide range of species differing in scale of study, life-history, function and the extent to which two or more may influence each other (e.g. rabbit grazing altering the floral composition which in turn alters the suitability of the landscape for other organisms such as bumblebees).

One feature which is common to all models based on ecological data is that the success of the model will depend on the quality of data used. For complex simulations (e.g. Crowley et al. 1987, Holst & Ruggle 1997), the amount of data required is enormous. This puts pressure on already stretched resources for autecology work. Taylor (1991) calls for the use 'of powerful and focused methodology' (in particular, experimentation) to directly describe movement rates and patterns rather than the use of crude observational data (e.g. extinction) to make inference about movements, but this may only solve part of the problem. What is needed is the development of a more theoretical approach which can be used to understand the important links between mechanisms and patterns. One such approach suggested by Wiens et al (1993) would be the judicious practice of reductionism, selecting for study those species which occupy key positions in gradients of life-history strategy, distribution, size etc.. Comparisons between such carefully chosen study species may provide powerful tools to develop generalisations and may provide a vehicle for the utilisation of much existing autecology data. However, resources directed towards selected species will still have to be considerable and carefully utilised where data are lacking.

A further consideration when looking at the development of spatial models is the political basis for their inception. At present spatial models, particularly those on a landscape scale, are often the result of a political need to answer questions about 'biodiversity'. The idea is that the use of a model will aid planning and implementation of management decisions and will allow the case for biodiversity to be understood. However, there may be little understanding of what biodiversity means. Politicians don't know, and scientists know enough to know that they can't say what is important! Therefore, it is up to scientists to tackle the problem of indicator species, and functional and general biodiversity, in order to create a widely accepted basis for measuring and describing the concept. In this way perhaps the spatial modelling effort can be targeted optimally.

Acknowledgements

The first author was supported by a grant from the Danish Strategic Environmental Research Programme within the Centre for Agricultural Biodiversity.

References

Alig, R. J. 1986. Econometric analysis of the factors influencing forest acreage trends in the southeast. *Forest Science* 32: 119-34.

A'Hara S., Harling, R., McKinlay, R. & Topping, C. J. submitted. Protocols for the storage, extraction and conditions for RAPD profiling of spider (Araneae) DNA. *J. Arachnol.*

Botsford, L. W. 1992. Individual state structure in population models. In: DeAngelis, D. L. & Gross, L. J. (eds.) *Individual models and approaches in ecology: populations communities and ecosystems*, pp. 213-36. Chapman & Hall, New York.

Burnham, B. O. 1973. Markov intertemporal land use simulation model. *Southern J. Agric. Econ.* 5: 253-58.

Caswell, H. 1978. Predator-mediated coexistence: a nonequilibrium model. *Amer. Nat.* 122: 127-154.

Constanza, R., Sklar, F. H. & White, M. L. 1990. Modelling coastal landscape dynamics: Process-based dynamic spatial ecosystem simulation can examine long-term natural changes and human impacts. *BioScience* 40: 91-107.

Cramp, S. (ed.) 1988. *Handbook of the birds of Europe, the Middle East and North Africa. The birds of the Western Palaearctic.* Vol. V. Oxford University Press, Oxford and NewYork.

Crowder, L. B., Rice, J. A., Miller, T. J. & Marschall, E. A. 1992. Empirical and theoretical approaches to size-based interactions and recruitment variability in fishes. In: DeAngelis, D. L. & Gross, L. J. (eds.) *Individual models and approaches in ecology: populations communities and ecosystems*, pp. 237-55. Chapman & Hall, New York.

Crowley, P. H. 1978. Predator-mediated coexistence: an equilibrium interpretation. *J. Theor. Biol.* 80: 129-44.

Crowley, P. H., Nisbet, R. M., Gurney, W. S. C. & Lawton, J. H. 1987. Population regulation in animals with complex life-histories: Formulation and analysis of a damselfly model. *Adv. Ecol. Res.* 17: 1-59.

Dawson, D. G. 1994a. Are habitat corridors conduits for animals and plants in a fragmented landscape? A review of the scientific evidence. *English Nature, Research Report 94,* Peterborough.

Dawson, D. G. 1994b. Narrow is the way. In: Duver, J. W. (ed.) *Fragmentation in Agricultural Landscapes. Proceedings of the third IALE(UK) conference.* 30-37.

DeAngelis, D. L. & Rose, K. A. 1992. Which individual based approach is most appropriate for a given problem. In: DeAngelis, D. L. & Gross, L. J. (eds.) *Individual models and approaches in ecology: populations communities and ecosystems*, pp. 167-87. Chapman & Hall, New York.

DeAngelis, D. L. Cox, D. C. & Coutant, C. C. 1979. Cannibalism and size dispersal in young-of-the-year largemouth bass: experiments and model. *Ecol. Modelling* 8: 133-48.

Den Boer, P. J. 1990. The survival value of dispersal in terrestrial arthropods. *Biol. Conserv.* 54: 175-92.

Drechsler, M. & Wissel, C. 1994. Ein stochastisches Modell für Metapopulationen – Analyse eines stochastischen Modells. *Verh. Ges. Ökol.* 23: 295-302.

Drechsler, M. & Wissel, C. 1995. Management-Hinweise für Metapopulationen – Analyse eines stochastischen Modells *Verh. Ges. Ökol.* 24:111-15.

Duelli, P. 1980. Adaptive dispersal and appetitive flight in the green lacewing *Chrysopa carnea. Ecol. Entomol.* 5: 213-20.

Duelli, P. & Obrist M. 1995. Comparing surface activity and flight of predatory arthropods in a 5 km transect.In: Toft, S. & Riedel, W. (eds.) *Arthropod natural enemies in arable land I. Acta Jutlandica* 70 (2): 283-93.

Duffey, E. 1963. A mass dispersal of spiders. *Transactions of the Norfolk and Norwich Nature Society* 20: 38-43.

Duffey, E. 1979. Aerial dispersal by linyphiid spiders from filter beds. *Br. arachnol. Soc. Secretaries Newsletter* 26: 3-4.

Dunning Jr, J. B., Stewart, D. J., Danielson, B. J., Noon, B. R., Root, T. L., Lamberson, R. H., & Stevens, E. E. 1993. Spatially explicit population models: current forms and future issues. *Ecol. Applications* 5: 3-11.

Ellis, S., & Seal, U. S. 1995. Tools of the trade to aid decision-making for species survival. *Biodiversity and Conservation* 4: 553-72.

Fahrig, L. 1990. Simulation methods for developing general landscape-level hypotheses of single species dynamics. In: Turner, M.G. & Gardner, R.H. (eds.) *Quantitative methods in landscape ecology. Ecological studies vol. 82.* pp. 417-42. Springer-Verlag, Berlin.

Folse, L. J., Packard, J. M. & Grant, W. E. 1989. AI Modelling of animal movements in a heterogenous habitat. *Ecol. Modelling* 46: 57-72.

Fritz, R. S., 1979. Consequences of insular population structure: distribution and extinction of spruce grouse populations. *Oecologia (Berl.)* 42: 57-65.

Fry, G. L. A. 1994. The role of field margins in the landscape. *Field Margins: integrating agriculture and conservation BCPC Monograph No. 58* pp.31-42.

Gilpin, M. & Hanski, I. (eds.) 1991. *Metapopulation Dynamics: Empirical and Theoretical Investigations.* London: Academic Press.

Gilpin, M. & Diamond, J. M., 1976. Calculation of immigration and extinction curves from the species-area-distance relation. *Proc. Nat. Acad. Sci., USA* 73: 4130-34.

Gilpin, M. & Diamond, J. M., 1981. Immigration and extinction probabilities for individual species: relation to incidence functions and species colonisation curves. *Proc. Nat. Acad. Sci., USA* 78: 392-96.

Gonzalez-Andujar, J.L. & Perry, J. N. 1995. Predictions of the control of the seedbank of *Avena sterilis*: the effect of spatial and temporal heterogeneity and of dispersal. *J. Appl. Ecol.* 32: 578-87.

Gross, J. E., Zank, C., Thompson Hobbs, N. & Spalinger, D. E. 1995. Movement rules for herbivores in spatially heterogeneous environments: responses to small scale pattern. *Landscape Ecol.* 10: 209-17.

Gruys, P: 1982. Hits and Misses. The ecological approach to pest control in orchards. *Entomol. Exp. Appl.* 31: 70-87.

Halley, J. M., Thomas, C. F. G. & Jepson, P. C. 1996. A model for the spatial dynamics of linyphiid spiders in farmland. *J. Appl. Ecol.* 33: 471-92.

Hanski, I., 1991. Single-species metapopulation dynamics: concepts, models and observations. *Biol. J. Linnean Soc.* 42: 17-38.

Hanski, I. 1983. Coexistence of competitors in patchy environments. *Ecology* 64: 493-600.

Hanski, I. & Gilpin, M., 1991. Metapopulation dynamics: brief history and conceptual domain. *Biol. J. Linnean* Soc. 42: 3-16.

Hanski, I. & Gyllenberg, M. 1993. Two general metapopulation models and the core-satellite species hypothesis. *Amer. Nat.* 142: 17-41.

Hastings, A. & Wolin, C. L., 1989. Within patch dynamics in a metapopulation. *Ecology* 70: 1261-66.

Holst, N. & Ruggle, P. 1997. Modelling the natural control of aphids. I. The metabolic pool model, winter wheat and cereal aphids. In: Powell, W. (ed.) *Arthropod natural enemies in arable land III. Acta Jutlandica* 72 (2): 195-206 (this volume).

Huston, M., DeAngelis D. L., & Post, W. 1988. New computer models unify ecological theory. *BioScience* 38: 682-91.

Huston, M., DeAngelis D. L. 1987. Size bimodality in monospecific plant populations: a critical review of potential mechanisms. *Amer. Nat.* 130: 168-98.

Jansen, M. J. W., Rossing W. A. H. & Daamen R. A. 1994. Monte Carlo estimation of uncertainty contributions from several independent multivariate sources. In: Grasman J. & Van Straaten, G. (eds.) *Predictability and nonlinear modelling in natural sciences and economics*, pp: 334-343. Kluwer, Dordrecht.

Jepson, P.C. & Sherratt, T.N. 1991. Predicting the long-term impact of pesticides on predatory invertebrates. *Proc. Brighton Crop Prot. Conf. - Weeds - 1991*, 991-19.

Jones, R. E. 1977. Movement patterns and egg distribution in cabbage butterflies. *J. Anim. Ecol.* 46: 195-212.

Jones, R. E., Gilbert, N., Guppy, M., Nealis, V. 1980. Long distance movement of *Pieris rapae. J. Anim. Ecol.* 49: 629-42.

Langton, C. G. 1990. Computations at the edge of Chaos: Phase transitions and emergent computation. *Physica D* 42: 12-37.

Lavers, C. P. & Haines-Young, R. H. 1996. Using models of bird abundance to predict the impact of current land-use and conservation policies in the flow country of Caithness and Sutherland, Northern Scotland. *Biol. Conserv.* 75: 71-77.

Law, A. M. & Kelton, W. D. 1982. *Simulation modelling and Analysis*. McGraw-Hill, New York.

Levins, R., 1969. Some demographic and genetic consequences of environmental heterogeneity for biological control. *Bull. Entomol. Soc. Amer.* 15: 237-40.

Levins, R., 1970. Extinction. In: Gesternhaber, M. (ed.) *Some mathematical problems in biology,* pp. 77-107. Providence, R. I., American Mathematical Society.

Lewis, D. M., 1995. Importance of GIS to community-based management of wildlife: lessons from Zambia. *Ecol. Applications* 5: 861-71.

Linden-Mayer, D. B. & Lacy, R. C. 1995. A simulation study of the impacts of population subdivision on the mountain brushtail possum *Tichosurus caninus* Ogilby (Phalangeridae: Marsupialia) in south-eastern Australia. I. Demographic stability and population persistance. *Biol. Conserv.* 73: 119-29.

Liu, J, Dunning, J. B. Jr. & Pulliam, H. R. 1995. Potential effects of a forest management plan on Bachman's Sparrows (*Aimophila aestivalis*): Linking a spatially explicit model with GIS. *Conserv. Biol.* 9: 62-75.

Lys, J-A. & Nentwig, W. 1991. Surface activity of carabid beetles inhabiting cereal fields. Seasonal phenology and the influence of farming operations on five abundant species. *Pedobiologia* 35: 129-38

MacLeod, A. 1994. Provision of plant resources for beneficial arthropods in arable ecosystems. PhD thesis, University of Southampton.

Mader, H. J. 1984. Effects of increased spatial heterogeneity on the biocenosis in rural landscapes. *Ecol. Bulletin* 39: 169-79.

Mader, H. J., Schell, C. & Kornacker P. 1990. Linear barriers to arthropod movements in the landscape. *Biol. Conserv.* 54: 209-22.

Maelfait, J-P. & De Keer, R. 1990. The border zone of an intensively grazed pasture as a corridor for spiders Araneae. *Biol. Conserv.* 54: 223-38.

MacArthur, R. M. & Wilson, E. O. 1967. *The theory of island biogeography.* Princeton University Press, Princeton.

Metz, J. A. J. & Diekmann, O. (eds.) 1986. *The dynamics of physiologically structured populations. Lecture notes in biomathematics 68.* Springer-Verlag, Berlin.

Murdoch, W. W., McCauley, E., Nisbet, R. M., Gurney, W. S. C. & de Roos, A. M. 1992. Individual-based models: combining testability and generality. In: DeAngelis, D. L. & Gross, L. J. (eds.) *Individual models and approaches in ecology: populations communities and ecosystems*, pp. 18-35. Chapman & Hall, New York.

Nachman, G. 1991. An acarine predator-prey metapopulation inhabiting greenhouse cucumbers. *Biol. J. Linnean Soc.* 42: 285-303.

Nisbet, R. M. & Gurney, W. S. C. 1982. *Modelling fluctuating populations.* Wiley, New York.

O'Neill, R. V., Johnson, A. R. & King, A. W. 1989. A hierarchical framework for the analysis of scale. *Landscape Ecol.* 3: 193-205.

Palmer, J. B. 1992. Hierarchical and concurrent individual based modelling. In: DeAngelis, D. L. & Gross, L. J. (eds.) *Individual models and approaches in ecology: populations communities and ecosystems*, pp. 188-207. Chapman & Hall, New York.

Perry, J. N. & Gonzalez-Andujar, J. L. 1993. Dispersal in a metapopulation neighbourhood model of an annual plant with a seedbank. *J. Ecol.* 81: 453-63.

Piatt, J. F. 1990. The aggregative responses of Common Murres and Atlantic Puffins to schools of capelin. *Studies in Avian Biology*, 14: 36-51.

Petersen, M. & Holst, N. 1997. Modelling the natural control of aphids II: The carabid *Bembidion lampros.* In: Powell, W. (ed) *Arthropod natural enemies in arable land III. Acta Jutlandica* 72 (2): 207-19 (this volume).

Phipps, M. J. 1992. From local to global: the lesson of cellular automata. In: DeAngelis, D. L. & Gross, L. J. (eds.) *Individual models and approaches in ecology: populations communities and ecosystems*, pp. 165-187. Chapman & Hall, New York.

Pulliam, R., 1988. Sources, sinks and population regulation. *Amer. Nat.* 132: 652-61.

Reh, W. & Seitz, A. 1990. The influence of land use on the genetic structure of populations of the common frog *Rana temporaria. Biol. Conserv.* 54: 239-49.

Royle, D.J., Rabbinge, R., & Fluckiger, C.R. (eds). 1988. Pest and disease models in forcasting, crop loss appraisel and decision-supported crop protection systems. *Bull. IOBC/WPRS* 11:48-62.

Ruxton, G.D. 1995. Foraging in flocks: non-spatial models may neglect important costs. *Ecol. Modelling* 82: 277-85.

Salthe, S. N. 1985. *Evolving hierarchical systems. Their structure and representation.* Columbia University Press, New York.

Sawyer, A. J. & Haynes, D. L. 1985. Simulating the spatiotemporal dynamics of the cereal leaf beetle in a regional crop system. *Ecol. Modelling* 30: 83-104.

Sawyer, A. J. & Haynes, D. L. 1986. Cereal leaf beetle spatial dynamics: simulations with a random diffusion model. *Ecol. Modelling* 33: 89-99.

Schneider, D. C. 1994. *Quantitative Ecology.* Academic Press Inc., San Diego, California.

Sherratt, T. N. & Jepson, P. J. 1993. A metapopulation approach to modelling the long-term impact of pesticides on invertebrates. *J. Appl. Ecol.* 30: 696-705.

Shugart, H. H. 1984. *Theory of forest dynamics: the ecological implications of forest succession models.* Springer-Verlag, New York.

Skirvin, D. J., Perry, J. N. & Harrington, R. 1997. A model describing the population dynamics of *Sitobion avenae* and *Coccinella septempunctata. Ecol. Modelling* (in press).

Sklar, F. H. & Costanza, R. 1990. The development of dynamic spatial models for landscape ecology: a review and prognosis. In: Turner, M.G. & Gardner, R.H. (eds.) *Quantitative methods in landscape ecology. Ecological studies vol. 82.* pp. 239-88. Springer-Verlag.

Stilmant, D., Hance, Th. & Noël-Lastelle, Ch. 1997. Discrimination of Belgian *Sitobion avenae* (F) populations by means of RAPD-PCR. In: Powell, W. (ed.) *Arthropod natural enemies in arable land III. Acta Jutlandica* 72 (2): 139-48 (this volume).

Sunderland, K.D. 1996. Studies on the population ecology of the spider *Lepthyphantes tenuis* (Araneae: Linyphiidae) in cereals. *Bull. IOBC/WPRS,* 19: 53-69.

Szentkiralyi, F. & Kozar, F. 1991. How many species are there in apple pest communities?: testing the resource diversity and intermediate disturbance hypothesis. *Ecol. Entomol.* 16: 491-503.

Taylor, A. D. 1991. Studying metapopulation effects in predator prey systems. *Biol. J. Linnean Soc.* 42: 305-23.

Taylor, L. R. 1961. Aggregation, variance and the mean. *Nature* 189: 732-35.

Taylor, L. R. 1965. A natural law for the spatial disposition of insects. *Proceedings of the XII International Congress of Entomology* pp.396-97.

Taylor, L. R. 1971. Aggregation as a species characteristic. In: Patil, G. P., Pielou, E. C. & Walters, W.E. (eds.) *Statistical Ecology* 1: 357-77.

Thiele H.-U, 1977. *Carabid Beetles in Their Environments.* Springer Verlag. Berlin.

Thomas, C. D. & Jones, T. M. 1993. Partial recovery of a skipper butterfly (*Hesperia comma*) from population refuges: lessons for conservation in a fragmented landscape. *J. Anim. Ecol.* 62: 472-81.

Thomas, C. D. & S. Harrison. 1993. Spatial dynamics of a patchily distributed butterfly species. *J. Anim. Ecol.* 61: 437-46.

Thomas, C.F.G. 1995. A rapid method for handling and marking carabids in the field. In: Toft, S. & Riedel, W. (eds.) *Arthropod natural enemies in arable land I. Acta Jutlandica* 70 (2): 57-59.

Thomas, C. F. G. 1997. Modelling dispersive spider populations in farmland. In: Powell, W. (ed.) *Arthropod natural enemies in arable land III. Acta Jutlandica* 72 (2): 79-85 (this volume).

Thomas, J.A., 1984. The conservation of butterflies in temperate countries: past efforts and lessons for the future. In: Vane-Wright, R.I. & Ackery, R.R. (eds.) *The biology of butterflies,* pp.333-54. Academic Press, London.

Topping, C. J. 1997. The construction of a simulation model of the population dynamics of *Lepthyphantes tenuis* (Araneae: Linyphiidae) in an agroecosystem. In: Powell, W. (ed.) *Arthropod natural enemies in arable land III. Acta Jutlandica* 72 (2): 65-77 (this volume).

Topping, C. J. & Sunderland K. D. 1994. A spatial population dynamics model for *Lepthyphantes tenuis* (Araneae: Linyphiidae) with some simulations of the spatial and temporal effects of farming operations and land-use. *Agric. Ecosystems Environ.* 48: 203-17.

Topping, C. J. & Sunderland, K. D. 1995. Methods for monitoring aerial dispersal by spiders. In: Toft, S. & Riedel, W. (eds.) *Arthropod natural enemies in arable land I. Acta Jutlandica* 70 (2): 245-56.

Topping, C. J., Sunderland, K. D. & Bewsey, J. 1992. A large improved rotary trap for sampling aerial arthropods. *Ann. Appl. Biol.* 121: 707-14.

Turner, M. G. 1987. Spatial simulation of landscape changes in Georgia: a comparison of 3 transition models. *Landscape Ecol.* 1: 29-36.

Turner, M. G. 1988. A spatial simulation model of land use changes in a piedmont county Georgia. *Applied mathematical Computation* 27: 39-51.

de Vasconcelos, M. J. P., Zeigler B. P. & Graham, L. A. 1993. Modeling multi-scale spatial ecological processes under discrete event system paradigm. *Landscape Ecol.* 8: 273-86.

Van Huizen, T. H. P. 1979. Individual and environmetal factors determining flight in carabid beetles. *Misc. Papers Landbouwhogeschool Wageningen* 18:199-211.

Van der Werf, W. 1995. How do immigration rates affect predator/prey interactions in field crops? Predictions from simple models and an example involving the spread of aphid-borne viruses in sugar beet. In: Toft, S. & Riedel, W. (eds.) *Arthropod natural enemies in arable land I. Acta Jutlandica* 70 (2): 295-312.

Vermeulen, R. & Opsteeg, T. 1993a. Movements of some carabid beetles in road-side verges. Dispersal in a simulation programme. In: Desender, K., Dufrêne, M., Loureau, M., Luff, M. L. and Maelfait, J-P (eds.) *Carabid beetles - Ecology and Evolution,* pp.393-98. Kluwer Academic Press, Dordrecht, The Netherlands.

Vermeulen, R. & Opsteeg, T. 1993b. Simulation of Carabid beetle movements. *Proceedings of the Second CONNECT Workshop on Landscape Ecology, 1993,* pp.67-69.

Wallin, H & Ekbom, B. 1988. Movement of carabid beetles inhabiting cereal fields: a field tracing study. *Oecologia (Berl.)* 77: 39-43

Weyman, G. S., Sunderland, K. D. & Fenlon, J. S. 1994. The effects of food deprivation on aeronautic dispersal behaviour (ballooning) in *Erigone* spp. spiders. *Entomol. Exp. Appl.* 73: 121-26.

Wiens, J. A. 1989. Spatial scaling in ecology. *Functional Ecology* 3: 385-97.

Wiens, J. A., Crawford, C. S. & Gosz, J. R. 1985. Boundary dynamics: a conceptual framework for studying landscape ecosystems. *Oikos* 45: 421-27.

Winder, L., Wratten, S.D. & Carter, N. 1997. Spatial heterogeneity and predator searching behaviour - can carabids detect patches of their aphid prey. In: Powell, W. (ed.) *Arthropod natural enemies in arable land III. Acta Jutlandica* 72 (2): 47-62 (this volume).

Woiwood, I. P. & Harrington, R. 1994. Flying in the face of change: the Rothamsted Insect Survey. In: Arleigh, R. & Johnson, A.C. (eds) *Long-term Experiments in Agriculture and Ecological Sciences*, pp.321-42. CAB International, Wallingford.

Wood, P. A. & Samways, M. J. 1991. Landscape element pattern and continuity of butterfly flight paths in an ecologically landscaped botanic garden, Natal, South Africa. *Biol. Conserv.* 58: 149-66.

Zeigler, B. P. 1990. *Object-oriented simulation with hierarchical modular models. Intelligent agents and endomorphic systems.* Academic Press, New York.

Zhou, G. & Liebhold, A. M. 1995. Forecasting the spatial dynamics of gypsy moth outbreaks using cellular transition models. *Landscape Ecol.* 10: 177-89.

THE COMMUNITY

Modelling natural control of cereal aphids. I. The metabolic pool model, winter wheat, and cereal aphids.

Niels Holst[1] & Patrick Ruggle[2]

[1]Department of Population Biology, Zoological Institute, University of Copenhagen
Universitetsparken 15, DK-2100 Copenhagen
[2]Zoological Institute, University of Basel, Rheinsprung 9, CH-4051 Basel

Abstract

The impact of natural enemies on a herbivorous pest cannot be studied without considering the interactions between the pest and its host plant. We investigated the population dynamics of cereal aphids (*Rhopalosiphum padi*, *Sitobion avenae*, and *Metopolophium dirhodum*) on winter wheat from early spring until harvest. As a method we used simulation modelling based on the concept of the metabolic pool and on the technique of object-oriented programming. In the model, aphid phenologies were closely linked to the phenology of the crop, and alate production was a major factor shaping aphid phenology. The results stress the importance of understanding the interactions between crop and cereal aphids, before trying to understand the effects of natural enemies. The object-oriented modelling approach made it possible to extend this model to several natural enemies, as explained in subsequent articles II to V in this volume.

Keywords: winter wheat, cereal aphids, simulation modelling, object-oriented, *Sitobion avenae, Metopolophium dirhodum, Rhopalosiphum padi.*

Introduction

Cultivation of cereals in Denmark requires control of several environmental factors, especially, soil nutrients, weeds, fungal diseases, and aphids. Outbreaks of aphid pests, including *Rhopalosiphum padi* (L.), *Sitobion avenae* (F.), and rarely *Metopolophium dirhodum* (Wlk.), occur every two or three years and are usually controlled using pesticides. Natural control, i.e. pest control by naturally occuring enemies, offers an ecologically sustainable alternative to chemical control. The potential of natural enemies to control cereal aphids has been reviewed by Vickerman & Wratten (1979) and Dixon (1987). In Denmark, all three aphid species are holocyclic. They over-winter as eggs and colonize their primary hosts early in the

Arthropod natural enemies in arable land · III The individual, the population and the community
W. Powell (ed.). *Acta Jutlandica* vol. 72:2 1997, pp. 195-206
© Aarhus University Press, Denmark, ISBN 87 7288 673 0

season. In spring or early summer they migrate from their primary hosts to the cereal fields when the preferred host plant organ has reached a suitable stage: *R. padi* feeds mainly on the culm (including leaf sheaths), *M. dirhodum* on the leaf blades, and *S. avenae* feeds initially on the flag leaf and then on the ear. During summer, the aphids leave the field as alates which are produced in response to host plant quality and crowding (Watt & Dixon 1981). These alates move to other summer hosts or their winter hosts depending on the time of the season. Natural control of cereal aphids is a process that overlies these complex life cycles. In this paper we present a simulation model of winter wheat and aphids, which forms the basis for studies on the effects of natural enemies on cereal aphid dynamics. The objectives of the model were (i) to summarise the existing knowledge on the system and (ii) to understand better the system dynamics.

Simulation models that include detailed sub-models of both host plant and cereal aphids have not been presented yet, and models including the dynamics of all three cereal aphid species are especially missing. The host plant has been either excluded from the model (Wiktelius & Petterson 1985) or simply modelled as a growth stage (Rabbinge et al. 1979, Carter et al. 1982, Holst & Ruggle in press). Several detailed models of winter wheat exist, but either they do not include aphids (e.g. Weir et al. 1984, Porter 1993) or aphids are simply entered as a forcing function to express a fixed aphid phenology (Rossing et al. 1989). The influence of crowding and crop growth stage on the appearance of alates has been quantified for *R. padi* (Wiktelius 1992), *S. avenae* (Watt & Dixon 1981), and *M. dirhodum* (Howard & Dixon 1992). These data have been included in simulation models of *S. avenae* (Rabbinge et al. 1979, Carter et al. 1982, Holst & Ruggle in press) and *R. padi* (Wiktelius & Petterson 1985), but only the effect of crowding on alate production is included in the latter model. The simulation model presented in this paper includes detailed sub-models of winter wheat and all three cereal aphid species. The sub-models are metabolic pool models (Gutierrez 1996), i.e. they are based on energy budgets. We estimated the parameter values for the three aphid species from literature data and the plant parameters from our own field data. Thus, we validated the aphid sub-models but not the plant sub-model against independent field data. We used an object-oriented approach (Yourdon 1994) to create a conceptual and technical modelling framework. The generality of this approach allows for an extension of this model to include the whole range of natural enemies, which are presented in subsequent papers in this volume: polyphagous predators (Petersen & Holst, Axelsen et al.), specific predators (Axelsen et al.), parasitoids (Ruggle & Holst), and fungal pathogens (Dromph et al.). For simplicity, and to show the commonalities in the model, we present all the modelling methods essential to its elements in the present paper.

Methods

Modelling approach

We used the metabolic pool approach (Gutierrez & Baumgärtner 1984, Gutierrez 1996) as a basis to develop the model. According to this approach, age-specific energy budgets represent population sub-models. The parameters of an energy budget are estimated from laboratory data at nearly optimal conditions and determine the demand rate (ΔD) for energy, or dry matter equivalents, of a species. Because field data are not usually used for parameter estimation, they can be used for validation. Populations are connected in a food-web using the Gutierrez-Baumgärtner (G-B) functional response (Gutierrez & Baumgärtner 1984, Schreiber & Gutierrez in press) for all trophic links, i.e. plant-light, herbivore-plant, predator-prey, parasite-host, and pathogen-host interactions. The G-B functional response takes into account the demand rates (ΔD) of consumers (herbivores, predators, parasitoids, pathogens), and the densities of consumers and resources (prey, hosts). Each trophic link is characterized by the search rate of the consumer, or from an alternative viewpoint, by the 'apparancy' (Schreiber & Gutierrez in press) of the resource. The search rate is an abstract description of the trophic interaction; it summarizes the complexities of behaviour and spatial heterogeneity. The functional response model determines the rate of supply (ΔS) to the consumer, and the ratio between ΔS and ΔD affects all processes in the system dynamically. The condition of a population is described in age-structured state variables, i.e., vectors characterising the state of the population divided into n age classes. Typical state variables are dry mass and number of individuals. Populations infected by parasites (including parasitoids and pathogens) have two-dimensional state variables (n_1 x n_2 matrices) holding all combinations of hosts (n_1) and parasites (n_2) in different age classes. Carruthers et al. (1986) and Stone & Gutierrez (1986) introduced two-dimensionally age-structured models. Recent models of the population dynamics of cereal aphid parasitoids (Holst & Ruggle in press, Ruggle & Holst 1997) adopted this modelling technique. Physiological development ("ageing") proceeds species- or stage-specific and was quantified in physiological time, e.g. day-degrees or non-linear scales of temperature-dependent development. The basic temperature-dependent time scale can optionally be modified according to $\Delta S/\Delta D$, nitrogen, humidity, or other factors. Age-distributed state variables are updated by increments on the corresponding physiological time scale with the 'distributed delay' procedure (Manetsch 1976, Vancikle 1977). By this, simulated development times have a spread (an Erlang or Gamma distribution) around the mean.

We started by developing a very generic modelling framework for metabolic pool models (Holst et al. in press), instead of solving specific modelling problems in

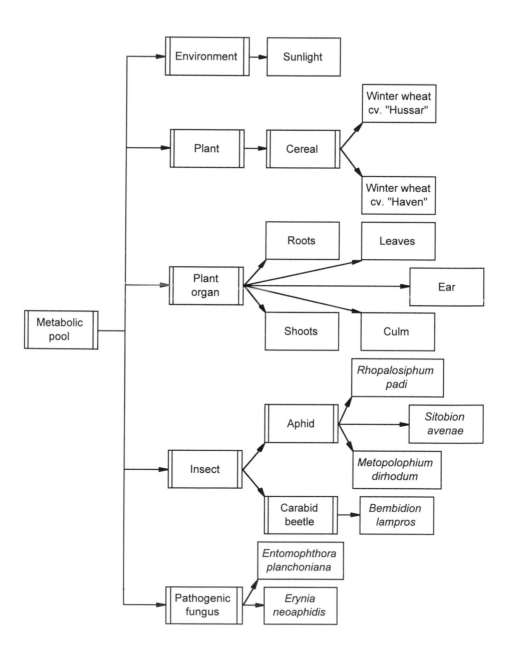

Fig. 1. Examples of classes (boxes with double vertical lines) and objects (simple boxes) derived (through arrows) from the metabolic pool class.

isolation. In our framework we used an object-oriented design (Yourdon 1994) and the C++ programming language (Stroustrup 1991). The framework is a 'workbench' for model development and consists of a general simulation engine, which carries out the simulation, and generic building blocks (*classes*, in object-oriented terms) of model components. The modelling framework can extend the set of classes available for model construction by deriving more specialized classes from the generic ones (for rules of class derivation, see Yordon 1994). For example, we used the generic metabolic pool class as a basis for deriving classes representing populations of different taxa, stages, and plant organs (Fig. 1). Other classes of the modelling framework represent energy budget components (e.g., age-specific fecundity and growth), trophical interactions (e.g., predation or parasitism), and driving variables (e.g, temperature) (Holst et al. in press). These classes can also be used as a basis for deriving classes with specific fucntionality. The resulting derivation trees (e.g, Fig. 1) reflect the functional similarities between model components. The model component classes are used as templates to make model *objects*, which are the actual entities in the running programme. Model objects have the functionality of their class but may have different parameter values, and each object has its own set of state variables. The Aphid class, for instance, is the basis for three aphid objects, one for each species (Fig. 1). Fig. 2 shows how objects of classes derived from the metabolic pool class can be combined to construct stage-structured plant and insect sub-models linked by trophic interactions. Once the model objects have been defined and connected in a trophic web, the simulation engine of the framework automatically carries out the simulation. Other examples of models built on this generic modelling framework can be found in this volume (Petersen & Holst, Axelsen et al., Ruggle & Holst, Dromph et al.).

Winter wheat-aphid model
The winter wheat-aphid model consists of five components: sunlight, winter wheat, and three aphid populations (Fig. 2). All units are on a per m^2 basis. A weather file supplies daily rates of sunlight and min-max temperatures. We modelled winter wheat in four growth stages each with four populations of plant organs: number of shoots, and dry mass of leaves (i.e., leaf blades), culm (i.e., stem and leaf sheaths), and ear. Thus, the winter wheat model is a hierarchy of metabolic pools: it is a metabolic pool (plant) of four metabolic pools (growth stages) which are in turn composed of four metabolic pools (plant organs). We calculated the crop area as the total area of all organs estimating for each organ an allometric relation between area and mass. The G-B functional response yielded the daily rate of photosynthetic production based on sunlight, crop area index, and the total energetic demand of plant organs. We split the photosynthates among plant organs proportional to their demands. Secondly, each plant organ and the aphids associated with it then shared

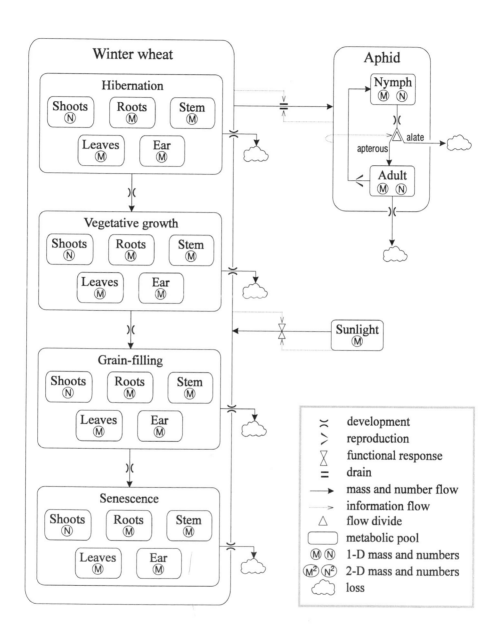

Fig. 2. The components and interactions of the simulated system.

this supply in proportion to their demands. Thus, the aphids acted as an extra sink, which is the meaning of the "drain" symbol in Fig. 2. The aphid model contains two stages, nymphs and adults, and holds both mass and number of individuals (Fig. 2). The proportion of alates among newly emerged adults depends on the density and the supply/demand ratio of the aphid population (Fig. 3; eq. 1),

$$P_{alate} = \left[1 + \exp\left\{-a\,\ln(N) - b\,\tfrac{\Delta S}{\Delta D}\right\}\right]^{-1} \qquad \text{(eq. 1)}$$

P_{alate} : proportion of alates
ΔS : daily supply (mg dry matter acquired)
ΔD : daily demand (mg dry matter needed for optimum performance)
a, b : constants

We derived eq. 1 from a function found empirically by Watt & Dixon (1981), and assumed that alate aphids leave the field as they emerge. Although these data were found for *S. avenae* only, we used the same function for all three aphid species. Parameters for the winter wheat model, including initial conditions on 1 January, were estimated by fitting model output to our field data, whereas aphid parameters were estimated from literature data exclusively (see Holst & Ruggle in press). The model ran from 1 January with aphids introduced at the date and density of their first

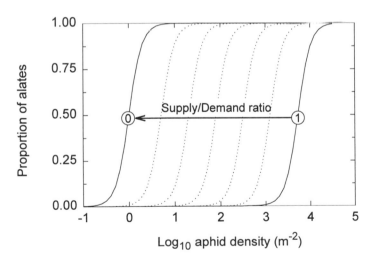

Fig. 3. The relationship between alate production and aphid density, which shifts from right to left as the supply/demand ratio decreases from one to zero.

detection in our field.

Field data
We collected field data in an unsprayed winter wheat field 20 km west of Copenhagen. Sixteen wheat plants were sampled every 14 days and split into tillers and plant organs of which we measured dry mass and area. Aphids were sampled once or twice a week from 5-20 tillers, depending on the expected average density, at 32 locations within the field. The aphids were washed off the tillers in the laboratory, identified and counted.

Simulation runs
We carried out three simulations: (1) winter wheat alone, (2) winter wheat and aphids, excluding production of alates, and (3) winter wheat and aphids, including production of alates.

Results

Winter wheat parameters were adjusted to fit the field data. Thus, there was a good correspondance between the first simulation and field observations (Fig. 4a), and we obtained reasonable estimates for plant parameters. In the second simulation, aphids were limited only by the prevailing temperatures and resource availability. This produced single-peaked phenologies for all three species: a steep exponential increase followed by a decline as resources decreased (Fig. 4b). The simulated populations peaked too late and their densities were much too high compared to the field observations. The third run showed a large impact of alate production on the aphid phenologies (Fig. 4c). The timing of population peaks corresponded better with field observations and population densities were more realistic, although still too high. In the declining phase, the model appeared to overestimate the production of alates, especially of *R. padi*, which led to an excessive population decline.

Discussion

Simulated and observed aphid densities corresponded surprisingly well, given that the aphid model was parameterized from literature data only. The model succeeded in reproducing the overall population trends but overestimated the actual population densities. The overestimation is probably due to the lack of aphid mortality factors in

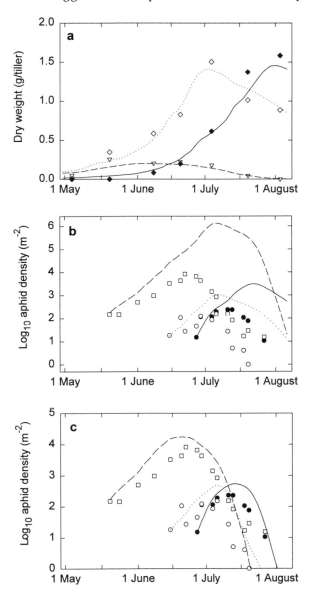

Fig. 4. Simulated (lines) and observed (symbols) winter wheat dry weight (a) and $\log_{10}$ aphid densities (b)-(c). Simulation of winter wheat without aphids (a) and of aphids, excluding (b) or including (c) alate production.

Symbols for (a): _ --- = leaves; à ×××× = culm; t ———— = ear.

Symbols for (b) and (c): ¨ --- = *R. padi*; m ×××× = *M. dirhodum*; • ———— = *S. avenae*.

the model and to an incomplete plant-aphid model. Alate production was a decisive factor in the timing and level of aphid population peaks. Additional simulations showed that the model is very sensitive to changes in the parameters of the alate model (eq. 1). Based on simulation studies of *S. avenae*, Rabbinge et al. (1979) and Carter et al. (1982) also concluded that alate formation is an important factor limiting the growth of cereal aphid populations. Carter et al. (1982) used a linear alate model, which they found not to be very sensitive to changes. This is in contrast to non-linear alate models (Rabbinge et al. 1979, Holst & Ruggle in press) like the one presented here, which are highly sensitive. We based our alate model for all three species (eq. 1) on rough extrapolations from a limited dataset on *S. avenae* (after Watt & Dixon 1981). More quantitative data on the mechanism of alate formation are needed. We included only the energetic demand of the aphids, but to model the plant-aphid interaction in full, the nitrogen demand of the aphids should be included as well (Rossing 1991). To model the aphid-plant interactions more realistically, including the nitrogen budgets of aphid and plant, we need a more advanced crop model than the one presented here.

We conclude that cereal aphid phenology is to a large extent determined by the phenology of the host plant. The relationship can be modelled with a simple growth stage representation of the plant. However, if the aim is a deeper understanding of aphid population dynamics then a detailed host plant model is needed. We used an object-oriented approach to develop the model, which imposed a consistent use of concepts throughout the project (Holst et al. in press). This encouraged us to find general solutions, rather than to develop a different model for every problem. The same approach is being used at The Danish Winter Wheat Project (Research Station Foulum), which is developing a detailed metabolic pool model of winter wheat, including effects of nitrogen and water stress. Due to the common object-oriented design, their winter wheat model and the aphid model presented here can be combined in the future. It is highly desirable, but unusual, that two complex simulation models developed at different institutions can be merged like this. The joint model serves as a framework for cross-institution team-work and, in the end, it will yield a more complete summary of the system processes. Several sub-models on aphid natural enemies are being developed within this framework (Axelsen et al., Dromph et al., Petersen & Holst, Ruggle & Holst; all in this volume) and will be integrated, together with the winter wheat and aphid sub-models, in one model incorporating the key elements of the system. We think that the effects of natural control need to be assessed with regard to all major interactions, and our approach appears to be very suitable.

Acknowledgements

We thank Trine Nielsen and Pernille Thorbek for their assistance in the field and laboratory work. The Department of Agricultural Sciences of the Agricultural University of Copenhagen provided the weather data. This work was supported by the Centre for Agricultural Biodiversity of the Danish Environmental Research Programme.

References

Axelsen, J.A., Ruggle, P., Holst, N. & Toft, S. 1997. Modelling natural control of cereal aphids. III. Linyphid spiders and coccinellids. In: Powell, W. (ed.) *Arthropod natural enemies in arable land. III.* 72 (2): 221-31 (this volume).

Carruthers, R.I., Whitfield, G.H., Tummala, R.L. & Haynes, D.L. 1986. A systems approach to research and simulation of insect pest dynamics in the onion agro-ecosystem. *Ecol. Modelling* 33: 101-21.

Carter, N., Dixon, A.F.G. & Rabbinge, R. 1982. *Cereal aphid populations, biology, simulation and prediction.* Pudoc, Wageningen.

Dixon, A.F.G. 1987. Cereal aphids as an applied problem. *Agric. Zool. Rev.* 2: 1-57.

Dromph, K., Holst, N. & Eilenberg, J. 1997. Modelling natural control of cereal aphids. V. Entomophthoralean fungi. In: Powell, W. (ed.) *Arthropod natural enemies in arable land III.* 72 (2): 247-58 (this volume).

Gutierrez, A.P. 1996. *Applied population ecology: a supply-demand approach.* John Wiley and Sons, New York.

Gutierrez, A.P. & Baumgärtner, J.U. 1984. I. Age-specific energetics models—pea aphid *Acyrthosiphon pisum* (Homoptera: Aphididae) as an example. *Can. Ent.* 116: 924-32.

Holst, N., Axelsen, J.A., Olesen, J.E. & Ruggle, P. (in press). An object-oriented implementation of the metabolic pool model. *Ecol. Modelling.*

Holst, N. & Ruggle, P. (in press). A physiologically based model of natural enemy-pest interactions. *J. Exper. Appl. Acarol.*

Howard, M.T. & Dixon, A.F.G. 1992. The effect of plant phenology on the induction of alatae and the development of populations of *Metopolophium dirhodum* (Walker), the rose-grain aphid, on winter wheat. *Ann. Appl. Biol.* 120: 203-13.

Manetsch, T.J. 1976. Time-varying distributed delays and their use in aggregate models of large systems. *IEEE Transactions on Systems, Man and Cybernetics* 6: 547-53.

Petersen, M.K. & Holst, N. 1997. Modelling natural control of cereal aphids. II. The carabid *Bembidion lampros.* In: Powell, W. (ed) *Arthropod natural enemies in arable land III.* 72 (2): 207-19 (this volume).

Porter, J.R. 1993. AFRCWHEAT2: A model of the growth and development of wheat incorporating responses to water and nitrogen. *European J. Agron.* 2: 69-82.

Rabbinge, R., Ankersmit, G.W. & Pak, G.A. 1979. Epidemiology and simulation of population development of *Sitobion avenae* in winter wheat. *Netherlands J. Pl. Pathol.* 85: 197-220.

Rossing, W.A.H. 1991. Simulation of damage in winter wheat caused by the grain aphid *Sitobion avenae*. 2. Construction and evaluation of a simulation model. *Netherlands J. Pl. Pathol.* 97: 25-54.

Rossing, W.A.H., Groot, J.J.R. & van Roermond, H.J.W. 1989. Simulation of aphid damage in winter wheat; a case study. In: Rabbinge, R., Ward, S.A. & Laar, H.H. (eds.) *Simulation and systems management in crop protection. Simulation Monographs*, pp. 240-61. Pudoc, Wageningen.

Ruggle, P. & Holst, N. 1997. Modelling natural control of cereal aphids. IV. Aphidiid and aphelinid parasitoids. In: Powell, W. (ed.) *Arthropod natural enemies in arable land. III*. 72 (2): 233-45 (this volume).

Schreiber, R.S. & Gutierrez, A.P. (*in press*). A supply/demand perspective of persistence in food webs: applications to biological control. *Ecology*.

Stone, N.D. & Gutierrez, A.P. 1986. Pink bollworm control in southwestern desert cotton. I. A field-oriented simulation model. *Hilgardia* 54: 1-24.

Stroustrup, B. 1991. *The C++ programming language*, 2nd ed. Addison-Wesley, New York.

Vansickle, J. 1977. Attrition in distributed delay models. *IEEE Transactions on Systems, Man and Cybernetics* 7: 635-38.

Vickerman, G.P. & Wratten, S.D. 1979. The biology and pest status of cereal aphids (Hemiptera: Aphididae) in Europe: a review. *Bull. Ent. Res.* 69: 1-32.

Watt, A.D. & Dixon, A.F.G. 1981. The role of cereal growth stages and crowding in the induction of alatae in *Sitobion avenae* and its consequences for population growth. *Ecol. Entomol.* 6: 441-47.

Weir, A.H., Bragg, P.L., Porter, J.R. & Rayner, J.H. 1984. A winter wheat model without water and nutrient limitations. *J. Agric. Sci.* 102: 371-82.

Wiktelius, S. 1992. The induction of alatae in *Rhopalosiphum padi* (L.) (Hom., Aphididae) in relation to crowding and plant growth stage in spring sown barley. *J. Appl. Entomol.* 114: 491-96.

Wiktelius, S. & Petterson, J. 1985. Simulations of bird cherry-oat aphid population dynamics: a tool for developing strategies for breeding aphid-resistant plants. *Agric. Ecosystems Environ.* 14: 159-70.

Yourdon, E. 1994. *Object-oriented systems design: an integrated approach*. Prentice Hall, London.

Modelling natural control of cereal aphids.
II. The carabid *Bembidion lampros*

Mette K. Petersen[1] & Niels Holst[2]

[1]Department of Plant Pathology and Pest Management, Danish Institute of Plant and Soil Science, Lottenborgvej 2, DK-2800 Lyngby.
[2]Department of Population Biology, University of Copenhagen, Zoological Institute, Universitetsparken 15, DK-2100 Copenhagen.

Abstract

Polyphagous predators are of special interest as natural enemies of cereal aphids, since they can be sustained in the field when aphid density is low. Their predation upon aphids early in the season can possibly prevent aphid outbreaks. We developed a simulation model to investigate the effect of the carabid *Bembidion lampros* on the cereal aphid *Rhopalosiphum padi* in winter wheat fields. The model is physiologically based. It simulates winter wheat growth and development, and the population dynamics of *R. padi* and *B. lampros*. The model presented here is preliminary and does not include any mortality factors for *B. lampros*. Simulations showed that *B. lampros* can control attacks of *R. padi* provided that (1) aphid arrival is well timed with predator activity; (2) the density of alternative food is not too high; and (3) *B. lampros* density is high enough. More conclusive results are expected when more detailed information becomes available, especially on mortality factors and food preferences of *B. lampros* and on availability of alternative food.

Key words: Polyphagous predator, Carabidae, *Bembidion lampros*, *Rhopalosiphum padi*, simulation modelling, aphid control.

Introduction

During the last two decades much work has been done to describe the abundance, distribution and function of polyphagous predators (carabids, staphylinids and spiders) in cereal fields (e.g. Sunderland 1975, Jones 1979, Sunderland & Vickerman 1980, Wallin 1985, Chiverton 1987, Wallin 1989, Winder 1990, Basedow et al. 1991, Ekbom et al. 1992, Bilde & Toft 1994, Lys 1995). Polyphagous predators are of special interest as natural enemies of pests, since they can be sustained in the field when pest density is low. If

Arthropod natural enemies in arable land · III The individual, the population and the community
W. Powell (ed.). *Acta Jutlandica* vol. 72:2 1997, pp. 207-219
© Aarhus University Press, Denmark, ISBN 87 7288 673 0

present early in the season, they have the potential to curtail a pest in its initial growth phase and thus prevent a pest outbreak. Whether this potential is realized depends on several factors: weather; food preference and search efficiency of the predators; and on the density and phenology of predators, pests and alternative food. In this paper we use simulation modelling to investigate the interaction between the carabid *Bembidion lampros* (Herbst) and the cereal aphid *Rhopalosiphum padi* (L.) in winter wheat.

The *B. lampros* and *R. padi* populations become active and migrate into the field at rates that are highly temperature-dependent. In the Danish climate, this results in *B. lampros* being present in the field before the arrival of *R. padi*. The precise timing of these migration processes will, however, vary from year to year. The aphid-controlling effect of *B. lampros* will depend on the co-occurrence of active, hungry predators and aphids. The presence of alternative food is a prerequisite for the sustenance of *B. lampros* at low aphid densities. On the other hand, alternative food may reduce the aphid-controlling effect of *B. lampros* if it is preferred over aphids or is available at a much higher density than the aphids. It has been shown that the density of carabid beetles can be increased by establishing overwintering sites within the field (Thomas et al. 1992). This could possibly increase the level of natural control of the aphids. However, what level of natural control could be expected given a certain increase in carabid density has not been determined. Ekbom et al. (1992) showed by simulation modelling that *B. lampros* in some cases could prevent an outbreak of *R. padi*. They did not, however, consider the possible effects of alternative food.

The model presented here is based on the physiology of the organisms involved. It simulates winter wheat growth and development, and the population dynamics of *R. padi* and *B. lampros*, in a field from early spring until harvest. No mortality factors of *B. lampros* are included in the model at this stage. Our modelling objective was to explore under which conditions *B. lampros* is able to control *R. padi*, i.e. prevent an aphid outbreak. We considered the following factors: (1) the period of aphid arrival, (2) presence or absence of alternative food and (3) the density of *B. lampros* adults.

Materials and methods

The carabid Bembidion lampros (Herbst)
The adult stage of *Bembidion lampros* is polyphagous. In cereals the beetles eat aphids and other food items (Sunderland 1975). The composition of their diet

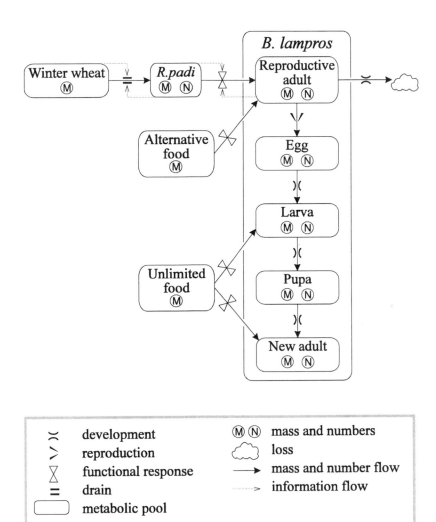

Fig. 1. Diagram of the *Bembidion lampros* model. For details of the winter wheat and aphid models, and of the symbolism used, see Holst & Ruggle (1997, this volume).

has, however, not been quantified. The adults forage on the ground and search for food mainly on the soil surface and at the crop stem bases (Chiverton 1988). The searching behaviour of the larvae is unknown, but larvae have been observed to eat *R. padi* (Petersen, unpublished data). *B. lampros* is univoltine.

Reproduction takes place in spring, and the new generation of adults emerges in late summer (Wallin 1989). The adults overwinter in aggregations in field edges and are sparsely distributed within the fields (Wallin 1989, Riedel 1991, Thomas et al. 1992). Activity starts in early spring, when the population disperses from the field edges into the field (Coombes & Sotherton 1986, Wallin 1985).

The model

The model is based upon the metabolic pool concept (Gutierrez & Baumgärtner 1984) using the object-oriented approach of Holst et al. (in press); see also Holst & Ruggle (1997). The model (Fig. 1) simulates the growth and development of winter wheat, and the population dynamics of *R. padi* and *B. lampros* by way of energy budgets. A common functional response model connects populations in the food web (Schreiber & Gutierrez in press); it takes into account the demands and search rates of the predators, and the densities of predators and prey. Temperature effects on development are included by using specific physiological time scales. Details of the winter wheat and aphid sub-models are given by Holst & Ruggle (1997). *B. lampros* is modelled as a sequence of life stages each holding mass and number of individuals (Fig. 1). The reproductive adults constitute the initial population in the model. They produce eggs at a rate which depends on their current age distribution and food intake. The food includes, in addition to *R. padi*, alternative food, which is available at a constant daily rate. We assumed that the search rate (per mg food) of *B. lampros* was the same for *R. padi* and for alternative food. The reproductive adults have a life expectancy that follows an Erlang distribution, wherein we assumed 100% survival of the subsequent life stages. The number of new adults that eventually emerge is thus equal to the number of eggs laid. The larval and new adult stages were modelled with an unlimited food resource.

For the *B. lampros* sub-model we estimated the following parameters:
* Temperature-dependent development of eggs, larvae, and pupae (Jensen 1990)
* Max. fecundity; set to 15 eggs per female (cf. Jones 1979 and Wallin et al. 1992)
* Dry mass of eggs (Petersen, unpublished data)
* Egestion, respiration, and conversion efficiency; assumed equal to those of the carabid *Notiophilus biguttatus* F. (De Ruiter & Ernsting 1987)
* Dry mass of adults and their fat reserves; estimated from field-collected adults (Petersen, unpublished data)
* Sex ratio; set to 1:1 (cf. Mitchell 1963)

* Average activity span of reproductive adults; estimated from pitfall trap catches (Petersen, unpublished data)
* Functional response to *R.padi* and alternative food; estimated from the simulation study of Ekbom et. al (1992)
* Day-degree model for the dispersal of adults into the field in the spring; estimated from pitfall trap catches (Petersen, unpublished data).

The spring dispersal of *B. lampros* was modelled using a function in which the immigration rate depended on day-degrees over 7°C accumulated from 1 March. In the field, the period and intensity of *R. padi* immigration depends on the density of overwintering *R. padi* and on spring weather. We modelled it more simply by assuming that the daily immigration rate follows a normal distribution with a standard deviation of 3.5 days around the peak immigration date. This means that nearly all (99.6%) of the aphids arrive within 20 days, which is in correspondance with the findings of Wiktelius (1982).

Simulations
We carried out nine simulations, varying only three parameters: date of *R. padi* peak arrival in the field, presence/absence of alternative food, and density of *B. lampros* adults in spring (Table 1). The total number of arriving *R. padi* was 10 per m² in all simulations. The model was driven by temperature data from 1994, from a weather station 20 km west of Copenhagen.

Results

The phenology of *B. lampros* is shown by simulation run 1 (Fig. 2). Under unlimited food conditions *B. lampros* attained the maximum net fecundity of 15 eggs per female resulting in maximum net reproduction.

Date and density for aphid peak and net reproduction of *B. lampros*, resulting from simulation runs 2-9, are shown in Table 1. Setting the density of alternative food to zero, the effect of *B. lampros* on *R. padi* was simulated with four peak arrival dates for *R. padi*, (runs 2-5; Fig. 3). In the case of earliest *R. padi* arrival, *B. lampros* consumed nearly all immigrant aphids and their offspring, which led to successful aphid control (run 2). Assuming later arrival of *R. padi* (1 or 15 May, 1 June; runs 3-5), the aphid population was only temporarily suppressed and peaked late June (run 3 and 4) or mid-July (run 5) respectively. By the latest arrival (run 5) the aphids partly escaped the predation

Table 1. Parameters for simulations and results of simulation runs. Aphid density m^{-2} was calculated assuming a density of 484 tillers m^{-2}.

Run No.	Date of *R. padi* peak arrival	Alternative food (mg/m^2/day)	No. of *B. lampros* present in spring (m^{-2})	Date of aphid peak	Peak aphid density (per tiller)	Net reproduction of *B. lampros* (per female)
1	No aphids	∞	5	-	-	15.0
2	1 May	0	5	5 May	0.03	0.0008
3	15 May	0	5	27 June	4.0	5.6
4	1 June	0	5	26 June	1.9	2.3
5	15 June	0	5	12 July	4.8	0.7
6	1 May	-	0	16 June	47.3	-
7	1 May	5	5	23 June	20.9	15.0
8	1 May	5	10	27 June	2.6	10.8
9	1 May	5	20	13 July	3.2	2.6

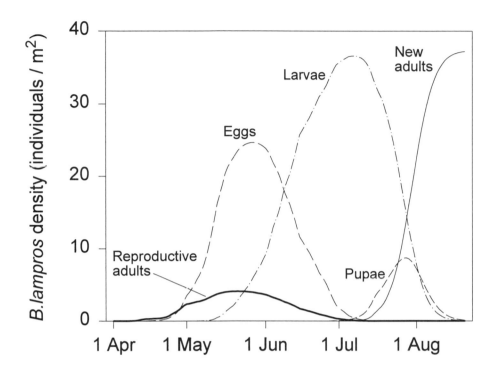

Fig. 2.   Phenology of *B. lampros* life stages with unlimited food (simulation 1).

pressure of reproductive *B. lampros*. The successful aphid control obtained from an early aphid arrival (run 2) caused a strong food limitation of reproductive *B. lampros*, which led to an extremely low net reproduction. A later arrival resulted in a higher net reproduction of *B. lampros* due to the larger aphid resource base (Table 1). However, reproduction decreased with later aphid arrival due to the prolonged initial period of *B. lampros* starvation.

Assuming an early arrival of *R. padi* (peaking 1 May), we simulated the control caused by *B. lampros* with (run 7) and without (run 2) alternative food available. When alternative food was available at a rate of 5 mg per m² per day (corresponding to about 40 adult *R. padi*), *R. padi* was released from predation (run 7; Fig. 4) compared to when alternative food was absent (run 2). However, *B. lampros* still roughly halved the peak number of *R. padi* and postponed the peak for a week (run 7) compared to when *B. lampros* was absent (run 6; Fig. 4

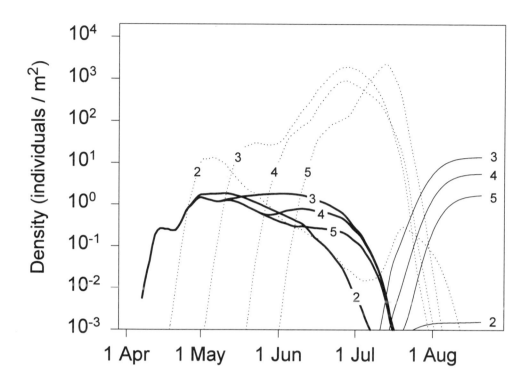

Fig. 3.   Effect of *R. padi* arrival period. Numbers refer to simulations 2-5 with peak arrival date: 1 May, 15 May, 1 June and 15 June, respectively. Reproductive *B. lampros* (thick lines) and the new generation of *B. lampros* (thin lines); *R. padi* (dotted lines).

and Table 1). The availability of alternative food resulted in optimal reproduction of *B. lampros* (Table 1). The amount of alternative food chosen (5 mg per m² per day) was the smallest amount that could produce a dramatic decline in the aphid-control exerted by *B. lampros*.

Runs 6-9 (Fig. 5) showed the potential of increasing the number of predators in the field (0, 5, 10, 20 per m²) with alternative food present. At a density of 10 or 20 *B. lampros* per m², *B. lampros* controlled the aphids. The aphid peaks, reduced and delayed, appeared in late June (run 8) or mid-July (run 9). The period of aphid suppression was prolonged with rising *B. lampros* density, but at the highest density (20 per m²; run 9) *B. lampros* over-exploited the aphid resource and net reproduction was severely reduced compared to that at lower densities (Table 1).

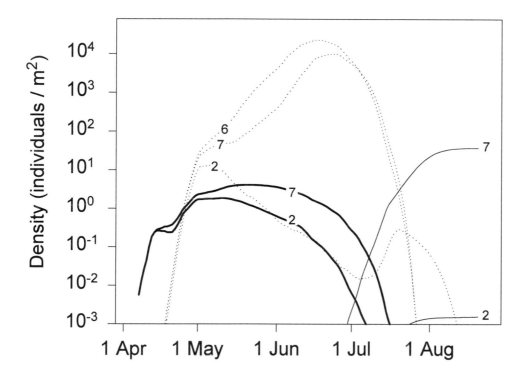

Fig. 4. Effect of presence/absence of *B. lampros* and alternative food. Numbers refer to simulations; alternative food absent (2), *B. lampros* absent (6) and alternative food present (7). Reproductive *B. lampros* (thick lines) and the new generation of *B. lampros* (thin lines); *R. padi* (dotted lines).

Discussion

By use of the model we have explored the potential of *B. lampros* to regulate populations of aphid pests. The presented model is preliminary and too optimistic in this respect since no mortality factors acting on *B. lampros* were incorporated. At this stage, model development is most useful to identify the parameters of importance for the population dynamics of *B. lampros* and in general the interactions between aphids and species of polyphagous predators.

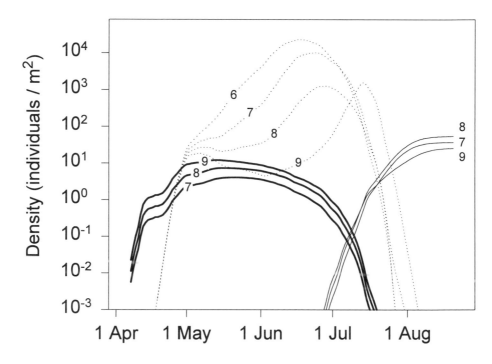

Fig. 5. Effect of initial *B. lampros* density with alternative food present. Numbers refer to simulations 6-9 with predator densities: 0, 5, 10 and 20 per m², respectively. Reproductive *B. lampros* (thick lines) and the new generation of *B. lampros* (thin lines); *R. padi* (dotted lines).

The simulations showed that the time of aphid arrival, as well as the availability of alternative food, is of importance for *B. lampros'* aphid control. An early arrival (run 2) resulted in a successful aphid control, whereas a late arrival (run 3-5) resulted in aphid populations with higher peaks. The initial density of migrating *R. padi* might also influence the control from *B. lampros*, as its searching strategy may change with changed aphid density. Variation of the initial aphid density is, however, not studied by the simulations presented here.

The availability of alternative food reduced the aphid-controlling effect but increased the net reproduction of *B. lampros*. There is scarce knowledge of which types of alternative food are available for polyphagous predators in the field and in which densities (Sunderland et al. 1997). Knowledge is also lacking of food preferences and search strategies of *B. lampros* and polyphagous predators in general under field conditions. Data to describe these conditions are needed to understand the population dynamics of polyphagous predators and their possible role as pest control agents.

Comparison of the present simulations with results from the model presented by Ekbom et al. (1992), in which alternative food was absent, indicates an over-estimation of the predation by polyphagous predators, when predator consumption of alternative food is neglected. The degree of over-estimation depends on the food preferences of the predator species, the density of *R. padi* and the availability of alternative food.

The simulations showed that the effect of *B. lampros* depends on the following factors: timing of aphid arrival, abundance of alternative food and initial density of *B. lampros*. These factors may also be of importance for other species of polyphagous predators. Variability in these factors can also explain why predation by *B. lampros* and other species of polyphagous predators is not successful in controlling aphids in every cereal field throughout each year.

Mortality factors on *B. lampros* are lacking in the current model. We plan to improve the realism of the model by making the following additions and corrections: 1) population dynamics of *B. lampros* throughout the year, including winter mortality as well as post-winter mortality before dispersal into the field, 2) adjustment of the reproductive success of *B. lampros* with rates of egg hatching and mortality during larval and pupal stages.

Incorporation of additional species of polyphagous predators in the model would provide a tool for studying the community of polyphagous predators in arable fields. Such a model could be used to simulate the dynamics of predator-prey interactions under different cultivation regimes, e.g. use of pesticides, crop rotation, amount of organic material in the soil, composition and density of alternative food. Inter- and intra-specific competition, by resource or interference competition, among polyphagous predators could be of importance but the subject has not been investigated in the field habitat.

An improved model of *B. lampros* is planned to be integrated with other sub-models of winter wheat, cereal aphids, parasitoids and entomopatogenic fungi, developed at the Centre for Agricultural Biodiversity. To estimate the economic value of polyphagous predators, a more detailed model of the crop-aphid interaction, including crop damage, would be needed. Since the timing of aphid arrival plays an important role, a model for aphid population development on the primary host and the emigration process might be needed as well. The model in its current form is far from being a tool for economic analyses, but it has proved to be a valuable tool for formulating hypotheses for further research.

Acknowledgements

The Department of Agricultural Sciences, Royal Veterinary and Agricultural University, Copenhagen, supplied the weather data. This work was supported by the Centre for Agricultural Biodiversity financed by The Danish Environmental Research Programme 1992-1996.

References

Basedow, T., Braun, C., Lühr, A., Naumann, J., Norgall, T. & Yanes, G.Y. 1991. Abundance, biomass and species number of epigeal predatory arthropods in fields of winter wheat and beets at different levels of intensity: Differences and their reasons. Results of a study at three intensity levels in Hesse, 1985-1988. *Zool. JB. Syst.* 118: 87-116.

Bilde, T. & Toft, S. 1994. Prey preference and egg production of the carabid beetle *Agonum dorsale*. *Entomol. Exp. Appl.* 73: 151-56.

Chiverton, P.A. 1987. Predation of *Rhopalosiphum padi* (Homoptera: Aphididae) by polyphagous predatory arthropods during the aphids' pre-peak period in spring barley. *Ann. Appl. Biol.* 111: 257-69.

Chiverton, P.A. 1988. Searching behaviour and cereal aphid consumption by *Bembidion lampros* and *Pterostichus cupreus*, in relation to temperature and prey density. *Entomol. Exp. Appl.* 47; 2: 173-82.

Coombes, D.S. & Sotherton, N.W. 1986. The dispersal and distribution of polyphagous predatory Coleoptera in cereals. *Ann. Appl. Biol.* 108: 461-74.

De Ruiter, P.C. & Ernsting, G. 1987. Effect of ration on energy allocation in a carabid beetle. *Functional Ecology* 1: 109-16.

Ekbom, B.S., Wiktelius, S. & Chiverton, P.A. 1992. Can polyphagous predators control the bird cherry-oat aphid (*Rhopalosiphum padi*) in spring cereals ? A simulation study. *Entomol. Exp. Appl.* 65: 215-23.

Gutierrez, A.P. & Baumgärtner, J.U. 1984. I. Age-specific energetics models-pea aphid *Acyrthosiphon pisum* (Homoptera: Aphididae) as an example. *Can. Ent.* 116: 924-32.

Holst, N., Axelsen, J.A., Olesen, J.E. & Ruggle, P. (in press). An object-oriented implementation of the metabolic pool model. *Ecol. Modelling*.

Holst, N & Ruggle, P. 1997. Modelling natural control of cereal aphids. I. The metabolic pool model, cereal aphids, and winter wheat. In: Powell, W. (ed.) *Arthropod natural enemies in arable land. III.* 72 (2): 195-206 (this volume).

Jensen, L.B. 1990. Effect of temperature on the development of the immature stages of *Bembidion lampros* (Coleoptera: Carabidae). *Entomophaga* 35; 2: 277-81.

Jones, M.G. 1979. The abundance and reproductive activity of common Carabidae in a winter wheat crop. *Ecol. Entomol.* 4: 31-43.

Lys, J.-A. 1995. Observation of epigeic predators and predation on artificial prey in a cereal field. *Entomol. Exp. Appl.* 75: 265-72.

Mitchell, B. 1963. Ecology of two carabid beetles, *Bembidion lampros* (Herbst) and *Trechus quadristriatus* (Schrank). I. Life cycles and feeding behaviour. *J. Anim. Ecol.* 32: 289-99.

Riedel, W. 1991. Overwintering and spring dispersal of *Bembidion lampros* (Coleoptera: Carabidae) from established hibernation sites in a winter wheat field in Denmark. *Behaviour and impact of Aphidophaga* 235-41.

Schreiber, R.S. & Gutierrez, A.P. (in press). A supply/demand perspective of persistence in food webs: applications to biological control. *Ecology.*

Sunderland, K.D. 1975. The diet of some predatory arthropods in cereal crops. *J. Appl. Ecol.* 12: 507-15.

Sunderland, K.D. & Vickerman, G.P. 1980. Aphid feeding by some polyphagous predators in relation to aphid density in cereal fields. *J. Appl. Ecol.* 17: 389-96.

Sunderland, K.D., Axelsen, J.A., Dromph, K., Freier, B., Hemptinne, J.-L., Holst, N.H., Mols, P.J.M., Petersen, M.K., Powell, W., Roggle, P., Triltsch, H. & Winder, L. 1997. Pest control by a community of natural enemies. In: Powell, W. (ed.) *Arthropod natural enemies in arable land. III.* 72 (2): 271-326 (this volume).

Thomas, M.B., Wratten, S.D. & Sotherton, N.W. 1992. Creation of island habitats in farmland to manipulate populations of beneficial arthropods predator densities and species composition. *J. Appl. Ecol.* 29 (2): 524-31.

Wallin, H. 1985. Spatial and temporal distribution of some abundant carabid beetles (Coleoptera: Carabidae) in cereal fields and adjacent habitats. *Pedobiologia* 28: 19-34.

Wallin, H. 1989. Habitat selection, reproduction and survival of two small carabid species on arable land: a comparison between *Trechus secalis* and *Bembidion lampros*. *Holartic Ecology* 12: 193-200.

Wallin, H., Chiverton, P.A., Ekbom, B.S. & Borg, A. 1992. Diet, fecundity and egg size in some polyphagous predatory carabid beetles. *Entomol. Exp. Appl.* 65: 129-40.

Wiktelius, S. 1982. Flight phenology of cereal aphids and possibilities of using suction trap catches as an aid in forecasting outbreaks. *Swedish. J. Agric. Res.* 12: 9-16.

Winder, L. 1990. Predation of the cereal aphid *Sitobion avenae* by polyphagous predators on the ground. *Ecol. Entomol.* 15: 105-10.

Modelling natural control of cereal aphids. III. Linyphiid spiders and coccinellids

J.A. Axelsen[1], P. Ruggle[2], N. Holst[3] & S. Toft[4]

[1]National Environmental Research Institute, Department of Terrestrial Ecology,
Vejlsøvej 25, P.O. Box 314, DK-8600 Silkeborg, Denmark
[2]Zoological Institute, University of Basel, Rheinsprung 9, CH-4002 Basel, Switzerland
[3]Department of Population Biology, Zoological Institute, University of Copenhagen,
Universitetsparken 15, DK-2100 Copenhagen
[4]Institute of Biological Researches, Department of Zoology, University of Aarhus, Building
135, DK-8000 Aarhus C, Denmark

Abstract

We present the results of a simulation study (metabolic pool model) of the relative importance of generalist and specialised predators for the control of aphids in winter wheat. The results support the hypothesis that predators can control aphids, and the simulation model is useful to quantify under which conditions control is best achieved by either of the two predator groups. The amount of alternative prey, aphid immigration rate and timing of immigration by aphid and specialised predator were identified as important factors for the control of aphids and the relative importance of the two predator groups.

Key words: winter wheat, cereal aphids, simulation modelling, specialised predators, generalist predators.

Introduction

The relative role of polyphagous predators (staphylinids, carabids and spiders), and stenophagous predators (coccinellids, syrphids, lacewings) as control agents for cereal aphid populations is not clear. The experimental removal of generalist predators can lead to a 16-fold increase in populations of *Rhopalosiphum padi* (L.) in the field (De Barro 1992). De Barro (1992) attributed this effect mainly to linyphiid spiders, because other predators like carabids were very rare. Similarly, Ekbom et al. (1992) found that the peak aphid levels were sensitive to predator density and suggested that polyphagous predators can prevent outbreaks of *R.*

padi. Recently, Toft (1995) and Bilde & Toft (1994) found that several species of spiders and the carabid *Agonum dorsale* (Pont.) are unable to satisfy their food demand on a diet of *R. padi* alone. These predators can consume only a very limited number of *R. padi*, which may have a chemical defence mechanism. Specialised aphid predators, such as coccinellids and syrphids, have high potentials to reduce populations of aphid pests. They can respond numerically to cereal aphids (Poehling 1988), but they usually arrive at cereal fields too late to prevent economic losses (Basedow 1982).

While it is obvious that both generalist and specialist predators have an impact on the population size of cereal aphids, it is not clear under which conditions predators are capable of controlling the aphids, and whether one group of predators may be more important than the other under certain circumstances. Most studies focus on one of the two predator groups or on just one taxonomic group. Therefore, it seems appropriate to assess the potential for aphid control by generalist and specialist predators acting simultaneously. We use a metabolic pool model driven by temperature and the supply/demand ratio of the organisms (Gutierrez et al. 1984, Graf et al. 1990). This model type is well suited to simulate interactions between multiple trophic levels and to incorporate the processes of predation and resource limitation. In our study, a linyphiid spider represents a generalist and the coccinellid *Coccinella septempunctata* L. the specialist predator. We used sensitivity analysis to evaluate how different conditions affect the efficiency of the two predators working simultaneously. The results showed that the amount of alternative prey for the generalist predator determines whether aphid control depends on the effects of the specialist alone or on the action of specialist and generalist together.

Methods

Holst & Ruggle (1997) (this volume) present the essential details of the metabolic pool model used here. Therefore, only a short description of the model incorporating spiders, coccinellids, aphids and alternative prey is given here.

The model system (Fig. 1) consists of the two predators, a linyphiid spider with size and reproductive properties resembling those of *Oedothorax apicatus* (Bl.) and temperature-growth relationships of *Erigone atra* (Bl.) (DeKeer & Malfait 1988) and the coccinellid *C. septempunctata*. The spider was simulated with a tolerance limit (1% of the predators dry weight per degree day) in the consumption of aphids, i.e. they could consume only a limited amount of aphids

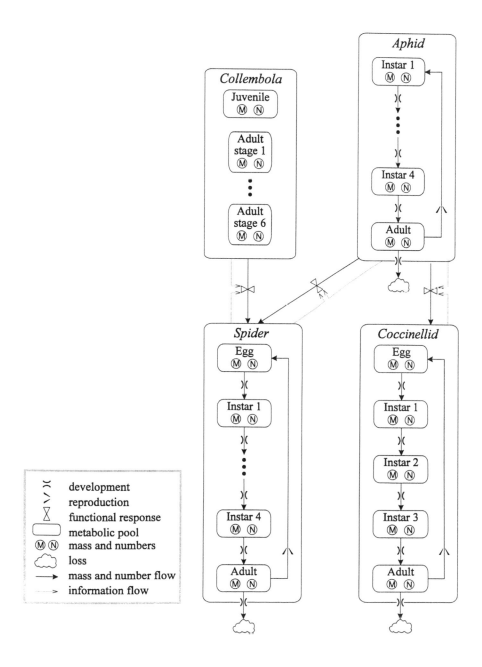

Fig. 1. The components and the interactions of the simulated system.

(Toft 1995, Bilde & Toft 1994), but all alternative prey was suitable. *Rhopalosiphum padi* (L.) represented the cereal aphids in the model. As alternative prey, an even stage distribution of collembola with the weight and growth characteristics of *Folsomia fimetaria* (L.) was used, and no chemical defence mechanism against spiders was assumed. The alternative prey was not eaten by the ladybirds. To mimic the slowing down of the aphid population increase due to the maturation of the winter wheat, we reduced the aphids survival rates from 1.0 to 0.999 and 0.998 $°D^{-1}$, which simulated the emigration of alate adults. The supply was also reduced to 90% and 75% of the demand at 900 and 1000 $°D$, respectively, which simulates the reduction in growth and reproduction as a consequence of reduced food quality of the crop late in the season. This relatively simple model reproduced the typical phenology of cereal aphids with a single peak after an exponential growth phase. Peak aphid population size (the habitat carrying capacity) was arbitrarily set to 20,000 individuals m^{-2} of each of the five stages. If the population exceeded this limit the survival was reduced to a level where the population increase was stopped. In the simulations, aphids arrived all at the same time on 1 June. If not stated elsewhere, the coccinellid arrived on 15 June, while the spiders were present in the field from the start of the simulation and had the possibility to initiate reproduction since alternative prey was present well before the arrival of the aphids. We feel that these assumptions are realistic for *R. padi* in Danish winter wheat fields. The simulations started 1 April and the climatic input came from hourly measurements of temperature 30 cm above the ground. The measurements were collected by an automatic weather station at Odum 15 km north or Aarhus, Denmark during 1991.

The search rate is a very important parameter in predator-prey interactions, and our model was sensitive to changes in this parameter. We fitted the search rate of the spider to make one individual catch 1.6 aphids per day at an aphid density of 100 m^{-2} and an average temperature of 15°C. Toft & Axelsen (unpublished data) determined this capture rate through fitting it to the data observed by Sunderland et al. (1986). We fitted the search rate of *C. septempunctata* to make one individual able to capture 32 aphids per day at an average temperature of 15°C and an aphid density of 1000 m^{-2}, which is close to the value reported by Honek (1985). The thermal threshold used for *C. septempunctata* was 13°C and was derived from the observations of Butler (1982), while the physiological parameters were taken from Baumgärtner et al. (1987) and Gutierrez et al. (1984).

We focused our sensitivity analysis on three parameters describing the conditions for the predators and on two responses describing the aphid

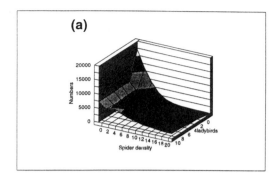

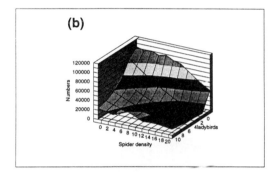

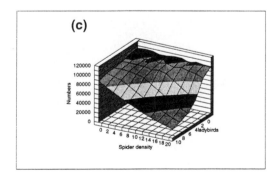

Fig. 2. Simulated peak density of aphids in relation to the density of spiders and coccinellids (all densities in individuals m^{-2}) at aphid immigration rates of 10 (a), 100 (b) and 200 (c) m^{-2} and high amounts of alternative prey for the spiders. The spiders were present in the field before the immigration of aphids and the coccinellids arrived 15 days after the aphids.

phenology. The conditions included (1) variable levels of initial aphid density (10, 100 and 200 aphids m^{-2}); (2) high and low levels of alternative prey, and (3) different times (Julian days 61, 68, 75 and 83) of coccinellid arrival to the field. To assess the efficiency of the predators, we used peak aphid density which is negatively affected by predation.

Results

At high levels of alternative prey for the spiders and at low aphid immigration rates (10 m^{-2}), peak aphid density was very sensitive to the number of spiders and less sensitive to the number of coccinellids (Fig. 2a). The impact of coccinellids was not improved through an increase in their density. On the contrary, there was a small negative impact of more than 2 coccinellids m^{-2}. This was due to intra and inter-specific competition leading to the starvation of

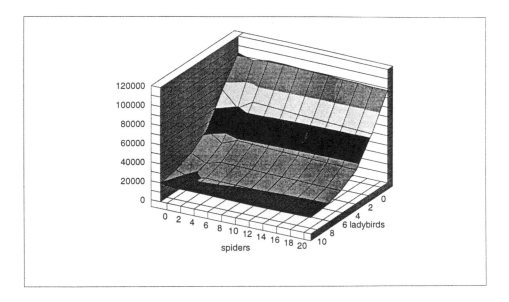

Fig. 3. Simulated peak aphid density in relation to the density of spiders and coccinellids (all densities in individuals m^{-2}) at an aphid immigration rate of 100 m^{-2} and low amounts of alternative prey for the spiders. The spiders were present in the field before the immigration of aphids and the coccinellids arrived 15 days after the aphids.

predators. At higher aphid immigration rates (100 and 200 m^{-2}), the role of coccinellids as aphid control agents became more apparent (Fig. 2b and 2c), as a strong reduction of the aphid could be achieved only in the presence of coccinellids. At the highest immigration rates, both predator types had to be abundant to reduce aphid numbers, and the coccinellids had the higher per capita effect (Fig. 2c). Peak aphid density reached the carrying capacity of the habitat, when densities of both predators were low and immigration levels of aphids high (Fig. 2c).

At low levels of alternative prey, the spiders were unable to contribute to aphid control, as peak aphid densities were almost unaffected by the number of spiders present (Fig. 3). At higher spider densities, they starved to death due to food limitations before the arrival of aphids, but at low densities there were a few survivors by the time of aphid arrival. Coccinellids had a large impact on the aphids and did not suffer from food limitation. Thus, the amount of alternative prey determined whether aphid control depends on the effects of the specialist alone or on the concerted action of specialist and generalist.

At the low aphid immigration rates, the time of coccinellid arrival hardly affected peak aphid populations levels (results not shown), which is not surprising since coccinellids had a low impact at low aphid densities (Fig. 2). Therefore it is only relevant to investigate the impact of coccinellid arrival time at higher aphid immigration rates. At an aphid immigration rate of 100 m^{-2}. and small amounts of alternative prey for the spiders, there was a clear relationship between time of coccinellid arrival and aphid control: the earlier the coccinellids arrive, the better the control of the aphid population (Fig. 4a). However, the difference between early and late arrival decreases at higher densities of coccinellids. Large amounts of alternative prey allow the spiders to survive and reproduce, which leads to a more complicated relationship between peak aphid density and coccinellid arrival time. Increasing the density of coccinellids beyond 6 m^{-2} reduced their impact on peak aphid population levels (Fig. 4b). When spiders got higher amounts of alternative prey they could exert serious intra- and inter-specific competition, leading to the starvation of either, or both, the coccinellid and the spider population. In our model they starved to death, but in nature they may leave to search for a better place.

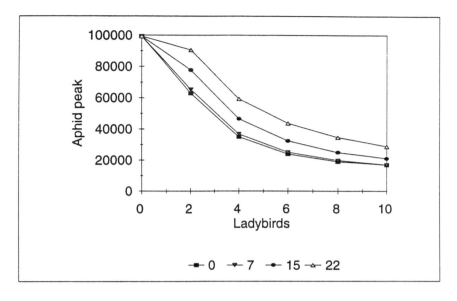

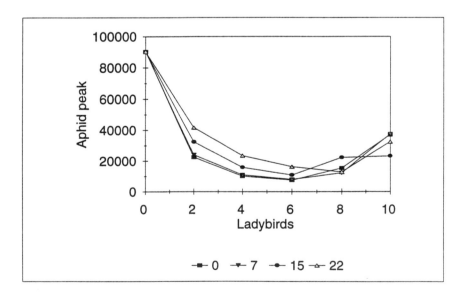

Fig. 4. The relationship between simulated peak density of aphids and the density of coccinellids immigrating 0, 7, 15 and 22 days after the aphids at low (a) and high (b) amounts of alternative prey. The number of immigrating aphids was 100 m^{-2}.

Discussion

The simulations presented in this paper suggest that predators are able to suppress the aphid population considerably. Assuming the spider and ladybird selected for the simulations are good representatives for polyphagous and specialised predators, respectively, the simulations indicate that the polyphagous predator is the most effective predator type at low aphid densities. At higher aphid densities, however, the specialised predators tend to be the most effective group. However, when the population of polyphagous predators is high, which may occur as a consequence of high densities of alternative prey, the two predator groups can both play an important role in keeping the aphid population at a low level. When the level of alternative prey and thus the number of polyphagous predators are low, the timing between the arrival of the specialist predator and the aphid is a crucial factor, especially at high aphid immigration rates. Under these circumstances the aphid control is improved by a relatively early arrival of the specialist. This result supports Basedow's (1982) conclusion that, to keep the aphid populations under control, coccinellids should arrive earlier than they actually do. He also concluded that the predator/prey ratio was a key factor controlling the production of coccinellid eggs and the survival of the larvae. This supports the result, that aphid control can be reduced due to strong intra-specific competition, as happened in simulations of low aphid densities.

The results of the simulations depend on crucial parameters in the model, namely predator search rate and spider tolerance limit to aphid diet. In the literature, there were no direct estimates on the search parameters of *C. septempunctata* or a linyphiid spider in winter wheat fields. This parameter is known to be crucial in predator-prey interactions (Axelsen et al. in press) and wrong estimates of the search rate can lead to erroneous conclusions. Our estimates of the search efficiency of the spider and the coccinellid reproduced observed field data (Sunderland et al. 1986 and Honek 1985, respectively) very well and were assumed to be realistic. However, the search rate is a product of ecological, behavioural and physiological attributes of a predator, and estimates tend to be specific to experimental conditions (e.g. geographic locations, predator strain, prey resource). The tolerance limit on aphid consumption is another parameter determining the efficiency of the spiders. In our simulations it was set to 1% of the spider dry weight $°D^{-1}$. A higher limit means that the prey is more suitable due to a weaker defence system or that the predator can cope better with the defence of the prey. Spiders and other polyphagous predators with higher tolerance limits than the species modelled here would be more efficient as aphid predators, especially at higher prey densities.

Rather than repeat how important predators can be for the control of cereal aphids, we presented simulations as a tool to identify the conditions (i.e., the combinations of predator densities, aphid immigration rates, and timing of arrival) required to obtain a certain level of control. The simulations could provide quantitative information on the dynamic factors required to control cereal aphid populations. Furthermore, the simulations have demonstrated that working with predation on aphid populations it is important to analyse both groups and also to include the availability of alternative prey for the generalists. Thus, our model results suggest both positive and negative interactions between the two types of predators, none of which have so far been documented by field studies.

References

Axelsen, J.A., Holst, N., Hamers, T. & Krogh, P.H. (in press). Simulations of the predator-prey interactions in a two species ecotoxicological test system. *Ecol. Modelling*.

Basedow, Th. 1982. Untersuchungen zur Populationsdynamik des Siebenpunktmarienkäfers *Coccinella septempunctata* L. (Col., Coccinellidae) auf Getreidefeldern in Schleswig-Holstein von 1976-1979. *Z. angew. Ent.* 94: 66-82.

Baumgärtner, J., Bieri, M. & Delucchi, V. 1987. Growth and development of immature life stages of *Propylea 14-punctata* L. and *Coccinella 7-punctata* L. [Col.: Coccinellidae) simulated by the metabolic pool model. *Entomophaga* 32: 415-23.

Bilde, T. & Toft, S. 1994. Food quality of cereal aphids to arthropod predators.

Butler, G.D. Jr. 1982. Development time of *Coccinella septempunctata* in relation to constant temperatures (Col.: Coccinellidae). *Entomophaga* 27: 349-53.

De Barro, P.J. 1992. The impact of spiders and high temperatures on cereal aphid (*Rhopalosiphum padi*) numbers in an irrigated perennial grass pasture in South Australia. *Ann Appl. Biol.* 121: 19-26.

DeKeer, R. & Maelfait, J.-P. 1988. Laboratory observations on the development and reproduction of *Erigone atra* Blackwall 1833 (Araneae, Erigoninae). *Bull. Br. arachnol. Soc.* 7: 237-42.

Ekbom, B.S., Wiktelius, S. & Chiverton, P.A. 1992. Can polyphagous predators control the bird cherry-oat aphid (*Rhopalosiphum padi*) in spring cereals? A simulation study. *Entomol. Exp. Appl.* 65: 215-23.

Graf, B., Baumgärtner, J. & Gutierrez, A.P. 1990. Modelling agroecosystem dynamics with the metabolic pool approach. *Mitteilungen der Schweizerischen Entomoligischen Gesellsachft* 63: 456-76.

Gutierrez, A.P., Baumgärtner, J.U. & Summers, C.G. 1984. Multitrophic models of predator-prey energetics. *Can. Ent.* 116: 923-63.

Holst, N. & Ruggle, P. 1997. Modelling natural control of cereal aphids. I. The metabolic pool model, winter wheat, and cereal aphids. In: Powell, W. (ed) *Arthropod natural enemies in arable land. III. Acta Jutlandica* 72 (2): 195-206 (this volume)

Honek, A. 1985. Activity and predation of *Coccinella septempunctata* adults in the field (Col., Coccinellidae). *Z. angew. Entomol.* 100: 399-409.

Poehling, 1988. Zum Auftreten von Syrphiden- und Coccinellidenlarven in Winterweizen von 1984-1987 in Relation zur Abundanz von Getriedeblattläusen. *Mitteilungen der Deutschen Gesellschaft für Allgemeine und Angewandte Entomologie* 6: 248-54.

Sunderland, K.D., Fraser, A.M. & Dixon, A.F.G. 1986. Field and laboratory studies on money spiders (Linyphiidae) as predators of cereal aphids. *J. Appl. Ecol.* 23: 433-47.

Toft, S. 1995. Value of the aphid *Rhopalosiphum padi* as food for cereal aphids. *J. Appl. Ecol.* 23: 433-47.

Modelling natural control of cereal aphids: IV. Aphidiid and aphelinid parasitoids

P. Ruggle[1] & N. Holst[2]

Department of Population Biology, Zoological Institute, University of Copenhagen, Universitetsparken 15, DK-2100 Copenhagen
[1]Current address: Zoological Institute, University of Basel, Rheinsprung 9, CH-4002 Basel, Switzerland
[2]Current address: Danish Institute of Plant and Soil Science, Flakkebjerg, DK-4200 Slagelse, Denmark

Abstract
We present an improved aphid-parasitoid model incorporating winter wheat, the three major cereal aphid species, and three common primary parasitoids. The improvements over an earlier model include: (i) a refined aphid-plant model producing more realistic aphid and parasitoid phenologies, (ii) emigration of adult parasitoids, depending on host availability, (iii) parasitoid preference for different aphid species, and (iv) host-feeding by aphelinids. These refinements of the model decreased parasitoid densities very much, but reduced aphid mortality due to parasitoids only slightly. The overall effects on the aphid populations remained small in comparison to other mortality factors. We point out some problems, which occur in the development of such detailed simulation models, and discuss possible reasons for the low effectiveness of parasitoids, namely parasitoid reproduction, size, preference, sex-ratio, and hyperparasitism.

Key words: simulation modelling, cereal aphids, parasitoids, emigration, preference, host-feeding.

Introduction

Some of the simulation models on cereal aphid-parasitoid interactions yield contradictory conclusions about the efficiency of parasitoids to control aphid populations (e.g., Carter & Dixon 1981, Rabbinge et al. 1984, Vorley & Wratten 1985). Most models lack essential, biological details (e.g., Rabbinge at al. 1979, Holz & Wetzel 1989) or depend on empirical data (e.g., Vorley & Wratten 1985)

*Arthropod natural enemies in arable land · **III** The individual, the population and the community*
W. Powell (ed.). *Acta Jutlandica* vol. 72:2 1997, pp. 233-245
© Aarhus University Press, Denmark, ISBN 87 7288 673 0

and are thus not independent of the field data they simulate. All these models were designed as predictive management tools (e.g. Carter et al. 1982, Holz 1991). A different approach was taken by Holst & Ruggle (in press) who used the biologically detailed metabolic pool model (Gutierrez & Wang 1977) to explain the population dynamics of three cereal aphids and their parasitoids. Their model parameters are derived from laboratory data and the simulations are independent of the field data except for daily weather data and initiation densities. The first simulations by Holst & Ruggle (in press) have a few shortcomings, however, which need improvements. The improvements presented here include: (i) a refined aphid-plant model producing more realistic aphid and parasitoid phenologies (see Holst & Ruggle, this volume), (ii) emigration of adult parasitoids, depending on host availability, (iii) parasitoid preference for different aphid species, and (iv) host-feeding by aphelinids. These improvements produced more realistic simulations of aphid and parasitoid densities and decreased aphid mortality due to parasitoids. The overall effect on the aphid populations remained small in comparison to other mortality factors.

Materials and methods

Aphid-parasitoid system
We simulated a system consisting of winter wheat, the three cereal aphids *Sitobion avenae* (F.), *Metopolophium dirhodum* (Wlk.), and *Rhopalosiphum padi* (L.), and the three common parasitoids *Aphidius rhopalosiphi* De Stefani-Perez, *Praon gallicum* Stary, and *Aphelinus abdominalis* Dalman. All three parasitoids attack all three aphid species (Table 1). We used life table data of *P. volucre* (Hal.) to model *P. gallicum*, because no information on the latter was available. The general biologies and abundancies of the two species are similar (Feng et al. 1992, Höller et al. 1993).

Improved aphid-plant model
To increase biological realism, we replaced the very simple photosynthate production model of Holst & Ruggle (in press) with a detailed wheat plant model and let the three different aphid species attack different plant parts. Thus, *R. padi*, *M. dirhodum* and *S. avenae* attacked the culm, leaves, and ear, respectively (for details, see Holst & Ruggle, this volume).

Table 1. Frequency-adjusted attack rates α (data from Höller et al. 1993)

	Aphidius rhopalosiphi	Praon gallicum	Aphelinus abdominalis	Sum
Frequency				
S. avenae	604	6	15	625
M.dirhodum	695	100	7	802
R.padi	89	8	1	98
Sum	1388	114	23	1525
Relative frequency				
S. avenae	0.435	0.053	0.652	
M. dirhodum	0.501	0.877	0.304	
R. padi	0.064	0.070	0.044	
Sum	1.000	1.000	1.000	
Adjusted α				
S. avenae	0.218	0.026	0.326	
M. dirhodum	0.250	0.439	0.152	
R.padi	0.032	0.035	0.022	
Sum	0.500	0.500	0.500	

Emigration

As shown in eq. 1, we related the current emigration rate (r_e) of adult parasitoids inversely to the supply/demand ratio (s/d) for oviposition and scaled it with the highest emigration rate (a) for *Aphidius* spp. reported in the literature (Vorley & Wratten 1987).

$$r_e = a * (1-s/d); d > 0 \qquad \text{(eq. 1)}$$

Because at the field temperature indicated by Vorley & Wratten (1987) *P. volucre* and *A. abdominalis* develop at about half the rate of *A. rhopalosiphi* (Ruggle, Zhili, Powell & Holst unpublished data), we reduced their *a* by 46%.

Preference

To incorporate preference into the functional response model, we used the 'apparency' index proposed by Schreiber & Gutierrez (in press). We computed the indices as frequency-adjusted attack rates (Table 1) from the relative abundance of the three cereal aphid parasitoids on the three different hosts (Höller et al. 1993) and used an equal total attack rate of 0.5 for all parasitoids. We assumed (best guess) equal densities for the three aphid species over all locations and years in study of Höller et al. (1993), because there were no data on the corresponding relative abundance of aphids The s/d incorporates the preference for all host-parasitoid interactions, which in turn affects the emigration rate of parasitoids.

Host-feeding
Within aphid parasitoids, host-feeding occurs only in aphelinids, and was observed in the populations of *A. abdominalis* found in Denmark (P. Ruggle personal observation). Because *A. abdominalis* can kill as many aphids by host-feeding as by parasitisation (Haardt 1991), we set demand for host-feeding equal to the oviposition demand. We also set the attack rates for host-feeding equal to the rates for oviposition, assuming that *A. abdominalis* is equally likely to parasitise or feed on each host species. Simulations of *A. abdominalis* were very sensitive to this assumption, but no data are available from literature on the likelihood of oviposition vs. host-feeding by *A. abdominalis* encountering different host/prey species.

Field data
Aphid and parasitoid data were collected from a winter wheat field 30 km east of Copenhagen (Ågerup, Denmark) during the summer of 1994. The collection of aphid data is described by Holst & Ruggle (in press). Parasitoid densities represent the sum of parasitised late instar nymphs (L_3-L_4) and adults, as determined by dissections, plus the number of mummies. Aphidiid parasitoids were pooled across genera, because the genera are distinguishable only in the first instar (O'Donnell & Mackauer 1989). Because they were about equally abundant in the field, we initialised the density of *A. rhopalosiphi* and *P. gallicum* with half of the total observed aphidiid density, and used the total observed aphelinid density to initialise *A. abdominalis*.

Results
Improved aphid-plant model
With the improved cereal aphid-wheat model, we were able to reduce extremely high parasitoid densities found in a previous simulation study (Holst & Ruggle in press). This effect was particularly strong late in the season (Fig. 1). Nevertheless, simulated parasitoid densities exceeded the observed densities by more than one order of magnitude during the second half of the season.

Emigration
The incorporation of emigration had very little effect on parasitoid densities (Fig. 2), because *s/d* remained near 1.0 during most of the season and high emigration rates would occur only 2-4 weeks after a very low *s/d*.

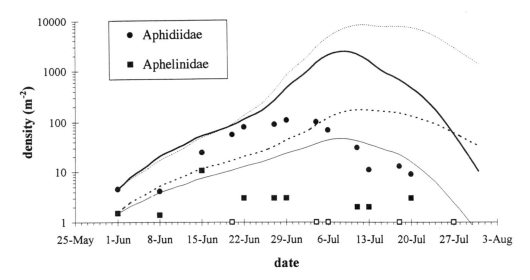

Fig. 1. Simulated (lines) and observed (symbols) densities of cereal aphid parasitoids. Dotted lines: previous simulation (Holst & Ruggle submitted); drawn lines: new aphid-wheat model (Holst & Ruggle this issue). Thick lines: Aphidiidae; thin lines: Aphelinidae; G = no Aphelinidae observed.

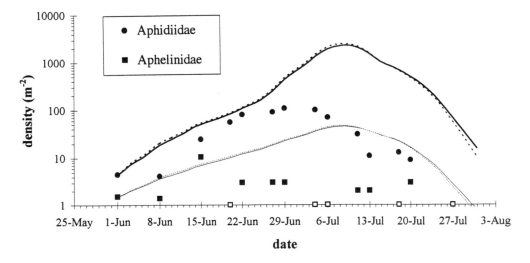

Fig. 2. Simulated (lines) and observed (symbols) densities of cereal aphid parasitoids. Dotted lines: previous simulation (Fig. 1.); drawn lines: model including emigration. Thick lines: Aphidiidae; thin lines: Aphelinidae; G = no Aphelinidae observed.

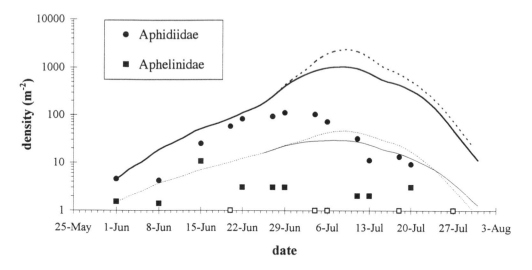

Fig. 3. Simulated (lines) and observed (symbols) densities of cereal aphid parasitoids. Dotted lines: previous simulation (Fig. 2.); drawn lines: model including parasitoid preference. Thick lines: Aphidiidae; thin lines: Aphelinidae; G = no Aphelinidae observed.

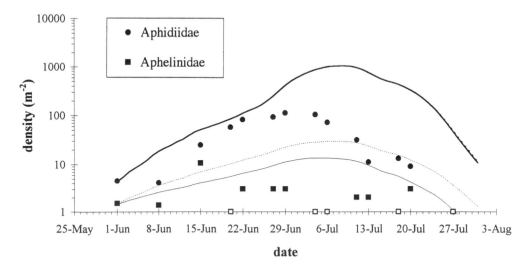

Fig. 4. Simulated (lines) and observed (symbols) densities of cereal aphid parasitoids. Dotted lines: previous simulation (Fig. 3.); drawn lines: model including host-feeding by *A. abdominalis*. Thick lines: Aphidiidae; thin lines: Aphelinidae; G: no Aphelinidae observed.

Preference

When we included the preference of parasitoids for different aphid species, we decreased peak aphidiid density by 56% and the corresponding density of aphelinids by 37% (Fig. 3). The difference was relatively small and represents the combined effects of preference and emigration.

Host-feeding

Host-feeding by *A. abdominalis* reduced its peak density by 53% (Fig. 4). Aphidiid densities were not affected. The inclusion of host-feeding in the model did not decrease the effect of *A. abdominalis* on the aphid populations despite its lower density in the field (Fig. 5). It is difficult, however, to interpret the phenology of aphelinids in the field, because they were not detected on several occasions, which may be due to our sampling plan being not designed particularly for aphelinids.

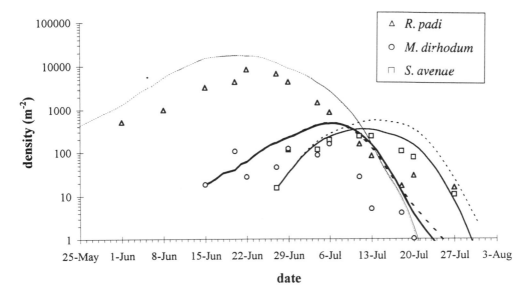

Fig. 5. Simulated (lines) and observed (symbols) densities of cereal aphids. Dotted lines: simulation without parasitoids; drawn lines: simulation with parasitoids. Thin line: *R. padi*; medium line: *M. dirhodum*; thick line: *S. avenae*.

The improvements of the model presented here decreased the extremely high parasitoid densities of the previous model by Holst & Ruggle (in press). Parasitoid densities, however, remained too high in the second half of the season (Fig. 4). This may be related to the over-estimation of aphid densities (Fig. 5), especially of *M. dirhodum*, which is the host preferred by aphidiids (Table 1). The addition of all new details into the model decreased peak population densities of parasitoids by a factor of 10, but their effect on peak aphid densities was decreased by a factor of 5 only. The overall effects, particularly on *R. padi*, remained small (Fig. 5).

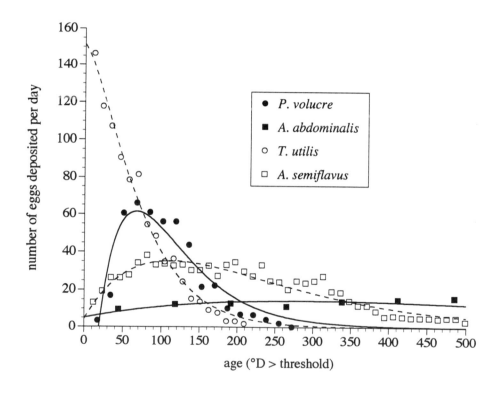

Fig. 6. Highest and lowest age-specific fecundity rates of parasitoids from the English grain aphid (closed symbols, drawn lines) and the spotted alfalfa aphid (open symbols, dashed lines). Data on *Praon volucre* from Stilmant (1994), on *Aphelinus abdominalis* from Haardt (1991), and on *Tryoxys utilis* and *Aphelinus semiflavus* from Force & Messenger (1964). Fitted function: $y = a*(x+c) / b^{(x+c)}$ (Bieri et al. 1983)

Discussion

Problems in the development of detailed simulation models

Problems that occur regularly in the development of such detailed simulation models include the lack of basic data and a danger of emphasising minor details. We were lacking basic data on the life history of *Praon gallicum* and had to use data on *P. volucre* instead. We also need better data on the preference, attack rates, and initial densities of different parasitoids. However, our modelling framework allows for any interesting details to be incorporated, if data become available at a later stage. We apparently over-emphasised the effects of parasitism on the aphid populations. Our objective, however, was to develop a strategic, explanatory model and not a tactical, predictive one. Thus, we intended to investigate the effects of parasitoids on cereal aphid populations and conclude now, that these effects are minor in comparison to the effects of the host plant and aphid crowding (Holst & Ruggle, this volume).

Reasons for the low effectiveness of cereal aphid parasitoids

Possible reasons for the low effectiveness of cereal aphid parasitoids include low parasitoid reproduction, small host and parasitoid size, unfavourable sex ratios, low preference for cereal aphids, and hyperparasitism. These factors are interdependent and may be most obvious in our data set with high density of *R. padi*, the smallest of the three cereal aphids. Even under reasonably good laboratory conditions, cereal aphid parasitoids show low reproduction compared to those of alfalfa aphids parasitoids (Fig. 6). The number of eggs deposited by individuals of *A. rhopalosiphi* during a constant time period is positively correlated with their size (Spearman rank order correlation coefficient: 0.5230, $P = 0.08$, $N = 12$, P. Ruggle unpublished; Fig. 7). Parasitoids emerging from *R. padi* are likely to be smaller than those emerging from the *S. avenae*, because the former is smaller in the field (P. Ruggle unpublished; Fig. 8). Because smaller host size may also decrease the sex ratio of parasitoids (review in Hardy 1994), we expect a lower ratio for parasitoids from *R. padi*, all factors other than size being equal. However, data on the sex ratio of adults emerging from field-collected aphids did not support that expectation. The 'preference' for *R. padi* by all parasitoids appears much lower that that for *S. avenae* or *M. dirhodum* (Table 1). This difference may be due to a behavioural selection by the adult female (preference in the strict sense) or to a combination of a size-dependent decrease in the fitness of parasitoids emerging from *R. padi*, because larger host size is generally correlated with higher fitness (review in Hardy 1994).

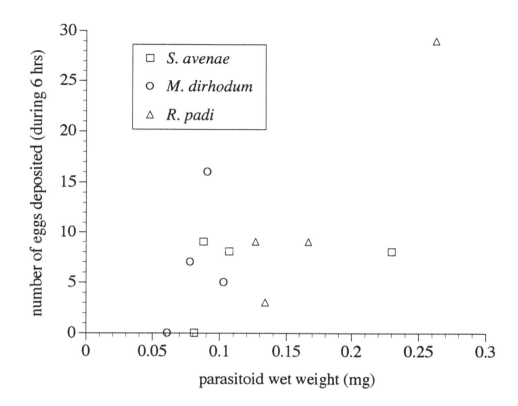

Fig. 7. Relationship between wet weight and oviposition rate of *Aphidius rhopalosiphi* on three different aphid hosts.

Parasitoids of cereal aphids have a wider host range than, for example, parasitoids of alfalfa aphids and may attack more suitable hosts among aphids not on cereals. Hyperparasitism potentially reduces the effectiveness of cereal aphid parasitoids (Höller et al. 1993), but the evidence is inconclusive, and the general importance of secondary parasitism is controversial (Horn 1989).

In general, the addition of more details into the model increased the level of realism and reduced the effect of parasitoids on their host population only slightly. Despite the fact that detailed simulation models with a high degree of

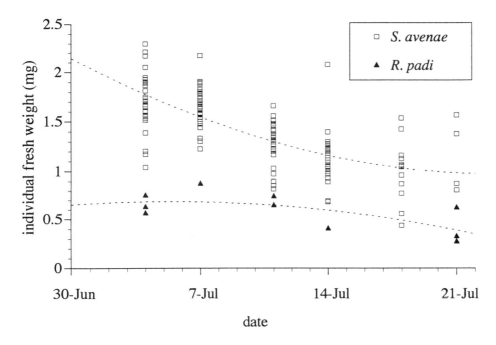

Fig. 8. Wet weights of field-collected cereal aphids (Ågerup, DK, 1995); The dashed lines (2nd degree polynomial fit) indicate that *S. avenae* is at least twice as heavy as *R. padi* at any given time.

realism are always limited by the availability of basic data, we can develop reasonable simulation models with the metabolic pool approach, because it is derived from first principles and thus very generic and flexible. We are able to integrate available knowledge and to explain the (in-) efficiency of parasitoids to control cereal aphid populations.

Acknowledgements

We thank Trine Nielsen and Pernille Thorbek for their assistance in the field and laboratory work. The Department of Agricultural Sciences of the Agricultural University of Copenhagen kindly supplied the weather data. This work was supported by the Centre for Agricultural Biodiversity of the Danish Environmental Research Programme.

References

Bieri, M., Baumgärtner, J.U., Bianchi, G., Delucchi, V. & Arx, R.V. 1983. Development and fecundity of the pea aphid *Acyrthosiphon pisum* Harris as affected by constant temperatures and pea varieties. *Mitteilungen der Schweizerischen Entomologischen Gesellschaft* 56: 163-71.

Carter, N. & Dixon, A.F.G. 1981. The 'natural enemy ravine' in cereal aphid population dynamics: a consequence of predator activity or aphid biology? *J. Anim. Ecol.* 50: 605-11.

Carter, N., Dixon, A.F.G. & Rabbinge, R. 1982. *Cereal aphid populations: biology, simulation and prediction.* Pudoc, Wageningen.

Feng, M. G., Johnson, J.B. & Halbert, S.E. 1992. Parasitoids (Hymenoptera: Aphidiidae and Aphelinidae) and their effect on aphid (Homoptera: Aphididae) populations in irrigated grain in southwestern Idaho. *Environ. Entomol.* 21: 1433-40.

Force, D. C. & Messenger, P.S. 1964. Fecundity, reproductive rates and innate capacity for increase of three parasites of *Therioaphis maculata* (Buckton). *Ecology* 45: 706-15.

Gutierrez, A. P. & Wang, Y. 1977. Applied population ecology: Models for crop production and pest management. In: Norton, G. A. & Holling, C.S. (eds). *Proceedings of the Conference on Pest Management*, pp. 255-80. New York: Pergamon Press.

Haardt, H. 1991. Untersuchungen zur Bedeutung von Aphelinus-Arten als Parasitoiden von Getreideblattläusen. PhD thesis, Christian-Albrechts-Universität, Kiel.

Hardy, I.C.W. 1994. Sex ratio and mating structure in the parasitoid Hymenoptra. *Oikos* 69: 3-20.

Höller, C., Borgmeister, C., Haardt, H. & Powell, W. 1993. The relationship between primary parasitoids and hyperparasitoids of cereal aphids: An analysis of field data. *J. Anim. Ecol.* 62: 12-21.

Holst, N. & Ruggle, P. 1997. Modelling natural control of cereal aphids: I. The metabolic pool model, winter wheat, and aphids. In: Powell, W. (ed.) *Arthropod natural enemies in arable land. III. Acta Jutlandica* 72 (2): 195-206 (this volume).

Holst, N. & Ruggle, P. (in press). A physiologically based model of natural enemy-pest interactions. *Exper. Appl. Acarol.*.

Holz, F. 1991. A model-based catalogue of case studies as a tool for decision making in the control of the cereal aphid *Macrosiphum [Sitobion] avenae* in winter wheat. *Bull. IOBC/SROP* 14: 35-41.

Holz, F. & Wetzel, T. 1989. Evaluation and utilization of a population model for the English grain aphid *Macrosiphum (Sitobion) avenae* (F.). *J. Appl. Entomol.* 108: 328-34.

Horn, D. J. 1989. Secondary parasitism and population dynamics of aphid parasitoids (Hymenoptera: Aphididae). *J. Kansas Entomol. Soc.* 62: 203-10.

O'Donnell & Mackauer, M. 1989. A morphological and taxonomic study of first instar larvae of Aphiniinae (Hymenoptera: Braconidae). *Syst. Entomol.* 14: 197-219.

Rabbinge, R., Ankersmit, G.W. & Pak, G.A. 1979. Epidemiology and simulation of population development of *Sitobion avenae* in winter wheat. *Netherlands J. Pl. Pathol.* 85: 197-220.

Rabbinge, R., Kroon, A.G. & Driessen, H.P.J.M. 1984. Consequences of clustering in parasite-host relations of the cereal aphid *Sitobion avenae*: a simulation study. *Netherlands J. Agric. Sci.* 32: 237-39.

Schreiber, S. R. & Gutierrez, A.P. (in press). A supply-demand perspective of persistence in food webs: applications to biological control. *Ecology*.

Stilmant, D. 1994. Differential impact of three *S. avenae* parasitoids. *Norwegian J. Agric. Sci.* Supplement No. 16: 89-99.

Vorley, W. T. & Wratten, S.D. 1985. A simulation model of the role of parasitoids in the population development of *Sitobion avenae* (Hemiptera: Aphididae) on cereals. *J. Appl. Ecol.* 22: 813-23.

Vorley, V. T. & S. D. Wratten, S.D. 1987. Migration of parasitoids (Hymenoptera: Braconidae) of cereal aphids (Hemiptera: Aphididae) between grassland, early-sown cereals and late-sown cereals in southern England. *Bull. Ent. Res.* 77: 555-68.

Modelling natural control of cereal aphids: V. Entomophthoralean fungi.

Karsten Dromph[1], Niels Holst[2] and Jørgen Eilenberg[1].

[1]Department of Ecology and Molecular Biology, Royal Veterinary and Agricultural University, Bülowsvej 13, DK-1870 Frederiksberg C, Denmark.
[2]Department of Population Biology, University of Copenhagen, Universitetsparken 15, DK-2100 København Ø, Denmark.

Abstract
The population dynamics of cereal aphids and entomophthoralean fungi are poorly understood. A detailed simulation model of this system is under development to investigate the relative importance of abiotic (temperature, humidity) and biotic (fungus inoculum level, host population dynamics) factors. In this paper the information needed to build a model of the interaction between cereal aphids and fungi is summarized, and the pertinent literature reviewed.

Key Words: simulation model, cereal aphid, entomophthoralean fungi, population dynamics, *Erynia neoaphidis.*

Introduction

Aphids are among the most important insect pests of agriculture in temperate regions. In Danish cereal fields, three species dominate: the English grain aphid, *Sitobion avenae* (F.), the birdcherry-oat aphid, *Rhopalosiphum padi* (L.) and the rose-grain aphid, *Metopolophium dirhodum* (Walk.) (Hansen 1995). Cereal aphids have many different natural enemies including predators, parasitoids and pathogens (Vickerman & Wratten 1979, Chambers et al. 1986, Wraight et al. 1993). The main groups of pathogens attacking aphids are fungi, and five genera of Entomophthorales are important for control of cereal aphids: *Conidiobolus, Entomophthora, Erynia, Neozygites* and *Zoophthora* (Latgè & Papierok 1988). A three years field study in Denmark showed that *Erynia neoaphidis* Remaud. & Henn., *Entomophthora planchoniana* Cornu and *Conidiobolus obscurus* (Hall & Dunn) Remaudiere & Keller were the most abundant, of which only the first two caused epizootics (Steenberg & Eilenberg 1995).

To try to model epizootics of entomophthoralean fungi in cereal aphid populations, it is necessary to understand the basic life cycle of the species involved, and quantify the single steps. In this paper information needed to build a population dynamics model of the aphid-fungus interaction are summarized and the pertinent literature reviewed, with the main emphasis on *E. neoaphidis* for which most information is available.

Life cycle of entomophthoralean fungi

Under satisfactory abiotic conditions an entomophthoralean primary conidium landing on a hydrophobic surface will adhere and germinate. On a susceptible host, a germ tube is produced which penetrates the cuticle. On other surfaces an actively discharged secondary conidium is produced which also has the potential to germinate and penetrate the cuticle of a host. The formation of higher order conidia can continue until the energy resources are exhausted (Zimmermann 1978).

Within the haemolymph of the host, the fungus multiplies in the form of hyphal bodies or protoplasts and after killing the host, the fungus starts to produce spores. These spores can either be produced non-sexually, as conidia projected actively from external conidiophores, or sexually as thick-walled resting spores inside the cadaver. The resting spores are adapted to persist for long periods during unfavourable climatic conditions such as low winter temperatures. These spores have to germinate to produce a primary conidium before infection can take place in the spring (Zimmermann 1978). *Erynia neoaphidis* is not known to produce resting spores (Wilding & Brady 1984c).

Model overview

At the moment, the model exists only as a sketch of components and their interactions (Fig. 1). The model is based upon the metabolic pool concept (Gutierrez & Baumgärtner 1984) using the object-oriented approach described by Holst et al. (in press). The aphid-fungus model is an extension of the winter wheat-aphid model of Holst & Ruggle (1997). In this model, the dynamics of aphid populations depends on host plant quality, on crowding, and on interactions with the fungus population. All components in the system depend on temperature and some (crop, fungus) depend on humidity as well. The fungus is modelled as a population consisting of two stages: sporulating cadavers and conidia. The cadaver stage includes the age-distributed mass and numbers of cadavers under decay. The cadavers produce conidia (resting spores are not

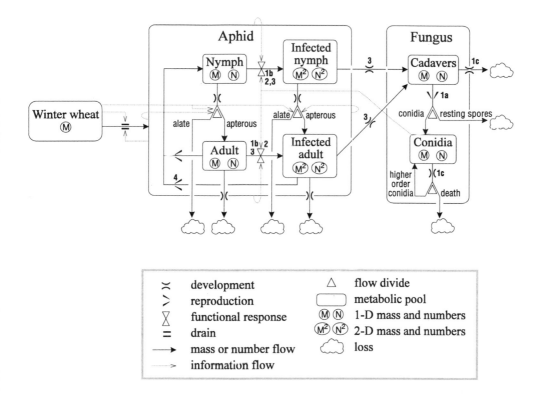

Fig. 1. Model of the interaction between winter wheat, cereal aphids and entomophthoralean fungi. Numbers 1a-c and 2-4 refer to the system components described further in the text and show where they enter the model. For details of the winter wheat model and of the symbolism used, see Holst & Ruggle (1997, this volume).

considered in the model) at a rate which depends on the current mass and its age distribution, and on climatic conditions. The conidium stage includes the age-distributed mass and number of conidia. Conidia have an average life span which is measured on a physiological time scale that depends on temperature and humidity. They can however prolong their life by producing higher order conidia, which in the model are simply put back into the conidium stage. This happens at a cost, however, which is subtracted from the mass (respiration). Below a certain mass the conidia die. The infection rate that converts healthy aphid nymphs or adults into infected ones depends on the density of hosts and conidia, and on the 'apparancy' of the hosts to the conidia as expressed by the Gutierrez-Baumgärtner functional response (Gutierrez & Baumgärtner 1984, Schreiber & Gutierrez in press). The apparancy parameter assumes many processes (1b, 2 and 3; see Fig. 1 and below) and must be estimated from trial runs of the model against field data.

System components

The available information is reviewed for the following components of the system: (1) inoculum level, (2) host density, (3) infection process and (4) pre-lethal effects of fungus infection. Since inoculum level is of primary importance (Soper & McLeod 1981), this component is divided further into: sporulation, transmission and conidia persistence. Fig. 1 shows how these components enter the model.

Inoculum level
Soper and MacLeod (1981) found in their study of the woolly pine needle aphid, *Schizolachnus piniradiatae* (Davidson), that the two most important factors for development of an epizootic were host density and inoculum level, especially the latter.

The level of inoculum present in a locality at a given time depends upon the number of conidia produced by cadavers, the transmission of conidia to and from the locality and the persistence of the conidia.

Sporulation
Sporulation from cadavers is an essential factor determining the total number of conidia present at a certain time. This was established by Wilding (1970) who showed that the daily pattern of conidia release from cadavers was correlated with the number of conidia present in the air.

Resting spores: Of the three species, only *C. obscurus* and *E. planchoniana* produce resting spores (Wilding & Brady 1984a-c). The formation of resting spores has only been studied for *C. obscurus*, where it is thought to be induced by unfavourable climatic conditions such as low temperature and poor nutrition (Papierok 1978, Latgé 1980).

Conidia: Aphids infected with *E. neoaphidis* normally die within a 4 hour period, 14 hours after dawn, and sporulation peaks 8-16 hours after death at 15°C (Milner et al. 1984). The total number of primary conidia produced by an aphid killed by *E. neoaphidis* is about 2-4 x 10^5 at 15-25°C, but is reduced significantly at temperatures below 10°C (Miller, 1981).

Sporulation is positively correlated with air humidity, peaking when the cadaver is in contact with liquid water (Wilding 1969), which is likely to be the reason why sporulation from cadavers peaks during the morning (Milner et al. 1984).

Total conidia production is probably positively correlated with cadaver size. This has been demonstrated for *Entomophthora muscae* (Cohn.) infecting the house fly, *Musca domestica* L. (Mullens & Rodriguez 1985). In addition, the stage or morph of the cadaver could affect the number of conidia produced independent of the effect of size.

The sporulation of *E. neoaphidis* continues for more than 48 hours but it peaks within the first 24 hours at temperatures between 10 and 25°C. At temperatures below 10°C no peak is seen (Milner 1981).

Transmission of infection
Conidia of insect pathogenic fungi can be spread in four possible ways.

(i) From cadaver to aphid. Transmission is directly from a sporulating cadaver to the surrounding healthy aphids (Coremans-Pelseneer et al. 1983).

(ii) From soil to aphid. Conidia from the genus *Conidiobolus* are abundant in the soil, from where they can be transferred to aphids on plants by rain splashes (MacDonald & Spokes 1981, Coremans-Pelseneer et al. 1983).

(iii) From air to aphid. Conidia of *E. planchoniana* and *E. neoaphidis* are carried over long distances with the wind. The conidia content of the air follows closely the release of conidia from cadavers. The concentration is therefore largest in the late summer when aphid populations peak (Wilding 1970, Coremans-Pelseneer et al. 1983). In addition, the conidia content of the air shows the same daily variation as the sporulation of individual cadavers. The peak number of conidia therefore occurs around dawn. Additional peaks occur later during the day, but only when temperature and especially relative humidity are optimal (Wilding 1970).

(iv) From vector to aphid. Very little is known about this method of transmission of conidia, but it has been shown in the laboratory that the seven-spot ladybird (*Coccinella septempunctata* L.) is able to carry conidia of *E. neoaphidis* passively from infected aphid colonies to healthy aphid colonies and initiate infection (Pell & Pluke 1994) In addition, parasitoids also have the potential to act as vectors.

Persistence of spores
(i) Climate. Several climatic factors such as temperature, humidity and sunlight

affect the survival of conidia. Persistence declines with increasing humidity. However the most unfavourable relative humidity (RH) was not 100% but 70-80%. This may be because this level of humidity is high enough to break the dormancy of the conidia but not sufficient for secondary conidia formation. At 100% RH the conidium will form higher order conidia and thereby exhaust its energy resources (Brobyn et al. 1987).

Sunlight probably affects the conidia in the same way as spores of hyphomycete fungi, where the spores are gradually killed by the UV-radiation. Hunt et al. (1994) found that the germination rate for *Metarhizium flavoviridae* Gams & Roszypal spores was affected by UV-radiation and that the damage was proportional to the radiation dose received.

In the field, conidia of *E. neoaphidis* can remain infective for up to 14 days on leaves. The rate of decline is highly dependent on the position of the conidia on the plant. Conidia on more exposed surfaces such as the upper side of leaves in the top of the plant persist only for 7 days, whereas conidia on the underside of leaves near the base of the plant stay infective longer. These differences probably occur because conidia at the base of the plant and on the underside of the leaves are more protected from some climatic factors such as rain and sunlight. At all positions, infectivity declines linearly with time (Brobyn et al. 1985). Under cool and dry conditions *E. neoaphidis* can survive for several months in cadavers, without any impact on the virulence of the conidia produced (Latteur et al. 1985).

(ii) Predation. Very little is known about predation on entomophthoralean spores. At present, investigations are being carried out at IACR-Rothamsted to elucidate the feeding responses of aphid predators, particularly coccinellids, to infected aphids and cadavers. These experiments show that under certain conditions both larvae and adults prey upon infected aphids, and even on sporulating cadavers (Pell & Pluke 1994, Roy & Pell 1995).

Host density
Aphid population density alone cannot be used for predicting epizootics, because aphids are often distributed contagiously in the field and the degree of aggregation is very important. In a less aggregated population the probability of sustaining an epizootic is higher, because of the reduced transmission of conidia directly from cadavers to healthy aphids sitting in the same colony (Soper & MacLeod 1981)

Infection process
After landing on the host, conidia have to germinate and penetrate the cuticle.

Both processes depend on the climate and on the host. The course of the infection further depends on the fungal strain involved.

Climate

Climatic factors such as temperature and relative humidity are important in the infection process. The optimum temperature for germination of both primary and secondary conidia of *E. neoaphidis* is 18-21°C (Morgan et al. 1995). The lethal time of aphids infected with entomophthoralean fungi is also temperature-dependent, decreasing with increasing temperature. *Erynia neoaphidis* takes 3-5 days to kill aphids at 20°C, while at 8°C it takes 12-15 days (Milner & Bourne 1983).

For infection with *E. neoaphidis* to occur at 20°C, at least 3 hours at 100 % RH is required and the required period increases as temperature decreases. At 10°C and high RH, the required time for infection is increased to 7 hours (Milner & Bourne 1983). In contrast, conidia of *E. planchoniana* germinate independently of RH, and appear therefore to be better adapted to relatively dry environments (Holdom 1986).

Host

Even within an insect species there can be a large variation in susceptibility to entomopathogenic fungi. In the pea aphid, *Acyrthosiphon pisum* (Harris), two distinct biotypes occur, one susceptible to *E. neoaphidis* and one almost resistant (Milner 1982).

The physiological condition of a susceptible host may also affect its infection with *E. neoaphidis*. Milner and Soper (1981) found that if aphids were starved for 24 hours after inoculation, fewer died as a result of the fungus.

The age of the aphid also plays a role in its susceptibility. Three to four days old adult *A. pisum* were less susceptible than both 0-1 day old and 6-7 days old adults (Lizen et al. 1985). Presumably the aphids are at their physiological peak at this time (nymph production also peaks at almost the same time) and the aphids therefore are better able to resist an infection.

Aphid instar at the time of inoculation may also have an effect. If nymphs receive the conidia just before they moult, the conidia will be lost together with the old cuticle (Latgé & Papierok 1988). However, when nymphs do become infected, the lethal time is the same as for adults for all nymphal instars (Schmitz et al. 1993). Studies have shown that alate morphs of *A. pisum* are up to 6 times more susceptible to *E. neoaphidis* infections than apterous morphs (Lizen et al. 1985).

Fungus

Characteristics of the fungus itself play an important role in its potential to infect the aphid. Different isolates may have different virulence towards aphids or have different host ranges (Milner & Soper 1981, Milner 1982, Yu et al. 1995). Within an isolate the different spore types may also show differences in virulence. This has been demonstrated for *E. muscae* in bioassays against *M. domestica*. In that case, the LC_{50} of the secondary conidia was 200 times lower than for the primary conidia (Bellini et al. 1992). It is not know if this is also the case for entomophthoralean species attacking aphids.

Prelethal effects of fungus infection

Infection with an entomopathogenic fungus may affect the host in several ways before killing it. Sugar-beet root aphids, *Pemphigus betae* Doane, infected with *Erynia nouryi* Remaud. et Henneb. climb up onto the leaves of the plant before dying (Harper 1958). Similarly, locusts and grasshoppers infected with *Entomophaga grylli* (Fres.) Batko, move to the top of the vegetation just before they die (Skaife 1925), whilst carrot flies, *Psila rosae* F., modify their oviposition behaviour when infected with *E. muscae*. Infected individuals lay their eggs in hedges and trees instead of on the carrots (Eilenberg 1987). It is possible that fungus infection may affect cereal aphids in other ways than just killing them, and this could be equally important in understanding the dynamics of the aphid - pathogen system. At present however very little is known about this.

Developmental time

Infection with *E. neoaphidis* has no significant effect on the duration of the single nymphal instars, except for the fourth apterous instar in which the duration is increased slightly by infection from 51.2 to 65.6 hours at 20°C (Schmitz et al. 1993).

Reproduction

Infection with *E. neoaphidis* has a great effect on the total reproductive capacity of an aphid if the infection happens shortly after the aphid has moulted from fourth instar to adult, because the daily reproduction rate peaks 5 days after the last moult (Schmitz et al. 1993). Infection by the hyphomycete fungus, *Beauveria bassiana* (Balsamo) Vuillemin, does not affect the reproductive rate of the Russian wheat aphid, *Diuraphis noxia* Kurdyumov, in any other way than does mortality due to age (Wang & Knudsen 1993). If infection occurs in one of the nymphal instars the aphid will die anyway before it has reproduced, even if the infection takes place in the last nymph instar (Schmitz et al. 1993).

Migration

Nothing is known on the possible effects of fungal infections on the migration behaviour of alates.

Discussion

At present both field and laboratory data is only available for *E. neoaphidis*. However, the information needed to develop a detailed model of *E. neoaphidis* epizootics is still incomplete. A more detailed set of data is still required for several factors, such as differences in susceptibility to *E. neoaphidis* among the three different species of cereal aphids. Other factors, such as interactions with natural enemies (both positive and negative) and prelethal effects of infection, are simply not understood, and must therefore be omitted from the model at this stage. Information on *E. planchoniana* is so incomplete that a simulation model could only be used as a tool to indicate the missing information. The relative importance of factors affecting the development of epizootics in cereal aphid populations in the field, for example aphid density and climatic factors such as humidity and temperature, is still not fully understood. Furthermore, the factors determining which pathogen species will be the dominant species at a specific locality at a given time are not known. A more detailed data set together with existing data from the literature will make it possible to develop a model of the interactions between aphids and *E. neoaphidis*. Hopefully, the model will answer some of these questions. We are in the first phase of developing a simulation model of fungal infections in cereal aphid populations. The modelling approach was chosen to integrate the existing knowledge on the subject and hence to guide our research. The model, though still vague, provides an outline for the literature survey and helps structure the information gathered.

Acknowledgements

This study was carried out as part of Centre for Agricultural Biodiversity supported by the Danish Environmental Research Programme.

References

Bellini, R., Mullens, B.A. & Jespersen, J.B. 1992. Infectivity of two members of the *Entomophthora muscae* complex (Zygomycetes: Entomophthorales) for *Musca domestica* (Diptera: Muscidae). *Entomophaga* 37: 11-19.

Brobyn, P.J., Wilding, N. & Clark, S.J. 1985. The persistence of infectivity of conidia of the aphid pathogen *Erynia neoaphidis* on leaves in the field. *Ann. Appl. Biol.* 107: 365-76.

Brobyn, P.J., Wilding, N. & Clark, S.J. 1987. Laboratory observations on the effect of humidity on the persistence of infectivity of conidia of the aphid pathogen *Erynia neoaphidis*. *Ann. Appl. Biol.* 110: 579-84.

Chambers, R.J., Sunderland, K.D., Stacey, D.L. & Wyatt, I.J. 1986. Control of cereal aphids in winter wheat by natural enemies: aphid specific predators, parasitoids and pathogenic fungi. *Ann. Appl. Biol.* 108: 219-31.

Coremans-Pelseneer, J., Villers, S. & Matthys, V. 1983. Entomophthorales found on wheat aphids, in soil and air on the same field. Four years compared results. *Med. Fac. Landbouww. Rijksuniv. Gent* 48: 207-213.

Eilenberg, J. 1987. Abnormal egg-laying behaviour of female carrot flies (*Psila rosae*) induced by the fungus *Entomophthora muscae*. *Entomol. Exp. Appl.* 43: 61-65.

Gutierrez, A.P. & Baumgärtner, J.U. 1984. I. Age-specific energetic models – pea aphid *Acyrthosiphon pisum* (Homoptera: Aphididae) as an example. *Can. Ent.* 116: 924-932.

Hansen, L.M. 1995. Aphids – the national pest in Denmark. *Danske Planteværnskonference. Sygdomme og skadedyr* 12: 115-28.

Harper, A.M. 1958. Notes on behavior of *Pemphigus betae* Doane (Homoptera: Aphididae) infected with *Entomophthora aphidis* Hoffm. *Can. Ent.* 90: 439-40.

Holdom, D.G. 1986. Moisture requirements and field occurrence of *Entomophthora planchoniana* Cornu. In: Baily, R. & Swincer, D. (eds.) *Pest Control: Recent Advances and Future Prospects,* pp. 368-74. S. Australian Govt. Printer, Adelaide.

Holst, N. & Ruggle, P. 1997. Modelling natural control of cereal aphids. I. Metabolic pool, winter wheat and cereal aphids. In: Powell, W. (ed.) *Arthropod natural enemies in arable land. III. Acta Jutlandica* 72 (2): 195-206 (this volume).

Holst, N., Axelsen, J.A., Olesen, J.E. & Ruggle, P. (in press) An object-oriented implementation of the metabolic pool model. *Ecol. Modelling*

Hunt, T.R., Moore, D., Higgins, P.M. & Prior, C. 1994. Effect of sunscreens, irradiance and resting periods on the germination of *Metarhizium flavoviride* conidia. *Entomophaga* 39: 313-23.

Latgé, J.P. 1980. Sporulation d'*Entomophthora obscura* Hall & Dunn en culture liquide. *Can. J. Microbiol.* 26: 1038-48.

Latgé, J.P. & Papierok, B. 1988. Aphid pathogens. In: Minks, A,K,W. & Harrewijn, P. (eds.) *World Crop Pests, Vol. 2B. Aphids, their Biology, Natural Enemies and Control,* pp. 323-35. Elsevier, The Netherlands.

Latteur, G., Lizen, E. & Oger, R. 1985. Influence de divers facteurs physiques sur la virulence des conidies de l'Entomophthorale *Erynia neoaphidis* Remaud. et Henn. Envers le puceron du pois, *Acyrthosiphon pisum* (Harris). *Parasitica* 41: 151-62.

Lizen, E., Latteur, G. & Oger, R. 1985. Sensibilite a l'infection par l'Entomophthorale *Erynia neoaphidis* Remaud. Et Henn. Du puceron *Acyrthosiphon pisum* selon sa forme, son stage et son age. *Parasitica* 41: 163-70.

MacDonald, R.M. & Spokes, J.R. 1981. *Conidiobolus obscurus* in arable soil: a method for extracting and counting azygospores. *Soil Biol. Biochem.* 13: 551-53.

Milner, R.J. 1981. Patterns of primary spore discharge of *Entomophthora spp.* From the blue green aphid, *Acyrthosiphon kondoi. J. Invert. Pathol.* 38: 419-25.

Milner, R.J. 1982. On the occurrence of pea aphids, *Acyrthosiphon pisum*, resistant to isolates of the fungal pathogen *Erynia neoaphidis. Entomol. Exp. Appl.* 32: 23-7.

Milner, R.J. & Bourne, J. 1983. Influence of temperature and duration of leaf wetness on infection of *Acyrthosiphon kondoi* with *Erynia neoaphidis. Ann. Appl. Biol.* 102: 19-27.

Milner, R.J. & Soper, R.S. 1981. Bioassay of *Entomophthora* against the spotted alfalfa aphid *Therioaphis trifolii* F. *maculata. J. Invert. Pathol.* 37: 168-73.

Milner, R.J., Holdom, D.G. & Glare, T.R. 1984. Diurnal patterns of mortality in aphids infected by entomophthoran fungi. *Entomol. Exp. Appl.* 36: 37-47.

Morgan, L.W., Boddy, L., Clark, S.J. & Wilding, N. 1995. Influence of temperature on germination of primary and secondary conidia of *Erynia neoaphidis* (Zygomycetes: Entomophthorales). *J. Invert. Pathol.* 65: 132-38.

Mullens, B.A. & Rodriguez, J.L. 1985. Dynamics of *Entomophthora muscae* (Entomophthorales: Entomophthoraceae) conidial discharge from *Musca domestica* (Diptera: Muscidae) cadavers. *Environ. Entomol.* 14: 317-22.

Papierok, B. 1978. Obtention in vivo des azygospores d'*Entomophthora thaxteriana* Petch, champignon pathogène de pucerons (Homoptères, Aphididae). *Compt. Rend. Acad. Sc. Paris, Série D,* 286: 1053-56.

Pell, J.K. & Pluke, R. 1994. Interactions between two aphid natural enemies, the entomophthoralean fungus Erynia *neoaphidis* and the predatory beetle *Coccinella septempunctata. Abstracts VI International Colloquium on Invertebrate Pathology and Microbial Control, Montpellier, France.* p. 114.

Roy, H.E. & Pell, J.K. 1995. Feeding behaviour of fourth instar *Coccinella septempunctata* larvae on *Acyrthosiphon pisum* aphids infected with *Erynia neoaphidis. Proceedings 28th Annual Meeting of the Society for Invertebrate Pathology, New York, U.S.A.* p. 53.

Schreiber, R.S. & Gutierrez, A.P. in press. A supply/demand perspective of persistence in food webs: applications to biological control. *Ecology*

Schmitz, V., Dedryver, C.H. & Pierre, J.S. 1993. Influence of an *Erynia neoaphidis* infection on the relative rate of increase of the cereal aphid *Sitobion avenae. J. Invert. Pathol.* 61: 62-68.

Skaife, S.H. 1925. The locust fungus, *Empusa grylli,* and its effects on its host. *S. African J. Sci.* 22: 298-308.

Soper, R.S. & MacLeod, D.M. 1981. *Descriptive epizootiology of an aphid mycosis.* USDA Technical Bulletin No. 1632.

Steenberg, T. & Eilenberg, J. 1995. Natural occurrence of entomopathogenic fungi on aphids on an agricultural field site. *Czech. Mycology* 48: 89-96.

Vickerman, G.P. & Wratten, S.D. 1979. The biology and pest status of cereal aphids (Hemiptera; Aphididae) in Europe: a review. *Bull. Ent. Res.* 69: 1-32.

Wang, Z.G. & Knudsen, G.R. 1993. Effect of *Beauveria bassiana* (Fungi: Hyphomycetes) on fecundity of the Russian wheat aphid (Homoptera: Aphididae). *Environ. Entomol.* 22: 874-78.

Wilding, N. 1969. Effect of humidity on the sporulation of *Entomophthora aphidis* and E.

thaxteriana. Trans. Br. Soc. 53: 126-130.

Wilding, N. 1970. Entomophthora conidia in the air-spora. *J. General Microbiol.* 62: 149-157.

Wilding, N. & Brady, B.L. 1984a *Conidiobolus obscurus.* CMI Descriptions of Pathogenic Fungi and Bacteria, No. 812.

Wilding, N. & Brady, B.L. 1984b. *Entomophthora planchoniana.* CMI Descriptions of Pathogenic Fungi and Bacteria, No. 814.

Wilding, N. & Brady, B.L. 1984c. *Erynia neoaphidis.* CMI Descriptions of Pathogenic Fungi and Bacteria, No. 815.

Wraight, S.P., Poprawski, T.J., Meyer, W.L. & Peairs, F.B. 1993. Natural enemies of Russian wheat aphid (Homoptera: Aphididae) and associated cereal aphid species in spring-planted wheat and barley in Colorado. *Environ. Entomol.* 22: 1383-91.

Yu, Z., Nordin, L., Brown, G.C. & Jackson, D.M. 1995. Studies on *Pandora neoaphidis* (Entomophthorales: Entomophthoraceae) infections of the red morph of tobacco aphid (Homoptera: Aphididae). *Environ. Entomol.* 24: 962-66.

Zimmerman, G. 1978. Zur Biologie, Untersuchungsmethodik und Bestimmung von Entomophthoraceen (Phycomycetes: Entompohthorales) an Blattläusen. *Z. angew. Entomol.* 85: 241-252.

Cereal aphid predation by the ladybird *Coccinella septempunctata* L. (Coleoptera: Coccinellidae) - Including its simulation in the model GTLAUS

By H. Triltsch[1] and D. Roßberg[2]

[1]Institute of Integrated Plant Protection, Kleinmachnow, Germany
[2]Institute for Technology Assessment in Plant Protection, Kleinmachnow, Germany
Present address: Federal Biological Research Centre for Agriculture and Forestry,
Kleinmachnow Branch, Stahnsdorfer Damm 81, 14532 Kleinmachnow, Germany

Abstract

Population development of *Sitobion avenae* (F.) on winter wheat in the presence and absence of adult *Coccinella septempunctata* L. was investigated in climate chambers under fluctuating temperatures, with daily averages of 17, 20, 22 and 25°C. The experiments indicated a strong temperature influence on the predatory effect of the ladybird. With rising temperature the increase in the rate of ladybird predation on aphids was much higher than the increase in aphid reproduction. The predator gained an advantage from a surprisingly small temperature increase.

The hunger level of the ladybird in the field was determined by comparing its field weight with the body weight under controlled conditions and with a food surplus. While the feeding condition was related to aphid presence and adult life stage, the hunger depended on temperature.

In the simulation model "GTLAUS" the aphid demand of the ladybird *C. septempunctata* is temperature-dependent. This demand was corrected by a parameter characterising the aphid-finding probability, which was assumed to increase with temperature (through increased beetle activity), and with aphid density. Finally the model compares the corrected aphid demand of the ladybird with the actual aphid presence.

Key words: aphid predation, *Coccinella septempunctata*, ladybird, simulation model, *Sitobion avenae*, wheat field.

Arthropod natural enemies in arable land · III The individual, the population and the community
W. Powell (ed.). *Acta Jutlandica* vol. 72:2 1997, pp. 259-270
© Aarhus University Press, Denmark, ISBN 87 7288 673 0

Introduction

Within computer models of predator-prey systems, the description of predator feeding rates is a basic problem. This paper presents investigations on the predation of cereal aphids by the adult ladybird *Coccinella septempunctata* L. and the utilisation of the results in the simulation model GTLAUS. GTLAUS is a discrete simulation model that is written in Turbo-Pascal and runs under DOS. The time step is one day. The model complex consists of three submodels: wheat ontogenesis, population dynamics of the cereal aphid *Sitobion avenae* (F.), and population dynamics and predation rate of aphid-specific predators, especially *C. septempunctata*. GTLAUS is not a prediction model for practical plant protection, but a tool for complex ecological studies in the tritrophic winter wheat – cereal aphid – predator system. A full description of the model is given by Roßberg & Freier (1993) and Freier et al. (1996b).

The first step in the investigations was climate chamber experiments aimed at quantifying the effect of adult ladybirds on aphid population development under controlled conditions. The daily predation rate on aphids was measured under conditions similar to those in the field, where the predator must actively search for food. This was necessary to get more realistic data than those obtained in feeding experiments with a food surplus and constant temperatures (e.g. Asgari 1966, Shands & Simpson 1972, Ghanim et al. 1984).

The second step was to estimate hunger levels of the ladybird in the field to get information on its success in aphid-finding, depending on environmental factors and aphid density.

The third step was to include a parameter characterising cereal aphid predation in the GTLAUS model on the basis of the daily ladybird food demand, corrected for the aphid-finding probability in relation to temperature and aphid density.

Climate chamber experiments: Predation under controlled conditions

Material and methods

The experiments took place under fluctuating temperatures (Table 1), with other conditions (daylength 16h, 15000 Lux, ca. 65 % RH) being equal. Each experiment included 10 pots with 15 or 16 stalks of winter wheat ("Orestis") infested with the cereal aphid *S. avenae* at growth stage DC 69 (end of flowering, see Stauss 1994) and half of them also containing one male adult *C. septempunctata*. The initial aphid density was 4 individuals (last instars or adults) per ear in 3 experiments and 8 per ear in 5 further experiments. A gauze netting above a wire frame on top of the pots prevented the insects from escaping. Population development of the aphid was ascertained by counting all individuals twice per week. Each experiment lasted until the end of aphid infestation, i.e. the death of all apterous aphids on the wheat plants.

Results

Table 1 shows the effects of increasing temperature on the cereal aphids and ladybird adults. With rising temperatures, crop development was faster and therefore the duration of aphid infestation became shortened, without any predatory impact. In the absence of the ladybird, the aphid population growth rate reached its maximum at 22°C and decreased rapidly at 25°C, whereas the

Table 1. Results of climate chamber experiments on the population development of *S. avenae* and predation by adult *C. septempunctata*

Temperature (average/ day/night) [°C]	Number of experiments/ Initial aphid density [aphids/ear]	Duration of aphid infestation without ladybird predation** [days]	Aphid reproduction rate without ladybird predation* [1/day]	Calculated ladybird predation rate* [aphids/day]
17 /20/11	1 /4	34	1.20	5.0
20 /23/14	3 /4,8,8	30 (30,29,30)	1.23 (1.20,1.18,1.34)	6.1 (8.2,6.8,3.4)
22 /25/16	2 /4,8	28 (28,28)	1.28 (1.30,1.27)	17.4 (14.3,20.4)
25 /27/19	2 /8,8	24 (22,26)	1.13 (1.13,1.13)	18.1 (23.9,13.2)

*   To calculate daily aphid reproduction and ladybird predation, the

    BOMBOSCH-equation (according to van Emden 1965) was used: $a_n = a_o q^n - k(q\dfrac{q^n - 1}{q-1})$

    (a...aphid density, n...number of days, q...daily aphid multiplication, k...daily predator feeding). This equation assumes an exponential aphid population growth and does not consider changes in the nutritional value of the plant for aphids, resulting from ripening. The predatory impact is given as a constant parameter.

** This duration means the period between wheat stage DC 69 and the death of all apterous aphids on the wheat plants.

feeding rate of the ladybird increased with higher temperatures over the whole range investigated.

A comparison of the number of aphids eaten daily by one ladybird adult with the daily number of aphid offspring from a given aphid population shows a temperature-dependent predator-prey ratio at which all of the offspring would be killed by the predator. This ratio would be 1:25 at 17°C, 1:27 at 20°C, 1:62 at 22°C and 1:151 at 25°C using data from Table 1. Obviously, the ladybird was better at controlling the aphids at higher temperatures. From 20 to 22°C the feeding rate of the predator increased much more than did the aphid multiplication, as revealed by the predator-prey ratios, and at 25°C the ladybird additionally profited from the decrease in aphid reproduction.

Wheat field: Ladybird hunger levels under field conditions

Material and methods

The "hunger" of the ladybird, as proposed by Frazer & Gill (1981), was used to assess prey-finding success. The individual body weight of beetles is correlated with the rate of food intake and can be used in determining beetle hunger (Frazer & Gill 1981). Ladybirds, 20 to 40 male and female adults of *C. septempunctata,* were sampled in the field and then weighed to estimate their field weight (W_{field}). They were weighed immediately after transportation to the laboratory in a cool box. A group of 16 beetles, 8 of either sex and of medium weight, were then chosen for the experiments. The ladybirds, kept individually with a food surplus of *S. avenae* at 20°C daily average (day/night=16h, 23°C/ 8h, 16°C; 65 %RH; 10 000 Lux), were weighed four times per day until no changes in their body weight occurred. Beetle sampling took place on two dates at each location: (1) at a hibernation site in March (inactive beetles), (2) in winter wheat in June (reproducing beetles) and (3) in spring wheat in July (new adult generation).

Under conditions of aphid surplus in the experiments, the body weight of the ladybirds increased within a few hours to a first maximum (W_{max}) and then reached a relatively constant level in the following days (W_{opt}). In accordance with Frazer & Gill (1981), who described hunger as feeding until first satiation, in our experiments the difference between the field weight and the first maximum of body weight under food surplus was defined as hunger (H). The difference between weight in the field and the long term average weight under food surplus, we termed food deficit (FD). Whereas hunger represents the actual demand for food, the food deficit is the result of a comparison between the field weight of a certain ladybird individual and its body weight under optimal food conditions, i.e. a complete satiation over a number of days.

$$H = W_{max} - W_{field} \quad [\Delta mg]$$
$$FD = W_{opt} - W_{field} \quad [\Delta mg]$$

To show differences of hunger and food deficit between sampling dates and sexes, we tested the two values separately using a t-test (P<0.05). Hunger and food deficit were then tested for any relationship with temperature and aphid density by calculating correlation coefficients.

Results

Food deficit was highest and hunger was lowest in ladybird adults from hibernation sites. This was true for both sexes. In reproducing ladybird adults

Table 2. Hunger and food deficit of field sampled adults of *C. septempunctata*

Life stage/ location	Date	Temp.* [°C]	Prey** density [aphids/ tiller]	Field weight [mg (SD)] Females	Males	Hunger [Δmg (SD)] Females	Males	Food deficit [Δmg (SD)] Females	Males
inactive/ hiber-	07.03.	6.7 (3.00)de	0.0	35.0 (4.08)	34.5 (5.79)	0.2 (0.29)a	2.0 (1.58)ab	12.7 (2.00)ef	8.8
nation site	13.03.	6.6 (3.19)bcd	0.0	37.6 (4.87)	27.7 (3.21)	0.8 (0.76)ab	1.7 (1.96)ab	15.9 (3.11)f	7.0
reprodu- cing/	06.06.	9.8	1.6	46.2 (3.72)	36.7 (1.47)	6.3 (3.97)cd	0.2 (0.64)a	3.6 (3.16)ab	0.0 (0.00)a
winter wheat	20.06.	19.2	3.6	53.6 (1.73)	32.4 (2.03)	8.9 (5.50)d	2.0 (1.23)bc	0.7 (1.16)a	1.0 (1.04)a
new ge- neration/	04.07.	25.7 (1.99)bc	3.5	39.5 (1.52)	31.5 (2.48)	9.1 (3.50)d	5.9 (2.33)cd	9.3 (3.58)cde	5.9
spring wheat	11.07.	18.6	2.0	41.1 (1.72)	37.2 (1.94)	8.4 (3.52)d	6.3 (2.02)cd	10.9 (6.83)def	4.7 (2.82)b

* Temperature is the daily average value of the previous day according to the nearest stationary weather station.

** Aphid density estimated by separate countings at the same time.

from winter wheat fields, food deficit was lowest and hunger differed significantly between the two sexes. The new adult generation as a whole showed high levels of food deficit and hunger (Table 2).

Ladybird hunger was significantly correlated with the temperature in the field ($r = 0.89$ in females, $r = 0.74$ in males and $r = 0.75$ both sexes together, $P<0.05$). Hunger was not correlated with food deficit for either sex ($r = -0.67$, $P=0.05$ in females; $r = 0.31$, $P=0.05$ in males). Especially in reproducing female ladybirds, hunger was much higher than food deficit. In this period, male adults had no food deficit and a low hunger level at an aphid density of 1.6 aphids/tiller. The females needed a higher aphid density (3,6 aphids/tiller) to compensate for their food deficit. Food deficit was related not only to aphid presence but also to the life stage of the adults.

Simulation: Predation in the model GTLAUS

The simulation of ladybird predation in the GTLAUS model is based on the daily food demand (D [aphids/day]) of adult ladybirds. We assumed that this demand depends only on temperature (T). We accounted for sex differences in the model by assuming a 1:1 sex ratio and using the average demand of the two sexes. From 10°C, i.e. close to the lower developmental threshold of immature ladybird stages (Butler 1982, Honek & Kocourek 1988, Frazer & McGregor 1992) food demand rises linearly up to 27°C above which it remains constant.

$$D = \begin{cases} T \leq 10°C: 0 \\ 10 < T \leq 20°C: 0.2(T\text{-}10) \\ 20 < T \leq 27°C: 0.4(T\text{-}20)+20 \\ T > 27°C : 50 \end{cases}$$

We corrected the demand (D) by multiplying it with the aphid-finding probability (P_{find}). It is extremely difficult to assess this probability, as also mentioned by Gutierrez et al. (1981). In our case, the results described above gave some information about the aphid-finding success of the ladybird under field conditions. We assumed that the probability of aphid-finding increases with temperature (T) (due to increased beetle activity), and with aphid density (ad). Hence, temperature has two effects on ladybird predation in the model, it determines beetle hunger (i.e. food demand) and beetle activity and thereby the chance of finding an aphid.

$$D_{cor} = P_{find} * D$$
$$P_{find} = f_1(ad) * f_2(T)/100$$

$$f_1 = \begin{cases} ad = [aphids/tiller] \\ ad \leq 1: 51.8\ ad \\ 1 < ad \leq 50: 0.8\ ad + 51 \\ ad > 50 : 91 \end{cases} \qquad f_2 = \begin{cases} T < 14°C: 0 \\ 14 \leq T < 26: (T-14)12 \\ T > 26: 1 \end{cases}$$

Finally, the model compares the corrected aphid demand (D_{cor}) of the ladybird with the aphid density. If aphid density is higher than the corrected demand of all predatory individuals, daily predation is equal to this demand. If aphid density is lower, the present number of aphids are eaten and it is then possible to distinguish a value of food scarcity.

We used our results from the climate chamber experiments to estimate the parameters of the equations for aphid demand and aphid-finding probability. Fig. 1 compares the simulated predation with data derived from the climate chamber experiments.

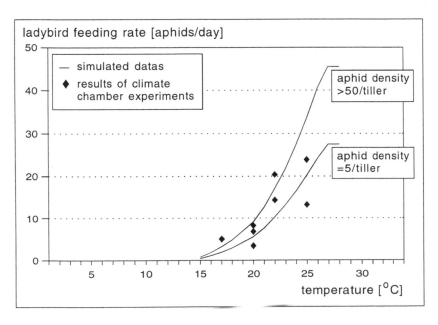

Fig. 1. Predation rate of *C. septempunctata* showing experimental data and simulations.

Discussion

In climate chamber experiments, three trophic levels – winter wheat, cereal aphids, and ladybird adults – were combined to investigate the predator-prey interaction under controlled conditions. In these experiments, predator and prey showed different responses to increasing temperatures. From 17 to 22°C, daily aphid multiplication increased by 110%, but daily ladybird predation increased by 300-500%. Cereal aphid reproduction was highest at 22°C and lowest at 25°C. Previously, Dean (1974) observed an optimum temperature of 22.5°C and the death of all larval aphids at 30°C. On the other hand, ladybird predation rose over the whole temperature range investigated. We used the BOMBOSCH-equation (van Emden 1965) to calculate aphid reproduction rate and ladybird predation rate. This equation describes predatory impact as a constant rate, but does not consider the host plant trophic level, i.e. it does not describe changes in the nutritional value of the plant for the aphids, resulting from ripening. Therefore, it is only used during the period of exponential aphid population growth between wheat stage DC 69 and DC 83.

In our experiments, we tried to simulate conditions relatively similar to those in the field, under which the predator must actively search for food. We chose to use fluctuating temperatures and potted wheat plants. Our calculated predation rates are comparable to those presented by Honek (1986). Honek (1986) estimated predation rates between 6.2 and 36.3 aphids/day for field collected ladybird adults on the basis of faecal production.

Because prey density changed during our experiments, it is difficult to attribute predation rates to prey density. We assumed a situation of food surplus, because aphid density was generally higher than 5 aphids/tiller. Nevertheless, different prey densities might be one reason for differences in calculated predation rates. The faster increase in ladybird predation rate compared to the increase in aphid reproduction rate is to a certain extend explainable by differences between the temperature requirements of aphid and ladybird. *Sitobion avenae* with its lower developmental threshold of 4.8°C (Campbell et al. 1974) had already nearly reached maximal reproduction at the starting point of our investigated temperature range. Searching activity of *C. septempunctata* starts at 13-17°C (Honek 1985) and temperatures in our experiments therefore corresponded to the period of increasing beetle activity. The observed advantage of predatory ladybirds at warmer conditions has already been described by Frazer & Gilbert (1976).

Under field conditions, aphid density is often significantly lower (Freier 1983, Freier et al. 1996a) than simulated in our experiments and there is still a lack of data on predation rates at low prey densities. Because little is known about the prey-finding success of ladybirds in the field (Gutierrez et al. 1981), we used a comparison between two values of beetle weight under food surplus, i.e. first maximum and average value, and beetle weight in the field to quantify hunger levels and to get information on prey-finding success. The reason for estimating two different parameters was differences in the curve of weight changes under food surplus between sampling periods. Though body weight of all ladybird adults increased within a few hours to a first maximum, this peak was relatively low in adults from hibernation sites compared to their weight over the following days, but was relatively high in adults from wheat fields.

The differences in food deficit between the three sampling periods show conformity with data concerning the gut contents of adult *C. septempunctata* (Triltsch 1996). Adults from hibernation sites as well as newly emerged adults had nearly empty guts, whereas aphid remains were present in 90% of reproducing ladybirds from winter wheat fields. The beetle hunger, which we understand as an actual demand for food, was not correlated with food deficit. However, hunger and food deficit in newly emerged adults were very similar, especially in adults from hibernacula, but in reproducing females the two values differed distinctly. It was possible to correlate hunger and field temperature, but our data base is relatively small and we do not consider possible physiological differences between the adults sampled at different periods. Nevertheless, it seems plausible that ladybirds sampled in March had little actual demand for food although their guts were empty, because temperatures were clearly below the activity threshold (see Honek 1985). Hodek (1967) stated that diapause in *C. septempunctata* terminates at the end of the year and after that the adults are still inactive only because of low temperatures. The higher activity threshold of the predator compared to its prey is easily understood as an adaptation to overcome the period of food scarcity.

Our method was limited in its suitability for reproducing females. Although we added egg weight to female body weight after each egg laying to compensate for weight loss due to reproduction, this loss in body weight was responsible for a relatively low average weight under food surplus compared to their first maximum weight. Egg laying was also responsible for large differences in body weight during the experiment.

Using analysis of production of faeces in adult ladybirds Honek (1986) found that *C. septempunctata* were hungry over long periods and on most host plants. Concerning the aphid-finding success of adult *C. septempunctata* in winter wheat fields, we found that males had no food deficit at a prey density of 1.6 aphids/tiller. The females needed a higher aphid density (3.6 aphids/tiller) to compensate for their food deficit, probably due to their reproductive demands and higher average body size. In periods of existing food deficit, temperature seems to define the degree with which the ladybirds strive to compensate for it.

In the GTLAUS model, ladybird predation is the important connection between the aphid submodel and the predator submodel. To build up predation in the model we used results from specific climate chamber experiments. Because our calculated predation rates are related only to relatively high aphid densities (>5 aphids/tiller) we assumed predation at lower aphid densities. As a whole prey-finding success in the field is not clear because we have limited indication at which aphid densities ladybird prey-finding is sufficient to fulfil their food demand. Another problem is predation at high temperatures (<25°C). No data are available, but it is likely that predation drops at very high temperatures (Skirvin 1995).

Description of aphid feeding of ladybird larvae in the model is incomplete. Probable differences in prey-finding between ladybird adults and larvae, especially younger instars, are not covered by experimental data. For this reason we only used the different predation rates based on data from Ghanim (1981).

The possibility of distinguishing a parameter of food scarcity by comparing aphid demand and aphid density is a special feature of the simulated predation in the GTLAUS model. This parameter influences other processes of the predator in the model, for instance time of larval development.

After complete build-up of the GTLAUS model it was validated with field data (Freier et al. 1996b). Because GTLAUS only describes the population dynamics of *S. avenae,* two years in which this species was dominant in winter wheat were chosen. Another critical point is the occurrence of other aphid predators in the field, which are not included in the model or are given as single regression equations. Scenario runs with GTLAUS in comparison with observed aphid densities in wheat fields showed that the model is able to simulate the general trend of aphid population development, i.e. growth or breakdown of population and end of infestation (Freier et al. 1996b).

Acknowledgements

The first author was funded by the BMBF.

References

Asgari, A. 1966. Untersuchungen über die im Raum Stuttgart-Hohenheim als wichtigste Prädatoren der Grünen Apfelblattlaus (*Aphidula pomi* DEG.) auftretenden Arthropoden. *Z. angew. Entomol.* 93: 51-60.

Butler, G. D. 1982. Development time of *Coccinella septempunctata* in relation to constant temperatures. *Entomophaga* 27: 349-53.

Campbell, A., Frazer, B.D., Gilbert, N., Gutierrez, A.P. & Mackauer, M. 1974. Temperature requirements of some aphids and their parasites. *J. Appl. Ecol.* 11: 431-38.

Dean, G.J. 1974. Effect of temperature on the cereal aphids *Metopolophium dirhodum* (WLK.), *Rhopalosiphum padi* (L.) and *Macrosiphum avenae* (F.) (Hom., Aphididae). *Bull. Ent. Res.* 63: 401-09.

van Emden, H. F. 1965. The effectiveness of aphidophagous insects in reducing aphid populations. In: Hodek, I. (ed.) *Ecology of Aphidophagous Insects*, pp. 227-35. Academia, Prague.

Frazer, B. D. & Gilbert, N. 1976. Coccinellids and Aphids: A quantitative study of the impact of adult ladybirds (Col.: Coccinellidae) preying on field populations of pea aphids (Hom.: Aphididae). *J. Entomol. Soc. Br. Columbia* 73: 33-56.

Frazer, B. D. & Gill, B. 1981. Hunger, movement and predation of *Coccinella californica* on pca aphids in the laboratory and in the field. *Can. Ent.* 113: 1025-33.

Frazer, B. D. & McGregor, R. R. 1992. Temperature dependent survival and hatching rate of eggs of seven species of Coccinellidae. *Can. Ent.* 124: 305-12.

Freier, B. 1983. Untersuchungen zur Struktur von Populationen und zum Massenwechsel von Schadinsekten des Getreides als Grundlage für ihre Überwachung, Prognose und gezielten Bekämpfung sowie für dic Entwicklung von Simulationsmodellen. Habilitationsschrift, Halle, Martin-Luther-Universität.

Freier, B., Möwes, M., Triltsch, H. & Rappaport, V. 1996a. Investigations on the predatory effect of coccinellids in winter wheat fields and problems of situation-related evaluation. *Bull. IOBC/WPRS* 19: 41-52.

Freier, B., Triltsch, H. & Roßberg, D. 1996b. GTLAUS – A model of wheat – cereal aphid – predator interaction and its use in complex agroecological studies. *Z. PflKrankh. PflSchutz.* 103: 543-54.

Ghanim, A. 1981. Untersuchungen über den Einfluß von Parasiten und Prädatoren auf die Entstsehung von Gradationen der Getreideblattläuse. Dissertation, Halle, Martin-Luther-Universität.

Ghanim, A.; Freier, B. & Wetzel, Th. 1984. Zur Nahrungsaufnahme und Eiablage von *Coccinella septempunctata* L. bei unterschiedlichem Angebot von Aphiden der Arten *Macrosiphum avenae* (FABR.) und *Rhopalosiphum padi* (L.). *Archiv Phytopathol. Pflanzensch.* 20: 117-25.

Gutierrez, A. P., Baumgärtner, J. U. & Hagen, K. S. 1981. A conceptual model for growth, development and reproduction in the ladybird beetle, *Hippodamia convergens* (Col.: Coccinellidae). *Can. Ent.* 113: 21-33.

Hodek, I. 1967. Bionomics and ecology of predaceous Coccinellidae. *Ann.Rev. Entomol.* 12: 79-104.

Honek, A. 1985. Activity and predation of *Coccinella septempunctata* adults in the field (Col., Coccinellidae). *Z. angew. Entomol.* 100: 399-409.

Honek, A. 1986. Production of faeces in natural populations of aphidophagous coccinellids (Col.) and estimation of predation rates. *Z. angew. Entomol.* 102: 467-76.

Honek, A. & Kocourek, F. 1988. Thermal requirements for development of aphidophagous Coccinellidae (Col.), Chrysopidae, Hemerobiidae (Neuropt.) and Syrphidae (Dipt.): some general trends. *Oecologia (Berl.)* 76: 455-60.

Roßberg, D. & Freier, B. 1993. Probleme bei der Erarbeitung von Simulationsmodellen für Nützlinge. *Berichte GIL*, 5: 232-36.

Shands, W.A. & Simpson, G.W. 1972. Insect predators for controlling aphids on potatoes. 8. Green peach aphid consumption by *Coccinella septempunctata* and *C. transversoguttata*. *J. Econ. Entomol.* 65: 1388-92.

Skirvin, D.J. 1995. Simulating the effect of climate change on *Sitobion avenae* F. (Hom., Aphididae) and *Coccinella septempunctata* L. (Col., Coccinellidae). PhD thesis, University of Nottingham.

Stauss, R. 1994. *Compendium of growth stage identification keys for mono- and dicotyledoneus plants.* unnummerierte Blattsammlung, Basel.

Triltsch, H. in press. Gut contents in field sampled adults of *Coccinella septempunctata* L. (Col., Coccinellidae). Proceedings of the IOBC-meeting „Aphidophaga 6", Gembloux. *Entomophaga*.

Pest control by a community of natural enemies

K.D. Sunderland[1], J.A. Axelsen[2], K. Dromph[3], B. Freier[4], J.-L. Hemptinne[5], N.H. Holst[6], P.J.M. Mols[7], M.K. Petersen[8], W. Powell[9], P. Ruggle[10], H. Triltsch[4], L. Winder[11] *

[1]Department of Entomological Sciences, Horticulture Research International, Wellesbourne, Warwick CV35 9EF, England
[2]Department of Terrestrial Ecology, National Environmental Research Institute, Vejlsøvej 25, DK-8600 Silkeborg, Denmark
[3]Department of Ecology and Molecular Biology, Royal Veterinary and Agricultural University, Thervaldsensvej 40, DK-1870, Frederiksberg C, Denmark
[4]Biologische Bundesanstalt für Land- und Forstwirtshaft, Institut für Integrierten Pflanzenschutz, Stahnsdorfer Damm 81, D-14532 Kleinmachnow, Germany
[5]Faculté des Sciences Agronomique de Gembloux, U.E.R. de Zoologie Générale et Appliquée, Passage des Déportes 2, B-5030 Gembloux, Belgium
[6]Research Center for Agricultural Biodiversity, University of Copenhagen, Universitetsparken 15, DK-2100 Copenhagen, Denmark
[7]Department of Entomology, Agricultural University, Binnenhaven 7, 6709 PD Wageningen, The Netherlands
[8]Department of Plant Pathology and Pest Management, Danish Institute of Plant and Soil Science, Lottenborgvej 2, DK-2800, Lyngby, Denmark
[9]Department of Entomology & Nematology, IACR-Rothamsted, Harpenden, Herts AL5 2JQ, England
[10]Zoological Institute, University of Basel, Rheinsprung 9, CH-4051 Basel, Switzerland
[11]School of Conservation Sciences, Bournemouth University, Fern Barrow, Poole, Dorset, England

*2nd to 12th author in alphabetical order

Abstract

The major characteristics of natural enemy communities, including biodiversity, trophic structure and guild structure are described. Interactions between predators, parasitoids and pathogens are reviewed in the context of their effects on the efficiency of pest control. Reference is made to models of pest population dynamics affected by a community of natural enemies, and suggestions are made concerning future research directions.

Key words: Biological control, natural enemy interactions, simulation models, metabolic pool models, biodiversity, generalists, specialists, predators, parasitoids, pathogens, guilds, pest dislodgement, foraging, hyperpredation, hyperparasitism

Arthropod natural enemies in arable land · III The individual, the population and the community
W. Powell (ed.). *Acta Jutlandica* vol. 72:2 1997, pp. 271-326
© Aarhus University Press, Denmark, ISBN 87 7288 673 0

Introduction

Many of the early studies and models of the effects of natural enemies on pest populations were on systems involving only two species, but such situations are probably very rare in nature (Hassell 1986). Outdoor crop pests are subject to attack by many species of predators, parasitoids and pathogens, and even glasshouse crops now contain a considerable diversity of natural enemies (Sunderland et al. 1992). It is, therefore, appropriate to consider pest control in the context of a community of natural enemies (Rosenheim et al. 1995). The complexity of these communities is probably advantageous for pest control because the great diversity of niches filled by natural enemies reduces the likelihood that the pest will find a refuge in time or space (Crawley 1992). However, a corollary of this diversity is that the interactions between different natural enemies of pests will reduce the efficiency of some species. Examples of such deleterious interactions are predators eating parasitised pests, interference between natural enemies foraging for pests, hyperpredation, hyperparasitism, and interspecific competition. On the other hand, a diversity of natural enemies may lead also to synergistic interactions between species, which could be of great benefit to pest control. Examples of potentially synergistic interactions are pest dislodgement by one natural enemy increasing the availability of the pest to others, dissemination of pathogens by foraging predators and parasitoids, and wounding of pests by predators or parasitoids increasing the probability of infection by pathogens. In the case of an assemblage of spider species, the positive interactions seemed to outweigh the interference effects, and pest control was enhanced by augmenting the spider community (Riechert & Bishop 1990), but many more examples of this sort are needed before generalisations are possible. A full understanding of pest population dynamics can probably only be gained from modelling the pest together with the community of its natural enemies, as some of the interactions may have counter-intuitive outcomes. A knowledge of these interactions is important also in relation to practical programmes of natural enemy enhancement in the field (Powell 1986).

The aim of this review is to summarise current knowledge of natural enemy interactions and their relevance to pest control. The major characteristics of natural enemy communities are discussed, then some of the interactions that can occur between predators, parasitoids and pathogens are examined. We have not attempted to be comprehensive (for example, some interactions, such as pathogen-pathogen interactions (Krieg 1971, Harper 1986), are omitted), but have concentrated mainly on interactions likely to impinge directly on the efficiency of pest control. We then describe the current position in the evolution of models of pest control by a community of natural enemies and, finally, suggest appropriate directions for future research.

Characteristics of natural enemy communities

Definition of community

There is no universally agreed definition for the concept of community. It can be defined at any size, scale or level within a hierarchy of habitats, and community ecology is the study of the community level of organisation rather than of a spatially and temporally definable unit (Begon et al. 1990). Definitions of community do, however, incorporate the concept of groups of species which are more or less coincident in space and time and which interact with each other (Askew & Shaw 1986, Anderson & Kikkawa 1986). In this review we will be concerned with groups of species of natural enemies that interact with each other (with consequences for pest control) during the cropping seasons of agroecosystems.

Size of natural enemy communities

Biodiversity

Agroecosystems are surprisingly rich in species. The total arthropod community of a typical temperate agroecosystem is estimated to contain at least 1500 - 3000 species (Nentwig 1995). Parasitoids, which account for one quarter of all insect species (Thompson 1984, Tscharntke 1992) are well-represented in agroecosystems. There were, for example, 59 species of parasitoid attacking herbivorous chalcids in UK grassland (Dawah et al. 1995), 61 associated with cynipid galls on oak in southern England (Memmott & Godfray 1993), 86 attacking Costa Rican leafminers (Memmott & Godfray 1993), 109 associated with the tortricid *Zeiraphera diniana* on conifers in the Central European Alps (Delucchi 1982), 41 associated with the hessian fly *Mayetiola destructor* (Askew & Shaw 1986) and 63 in alfalfa at one site over two years (Pimentel & Wheeler 1973). Numerous species of predator are also found in crops, e.g. 216 species in alfalfa (Pimentel & Wheeler 1973) with at least 56 species in the vegetation layer alone (Wheeler 1977), *ca.* 400 in UK cereals (Potts & Vickerman 1974, Sunderland & Chambers 1983) and *ca.* 600 - 1000 in USA cotton and soybean (Whitcomb & Bell 1964, Gross 1987). Usually 10-40 species of ground beetles (Carabidae) are active in a habitat in the same season (Lövei & Sunderland 1996). Spiders, which exist in virtually all terrestrial environments and are the dominant insectivores in some ecosystems (Thompson 1984), were represented by 60 species in apple orchards (Bogya & Mols 1996) and by 88 species in barley fields in Denmark, which is comparable to the richest natural Danish biotopes known (Toft 1989). Spider species richness can also be very high in other crops, with 112 species in USA alfalfa (Howell & Pienkowski 1971) and more than 600 species in USA field crops as a whole (Young & Edwards 1990).

Entomopathogenic nematodes do not appear to be particularly species-rich, but they are morphologically conservative and it is likely that many more species will be discovered by molecular techniques (Hominick et al. 1996). Although species diversity of entomopathogenic fungi is high in the tropics (Evans 1982), and approximately 700 species have been described globally (Charnley 1989), less than ten species were recorded from agricultural soils in the UK (Chandler et al., in press) and Finland (Vanninen et al. 1989). However, different isolates of the same species can differ in specificity and the host range of a single species can be very large; *Beauveria bassiana*, for example, is known to attack more than 700 species of host (Goettel et al. 1990). Most of the species from temperate regions are asexual. These species show significant intraspecific heterogeneity within ecological zones, which can affect host preference and virulence, but the ecological role of such diversity, and its extent within agrohabitats, are unknown (D. Chandler, personal communication). Several species of entomopathogenic fungi are known to be associated with cereal aphids, but only three species belonging to the Entomophthorales commonly cause epizootics (Dean & Wilding 1971, Feng et al. 1991, Steenberg & Eilenberg 1995).

Factors affecting parasitoid species richness
The literature on the determinants of species richness of predators and pathogens in agricultural communities is rather sparse, so discussion is confined to parasitoids.

For a range of annual crops, the number of parasitoid species per specific pest host in any area varies from 2-14. Nearly 50% of pest species have more than 10 parasitoid species each (Altieri et al. 1993). The mean number of parasitoid species per host is 2.6 for Hemiptera hosts, 4.6 for Coleoptera, 5.2 for Diptera, 7.0 for Hymenoptera and 9.4 for Lepidoptera (Hawkins & Lawton 1987, Price 1994). Species richness decreases with increase in intensity of agroecosystem management and is greater on tree crops than annual crops (Altieri et al. 1993), being 4.7 species per host on grasses and herbs, 6.1 on shrubs and 7.8 on trees (Price 1994). Increased species richness on large and spatially complex plants may be due to polyphagous parasitoids finding more hosts on such plants (Hawkins & Lawton 1987). Parasitoid species richness increases with plant succession and early colonisers have high fecundity and low competitive ability, whereas later colonists have low fecundity and high competitive ability (Price 1973, 1994, Askew & Shaw 1986). Although many factors, such as habitat diversity (Hirose 1994), habitat disruption (Miller 1980), plant geographical range (Price 1994), intraspecific plant variation (Craig 1994), plant toxins (Price 1994), plant gall size (Stiling & Rossi 1994) and gall distribution (Tscharntke 1992), and the predictability of host abundance (Mills 1994b, Sheehan 1994), influence parasitoid species richness, the feeding site of

the host emerges as the predominant factor (Hawkins 1993, Mills 1993,1994a).
In a study of 285 species of herbivore belonging to 42 families, Hawkins &
Lawton (1987) found that host feeding site accounted for a fifth of the variance in
species richness. Exposed hosts and fully concealed hosts (e.g. root feeders)
were not attacked by as many parasitoid species as were intermediately
concealed hosts, such as leaf-miners (Hawkins & Lawton 1987).

Biodiversity in relation to pest control
In general, pests are more likely to be controlled below the economic threshold
when natural enemy biodiversity is high. From an analysis of small-scale
experiments and a review of the literature pertaining to a range of crops, Riechert
(1992) concluded that the presence of a diverse assemblage of spider species at
the time of pest invasion may, under suitable conditions, prevent pest numbers
from ever building up to damaging levels. Percentage parasitism and subsequent
pest control is often positively correlated with parasitoid species richness (Force
1974, Redfearn & Pimm 1992, Waage & Mills 1992, Hawkins 1993, Hochberg
& Hawkins 1994), although the relationship can be variable, both in theoretical
models and in actual cases (Miller & Ehler 1990, Mills 1994b). An extreme case
running counter to the general trend was reported by Frazer & Van den Bosch
(1973) where the addition of one species of parasitoid to walnut in California
resulted in greatly improved control of walnut aphid but a subsequent loss of 15
species of predator from the system.

Pest control is likely to be enhanced by maximising the proportion of the
pest's life cycle that is exposed to natural enemies (Waage & Mills 1992). To
achieve this, a rich diversity of natural enemies with different phenologies is
required, and this may be encouraged by habitat diversification. For example,
parasitoid species richness and percentage parasitism of pests is greater in
complex (e.g. intercropping, cover crops, living mulches) than in simple cropping
systems (Altieri et al. 1993). Conversely, farming practices that reduce natural
enemy diversity could reduce trophic complexity, which would cause pest
outbreaks to occur faster (Redfearn & Pimm 1987). Pest control on a mosaic of
crops is favoured by having a diversity of natural enemies because the efficiency
of any one species varies with conditions such as density of vegetation (Raw
1988). Natural enemies can boost overall invertebrate biodiversity by regulating
the abundance of species that would otherwise competitively exclude other
species (Kitching 1986, Lasalle & Gauld 1993). Rich invertebrate biodiversity is
advantageous in providing natural enemies with mixed diets and alternative
foods/hosts, enabling them to increase rapidly in the crop and be present when
pests arrive (Axelsen et al. 1997).

Structure of natural enemy communities

Trophic structure

Communities have 2-5 trophic levels, with 3-4 being the norm (Begon et al. 1990). It is a general feature of communities to have a fairly constant ratio of prey to natural enemies (Jeffries & Lawton 1984), i.e. the proportion of top natural enemies to basal and intermediate species is reasonably constant in all food webs (Pimm et al. 1991). This may result from competition for "enemy-free space" culminating in limits to the number of victims able to coexist with different species of predators (Jeffries & Lawton 1984). Enemy-free space was defined by Jeffries & Lawton (1984) as "ways of living that reduce or eliminate a species' vulnerability to one or more species of natural enemies". Competition for enemy-free space can be a driving force behind niche differentiation by phytophages that need to develop refuges from shared natural enemies (Lawton 1986, Schoener 1986); e.g. caterpillars of *Malacosoma californicum* feed early in the season, even though leaves are less suitable then, and this behaviour may have evolved in response to heavy parasitism by Tachinid flies, which occurs later (Lawton 1986). In a recent review, Berdegue et al. (1996) concluded that enemy-free space may be important in moulding the niches of arthropods.

Numerous interactions between trophic levels are theoretically possible, but their patterns in nature are simpler than would be expected by chance (Redfearn & Pimm 1987). Nevertheless, dynamic interactions can extend across many trophic levels. Plant resistance can have an impact at the fourth trophic level; for example a scelionid egg parasitoid (*Telenomus podisi*) had reduced fecundity on eggs of the predatory bug *Podisus maculiventris* which was fed on a noctuid caterpillar (*Pseudoplusia includens*) reared on resistant compared with susceptible soybean (Hare 1992). Food web modelling has revealed various properties of community trophic dynamics. Species at higher trophic levels are thought to recover from disturbance less rapidly than those below (Pimm & Kitching 1987) and trophic loops (e.g. A eats B, and B eats A) are expected to be rare (Pimm et al. 1991). However, Polis et al. (1989) demonstrated that, in nature (as opposed to in models), trophic loops (termed by Polis et al. 1989 as symmetric intra-guild predation) are surprisingly common, especially if predation on different life stages of a species is taken into account. For example, loops were observed between 7 out of 12 species pair combinations of spiders (Polis et al. 1989).

Modelling suggests that parasitoid trophic webs are more stable than other webs and that omnivory, defined as feeding on more than one trophic level (Begon et al. 1990), is more likely in parasitoid trophic webs; this was supported by an analysis of data from 19 food webs (Pimm & Lawton 1978, Lawton 1994). Parasitoid webs are also thought to gain stability from compartmentalisation

(Memmot & Godfray 1993), i.e. the web has several clusters of specialist parasitoid species which are linked by a relatively small number of polyphagous parasitoid species.

Generalists and specialists
Early research into biological pest control tended to focus mainly on specialist natural enemies, but now there is an appreciation of the value of generalists, to some extent in their own right, but especially in combination with specialists. Generalist natural enemies, rather than specialists, appear to be dominant in fluctuating unpredictable environments, such as agriculture (Ehler & Miller 1978). However, in the particular case of parasitoids, early successional plant communities have a greater proportion of specialist parasitoids and late successional communities, a greater proportion of polyphagous parasitoids (Price 1994). Polyphages can be more successful than stenophages in controlling pests (Watt 1965). For example, there is evidence from cotton in California, that noctuid pests can be controlled predominantly by generalist natural enemies without significant help from specialists (Ehler & Miller 1978). More generalist than specialist parasitoids were found to attack Lepidoptera in the USA during the Spring (Sheehan 1994).

Most natural enemy communities contain both generalists and specialists and they tend to be complementary in terms of pest control. Biological control of woolly apple aphid (*Eriosoma lanigerum*) in apple orchards of the temperate zones of Europe is accomplished by a combination of specialists (the parasitoid *Aphelinus mali* plus the coccinellids *Exochomus quadripustulatus* and *Coccinella 7-punctata*) and a generalist predator, the earwig *Forficula auricularia* (Mols 1996, Mols & Diederik 1996). Early-season generalists and specialists may reduce the rate of pest increase, allowing the later-arriving specialists to keep pests below the economic threshold. In European cereal crops, for example, early-season synchrony between colonising aphids and their specialist parasitoids is thought to be the key to aphid control (Wratten & Powell 1991). Early-season parasitism and predation by polyphagous predators reduces the growth rate of the aphid population, allowing later-arriving specialist predators to impose effective control. In some cases generalists may be effective at low pest density whilst specialists are more effective at high host density. *Calosoma sycophanta* is a specialist carabid beetle predator of caterpillars and pupae and is only active against the gypsy moth *Lymantria dispar* when moth populations are large; generalist predators cause significant mortality when moth are rare (Weseloh 1990). Another example of complementary effect is that specialists searching on the crop can cause dislodgement of pests which are then attacked by ground-based generalist predators (see later section on "Dislodgement of pests by natural enemies").

Some aspects of agroecosystems favour specialists whilst others favour generalists. Compared with specialists, generalists tend to be more susceptible to the deleterious effects of food plant allelochemicals sequestered in the host or prey (Faeth 1994). Endophytic hosts are more likely to support generalist parasitoids and exophytic hosts are more likely to support specialists (Jones et al. 1994). Generalists are not usually host plant specific, but the species of specialist natural enemies of a given pest can vary greatly from crop to crop, partly because of host location cues emanating from the host plant (Waage & Mills 1992).

Guild structure
A guild is defined, after Root (1967), as "a group of species that exploit the same class of environmental resource in a similar way". A species can belong to more than one guild. The guild concept is useful in dividing complex biological communities into functional units (Adams 1985). Moran & Southwood (1982) found guild analysis to be effective in analysing arthropod communities in trees. Guild structure is expected to be more predictable and stable than species composition (Hawkins & MacMahon 1989) and community organisation would be misconstrued by presuming particular roles for species based on taxonomic affiliations alone (Blaustein & Margalit 1994). Ecosystems may contain many functionally redundant species, i.e. all capable of performing the same role in the ecosystem. So guilds might remain fairly stable in relation to resource availability whilst individual species vary due to factors (such as weather, natural enemies and environmental conditions) unrelated to resource availability (Hawkins & MacMahon 1989). Guild structure can be affected by intrinsic factors, such as intraspecific competition, interspecific competition and intraguild predation, and by extrinsic factors, such as spatial structure of the host/prey population, habitat disturbance and host plant structure (Ehler 1992).

In nearly all guilds studied, several dimensions, usually three, are needed to segregate species (Schoener 1986). Schoener (1974) considered that predator species tend to be segregated most in relation to diel activity cycles (rather than in relation to other factors such as food or habitat). This may be because prey availability varies on a diel cycle, whereas, for example, leaf supply for a herbivore usually does not (Schoener 1986). However, seasonal separation of peak activity was the main factor allowing intraguild coexistence of species of Thomisidae and Oxyopidae spiders on flower heads (Turner & Polis 1979).

Authors have used various definitions to delineate parasitoid guilds. One definition is that parasitoid species which attack the same stage of the same host species belong to the same guild (Miller & Ehler 1990, Ehler 1994). The parasitoid guilds of Mills (1994a), on the other hand, require more information and are defined in terms of (i) the host stage attacked, (ii) whether the parasitoid is an endo- or an ectoparasitoid and, (iii) whether it is an idiobiont or a koinobiont. The terms idiobiont (parasitoids that kill the host rapidly) and koinobiont (parasitoids that kill the host after a protracted period) were created by Askew & Shaw (1986). Idiobionts are often generalists and ectoparasitoids, whilst koinobionts are often specialists and endoparasitoids. For tortricoid hosts, Mills' definition provided 11 guilds, and this guild analysis greatly helped the determination of the role of various biotic factors (e.g. host larval feeding site) in limiting the utilization of hosts by these parasitoids (Mills 1993). Guild analysis also allowed Mills (1994a) to determine the factors affecting species richness of the parasitoid communities attacking weevils. Hoffmeister & Vidal (1994) found no relationship between total species richness of parasitoids attacking fruit flies (Tephritidae) with any of the variables measured. However, analysis of species richness at the guild level showed it to be affected by the host feeding site and the taxonomic status and degree of polyphagy of the parasitoids. A small degree of temporal stability, rather than interspecific competition, was found to be the main factor affecting intra-guild species composition amongst the parasitoids attacking the lepidopteran pest *Spodoptera praefica* in USA alfalfa (Miller 1980). Hawkins & Mills (1996) suggest that the number of parasitoid guilds that hosts support is very variable, both among host species and among populations within host species, and that parasitoid communities are undersaturated with species, i.e. about 70% of the potential parasitoid guilds that hosts could support are actually present. The reason for this is not clear, but it does not appear to be due to inter-guild competition (Hawkins & Mills 1996).

Very few studies have attempted to construct food webs describing the interactions between guilds of hosts and parasitoids at a single locality (Hassell & Godfray 1992); a recent exception is described by Dawah et al. (1995). Although guild analysis is clearly useful, the guild structure of the vast majority of natural enemies remains undescribed.

Interactions between natural enemies

Sometimes mortality caused by an entomophage contributes to reduction of pest density only if the mortality is irreplaceable (Thompson 1955) i.e. only if the pest is not under imminent control by another natural enemy. Little attention has so far been paid to this limitation in biological control (Jones 1982). Life Table analysis assumes that mortality factors operate independently of each other and is not equipped to deal with interaction effects between mortality factors (Kidd & Jervis 1996). Simulation models also do not currently incorporate many of the intricacies of natural enemy interactions, largely because the need to do so has not been demonstrated. Indeed, there is currently very little insight or experimental knowledge of interactions between all the natural enemies of a given pest and of the way these interactions affect its density (Miller & Ehler 1990). The relative importance for biological control of qualitative versus quantitative aspects of natural enemy interactions remains undetermined, but initial studies suggest that qualitative aspects may be of greater significance (Heinz & Nelson 1996). Knowledge of interactions can guide IPM decisions; for example, the microbial insecticide *Nosema pyrausta* and the predator *Chrysoperla carnea* are compatible (Sajap & Lewis 1989), but other natural enemy combinations may not be. Although the field of natural enemy interactions is largely unexplored, an indication of the small amount of information currently available is set out in this section, organised mainly under natural enemy headings. The likely relevance of these interactions to pest control is indicated.

Dislodgement of aphids (and other pests) by natural enemies

In many crops, natural enemies are stratified vertically and the number of species that climb crop plants to attack pests is often very limited. The majority of natural enemy species are confined to the ground. Therefore, any climbing natural enemy that dislodges pests (for example by triggering production of an alarm pheromone) will increase the pest's encounter rate with ground based natural enemies to a greater degree than that resulting simply from random wandering and dislodgement by weather and other factors. This synergistic interaction would be further enhanced if a few falling pests trigger facultatively climbing predators to climb the plant (Loughridge & Luff 1983). Since the degree of aggregation of pests is a very important factor affecting epizootics by entomogenous fungi, changes to the pest's distribution caused by searching predators and parasitoids could significantly influence pest control by these pathogens. For example, *Entomophthora* epizootics amongst *Schizolachnus piniradiatae* aphids on pine were favoured by a more even distribution of aphids (Soper & Macleod 1981), which could result if colonies were disrupted by foraging natural enemies. The level of infection of the melon and cotton aphid, *Aphis gossypii,* by the fungus *Verticillium lecanii* was enhanced by spraying synthetic aphid alarm pheromone onto the chrysanthemum plants, thereby increasing aphid movements and subsequent contacts with fungal spores on the plant surface (Hockland et al. 1986). On the negative side, this behaviour could be detrimental to plant health if the pests are transmitting plant pathogens, such as Barley Yellow Dwarf Virus spread by cereal aphids, especially since alatae are more sensitive to alarm pheromone than apterae (Nault et al. 1973). Bean Yellow Mosaic Virus transmitted by the aphid *Acyrthosiphon pisum* was spread four times further in the presence of coccinellid beetles (Roitberg & Myers 1978b).

Overall aphid dislodgement rates (due to all causes)

Aphids could appear on the ground as a result of random wandering (Dean 1973), by falling during a moult, by being brushed off by wind, rain (Mann et al. 1995) or farm machinery or by being dislodged by predators and parasitoids. Further interactions with a range of factors, including crop cultivar (Bhuiyan & Wratten 1994), are likely. Overall dislodgement rates can be very high but the percentage contribution of natural enemies to these recorded rates is not currently known.

The number of aphids found on the soil in winter wheat crops, as a percentage of the total number of aphids present in the crop field, is very variable, but can exceed 50% on some occasions (Griffiths et al. 1985; Sunderland et al. 1986a, Kennedy 1990). It appears that a greater proportion of the aphid population is found on the soil surface at lower aphid densities, but the reasons for this relationship are unknown (Sopp et al. 1987). Large numbers of aphids may leave, or be dislodged from, cereal plants each day (e.g. 26% of the population in a study by Watson (1983) quoted in Dixon (1987)).

Aphid responses to alarm pheromone

Montgomery & Nault (1977) found that 8 out of the 14 species of aphid subjected to alarm pheromone were more likely to drop off the host plant than to walk away. Dropping was especially prevalent amongst species feeding on Gramineae. Aphid alarm pheromones are usually also effective interspecifically (Nault et al. 1973) and this increases the likelihood of pheromone-induced dislodgement. Percentage dislodgement rates in response to alarm pheromone vary within and between clones, (Roitberg & Myers 1978a, Muller 1983) and in relation to aphid size, host plant quality and geographical region (Roitberg & Myers 1978a, Stadler et al. 1994). For pea aphid, *Acyrthosiphon pisum*, a combination of vibration and alarm pheromone caused the greatest response (Clegg & Barlow 1982), and this combination of stimuli caused more than 90% of first instars to abandon the plant (Roitberg & Myers 1978a).

Pest dislodgement by parasitoids

The impact of parasitoids on pests is not confined to parasitism, but also includes dislodgement, which can expose the pest to predation (Van Driesche 1983, Longley & Jepson 1996). Of the aphids and hymenopteran parasitoids, *Myzus persicae* and *Neomyzus circumflexus* are dislodged by *Diaeretiella rapae* (Klingauf 1967), *Acyrthosiphon pisum* by *Aphidius ervi* and *Aphidius smithi* (Bueno et al. 1993), *Metopolophium dirhodum* by *Aphidius rhopalosiphi* (Gowling & Van Emden 1994) and *Sitobion avenae* by *Aphidius* spp. (Longley & Jepson 1996). Up to 63% of *Schizaphis graminum* were dislodged by *Lysiphlebus testaceipes* (Hight et al. 1972, Ruth et al. 1975). The phenomenon is not confined to aphids but also occurs in other groups of pests, such as Lepidoptera. Larvae of green cloverworm, *Plathypena scabra*, jump off soybean if disturbed by the braconid parasitoid *Diolcogaster facetosa* (Yeargan & Braman 1986) or by the ichneumonid hyperparasitoid *Mesochorus discitergus* (Yeargan & Braman 1989). Dislodged caterpillars may sometimes remain attached to the plant by a silken safety line, enabling them to climb back later.

There is more than one mechanism of aphid dislodgement by hymenopteran parasitoids. Early instar apterae may be physically knocked off the plant by thrusts from the parasitoid's ovipositor, and parasitised alatae fly very readily if approached by a parasitoid for the second time (Hight et al. 1972). When an aphid is stung, other aphids in the vicinity drop off the plant (Goff & Nault 1974, Ruth et al. 1975) and a chain reaction of falling apterae can be triggered (Hight et al. 1972), presumably due to sequential release of alarm pheromone. Dislodgement can also occur in the absence of stinging; Tamaki *et al.* (1970) showed that *A. pisum* left the host plant as a result of attacks by *A. smithi* even though they were unable to parasitize (the ovipositor valvulae having been experimentally amputated).

In some cases, aphids dislodged by parasitoids move a greater distance from the original host plant than aphids that are randomly dispersing (Tamaki et al. 1970) and they may take longer to re-climb plants (Gowling & Van Emden 1994). These behaviours contribute to an increase in the time spent on the ground, which, in turn, entails a greater risk of predation.

Pest dislodgement by predators
Aphids can be dislodged by larvae of Coccinellidae, Chrysopidae, Syrphidae and Chamaemyiidae, and by Nabidae, Reduviidae and Anthocoridae (Klingauf 1967, Nault et al. 1973), but there is usually only a weak response to larvae of *Aphidoletes aphidimyza* (Cecidomyiidae) (Klingauf 1967).

Aphid species vary in the strength of their response to adult coccinellids. For example, *Acyrthosiphon pisum* and *Therioaphis maculata* drop readily but *Brevicoryne brassicae* and *Aphis fabae* are very sedentary and do not (Roitberg et al. 1979). When the response is very strong, it can result in more dislodgement than predation by coccinellids. McConnell & Kring (1990) recorded that *Schizaphis graminum* were five times more likely to be dislodged than consumed by adult *Coccinella 7-punctata*. Although *A. pisum* readily falls off the plant when confronted by an adult *Coccinella californica* (Roitberg & Myers 1978b, 1979), the response is moderated on high quality plants and in hot, dry environments (Dill et al. 1990), probably reflecting selection pressures in relation to the ratio of risks to fitness-promoting advantages. This aphid species usually drops off the plant when it is encountered by an adult *Adalia bipunctata*

(Coccinellidae), but it backs away from larvae of *Metasyrphus corollae* (Syrphidae) (Brodsky & Barlow 1986). It falls off the plant in response to *Anthocoris nemorum*, and the incidence of dislodgement becomes greater with increase in the predator-prey size ratio (Evans 1976). Similarly, *Eucallipterus tiliae* were increasingly more likely to jump off lime leaves with increase in size of an approaching *A. bipunctata* larva (Wratten 1976).

McClure (1995) described an unusual case of biological control of an adelgid (*Adelges tsugae*) by dislodgement from hemlock trees in Japan. Arboreal non-predatory oribatid mites (*Diapterobates humeralis*) eat the woolly filaments that protect the oviscac of this pest. A high proportion of eggs fall to the ground and die of desiccation or are eaten by ants and spiders. The mite is considered to be an important natural enemy of *A. tsugae* throughout the hemlock growing areas of Japan.

Dislodgement of aphids by predators such as coccinellids (Evans 1991) and anthocorids (Dixon & Russel 1972) can exert a considerable influence on aphid distribution, with potential implications for disease transmission and for aphid control by natural enemies that respond to pest aggregations. The distance that aphids move away from the source plant after dislodgement by coccinellids is greater for larger than for small aphids and greater at higher than at low aphid densities (Roitberg et al. 1979).

Caterpillars, such as larvae of the Cinnabar Moth, *Tyria jacobaeae* (Dempster 1971), may also drop from the leaf when disturbed by predators. Larvae of green cloverworm *Plathypena scabra* jump off soybean if disturbed by the heteropteran predators *Podisus maculiventris* and *Orius insidiosus* (Yeargan & Braman 1986). Larvae of the Fall Webworm, *Hyphantria cunea*, fall off the tree when touched by a *Polistes* wasp, and may then be eaten on the ground by tree sparrows (Ito & Miyashita 1968). In apple orchards, 34% of *Spodoptera littoralis* larvae left branches in response to disturbance by the spider *Chiracanthium mildei* (Mansour et al. 1981) and tortricid larvae are very often dislodged by earwigs entering leafrolls, but the impact of this behaviour on mortality of the larvae has not yet been quantified (P.J.M. Mols personal communication). In taro fields 4% of *Spodoptera litura* caterpillar mortality was due to spider predation and 38% due to mortality associated with dislodgement (Nakasuji et al. 1973). Herbivores forced to disperse from their host plants after disturbance by spiders die from failing to locate a suitable feeding site. (Riechert 1992).

Ground-based predators consuming fallen aphids
A high proportion of predators in agroecosystems are ground-based. For example, 60% of the 279 species of predator in UK cereals are confined to the ground zone (Sunderland et al. 1986a, 1988). Spider communities are often

vertically stratified (Turnbull 1973, Sunderland et al. 1986b) but even web-spiders on the ground obtain a proportion of their food falling from above.

The carabid beetles, *Pterostichus melanarius* and *Pterostichus madidus* rarely climb plants, yet often contain aphid remnants in their guts. There was a significant increase in the percentage of *P. madidus* containing aphids after spraying potatoes with an insecticide which caused the aphids to fall from the plants (Dixon & McKinlay 1992). *Agonum dorsale* obtained most of the aphids in its diet by foraging on the ground (Griffiths et al. 1985). Even predators that do climb may spend a large proportion of their time searching on the soil surface (e.g. the carabid *Demetrias atricapillus* (Winder 1990) and coccinellids (Ferran & Dixon 1993)). Approximately one third of the cereal aphid prey of *Coccinella 7-punctata* in wheat was captured on the ground (Ferran et al. 1991).

Winder (1990) measured aphid falling and re-climbing rates in winter wheat plots where polyphagous predator density had been manipulated, and concluded that predators on the ground reduced the rate at which aphids returned to the crop canopy. Winder et al. (1994) estimated that more live aphids than dead aphids fall to the ground in winter wheat crops, but the availability of live aphids to ground predators depends on how rapidly they re-climb the plants. Aphids dislodged by natural enemies may spend longer on the ground than aphids voluntarily dispersing between plants, but good data are lacking. It was suggested from analytical and simulation modelling that residence time of aphids on the ground could be a very important factor influencing their consumption rate by the carabid *Agonum dorsale* (Winder, Wratten & Carter 1997).

Predator-predator interactions
Interference with foraging
Predators can, in theory, suffer reduced foraging efficiency at high density as a result of mutual interference. Most information relates to intraspecific interactions (e.g. amongst carabid beetles (Griffiths et al. 1985, Hance 1987)), but there are also some records of interspecific agonistic behaviour in the field. Lys (1995), for example, observed artificial baits by day and night in a wheat field and recorded four species of carabid that were very aggressive towards other smaller species of predator at the baits, chasing them away or biting them. Interspecific interactions amongst the Coccinellidae did not appear to greatly influence foraging efficiency; *Acyrthosiphon pisum* aphids on beans were reduced equally by heterospecific and conspecific pairs of ladybird larvae (*Coccinella 7-punctata* and three species of *Hippodamia*) (Evans 1991). In general, intraspecific effects seem to be greater than interspecific effects in the Coccinellidae. Hemptinne, Dixon & Coffin (1992) found that adult female *Adalia bipunctata* attempted to leave an area when they encountered conspecific females or larvae and they showed a marked reduction in oviposition rate in the

presence of conspecific larvae but not in the presence of other species of aphidophagous ladybirds. The money spider *Lepthyphantes tenuis* accepts webs constructed by related species and contests (mainly intraspecific) over web ownership are very common (Samu et al. 1996). Interspecific web take-over involving related species with similar web structure can be common in some habitats and this behaviour sometimes results in hyperpredation (see below) (Toft 1988).

Predators may also have a detrimental effect on the availability of suitable oviposition sites for other species of predator. Coccinellids laid fewer eggs on plants where syrphid larvae were already present, compared with predator-free plants (Hemptinne et al. 1994).

Hyperpredation

Many predators include other predators amongst their prey. This behaviour can help to buffer a natural enemy community from rapid changes in predator species diversity resulting from changes in herbivore prey availability. Over a longer period it may promote predator guild diversification (Polis et al. 1989), which would benefit pest control by increasing the number of pest species or pest life stages that are exposed to predation, but in the shorter term it can reduce the efficiency of pest control (Polis et al. 1989, Cloutier & Johnson 1993, Rosenheim et al. 1993, 1995) or contribute to displacement of some species of predators from crops (Macrae & Croft 1993). A small sample of the literature on hyperpredation is given below, using the Arachnida as an example (for references to hyperpredation involving some other groups of predators, see Rosenheim et al. 1995 and Sunderland 1996).

Some spiders specialise in attacking other spider species or in eating their eggs (Nyffeler et al. 1990, Jackson & Pollard 1996), but polyphagous spiders are also often arachnophagous. The wolf spider genus, *Pardosa*, which themselves often eat other spider species (Edgar 1969), formed one tenth of the prey of ground-living crab spiders (*Xysticus*) in cultivated meadows (Nyffeler & Benz 1979), and *Oxyopes salticus* killed various spider species but was itself killed by lycosid spiders in Texas peanut fields (Agnew & Smith 1989). Polis et al. (1989) reported that heterospecific spiders formed 1-49% of the diet of 27 species of spiders. *Amblyseius andersoni* was a strong hyperpredator of three other species of Phytoseiidae mites under laboratory conditions (Zhang & Croft 1995). Populations of the predatory mite *Phytoseiulus persimilis* were reduced by *Amblyseius degenerans*, even in the presence of ample *Tetranychus pacificus* prey, and there was mutual predation of eggs and larvae by adults of these two phytoseiid predators (Yao & Chant 1989). Predation of spiders by spiders can also be a major factor in keeping spider populations below their carrying capacities (Polis et al. 1989). Although spiders have a large armoury of anti-predator devices (Cloudsley-Thompson 1995), they still fall prey to a wide range of invertebrate predators. Lacewing larvae (Chrysopidae), for example, will gorge themselves on immature oxyopid spiders (Agnew & Smith 1989). Sphecidae wasps hunt Araneidae and Theridiidae to provision their nests but are, in turn, sometimes killed by web spiders (Dean et al. 1988). Some species of Carabidae, Staphylinidae and Nabidae were voracious predators of spiders in the laboratory and the density of immature spiders was greater in field plots of winter wheat where carabid density had been reduced (Sunderland et al. 1994). Guillebeau & All (1989) recorded assymmetrical mutual hyperpredation between the spider *Oxyopes salticus* and *Geocoris* spp. bugs, the former being more likely to reduce populations of the latter. Coccinellids will eat spider eggs (Whitcomb 1974 in Guillebeau & All (1989)), but some species of spiders also eat coccinellids (Frazer & Gilbert 1976). It can be common for spiders to kill other natural enemies. For example, *Pardosa lugubris* killed anthocorid bugs and opilionids (Edgar 1969). The proportion of beneficials in the diet of spiders is quite variable; Hymenoptera and Chrysopidae were rarely caught in spiders' webs in Swiss winter wheat fields (Jmhasly & Nentwig 1995), but the ratio of beneficials to pests in the diet of *Peucetia viridans* in various fields in Florida was 4:1 (Randall 1982) and natural enemies accounted for half of the diet of spiders in Texas peanut fields (Agnew & Smith 1989). Although hyperpredation by spiders is common in US cotton, the net effect of spider predation is considered to be positive (Nyffeler et al. 1994).

On occasion, predation on natural enemies can be so intense that it has a deleterious effect on pest control. For example, anthocorid bugs eating Braconidae parasitoids allowed an 180% increase in the pyralid caterpillar host of the braconid (Polis et al. 1989). Heteropteran bugs feeding on lacewing larvae (*Chrysoperla carnea*) in USA cotton released aphids (*Aphis gossypii*) from their former level of satisfactory biological control (Rosenheim et al. 1993). There are also cases of hyperpredation being associated with improved pest control (Rosenheim et al. 1995).

Competition

Competition between predator species is most likely to occur within a guild. A top predator, however, may forstall competitive exclusion between two or more lower predators, by reducing them to levels where competition is weak or absent (exploiter-mediated coexistence). This effect would be amplified if the top predator switches to concentrate on species of lower predators that become especially abundant (Begon et al. 1990).

In the laboratory, predator size, degree of polyphagy and prey density all affected the outcome of interspecific competition between immature phytoseiid mites (Zhang & Croft 1995). Evidence for interspecific competition in the field, however, is difficult to obtain. Four out of five field experiments with spiders failed to detect competition, but it has been detected between some orb-weaver species (Schoener 1983). For wandering spiders in a forest, niche overlap was less between species pairs that were taxonomically close or similar in size and this may have resulted from interspecific competition (Uetz 1977). Toft (1990) observed web take-overs between related spider species and described this as a "producer-scrounger" relationship, rather than direct competition for a shared resource. Evidence for interspecific competition between carabid adults is in general still inconclusive, but competition is more likely between larvae (Lövei & Sunderland 1996). Currie, Spence & Niemela (1996) noted that intraguild predation between the carabids *Pterostichus melanarius* and *Pterostichus adstrictus* was exacerbated when food was scarce. They considered that competition for food could amplify interspecific predation through complex interactions with foraging behaviour.

Parasitoid-parasitoid interactions
Interference with foraging
Mutual interference between searching individual parasitoids at high parasitoid density could decrease their searching efficiency, but density-dependent interference may help to stabilise host-parasitoid population dynamics (Godfray 1994).

It is quite normal for several parasitoids to search a patch simultaneously. For example, adult parasitoids of bark beetles (Scolytidae) often encounter each other and show agonistic behaviour. When adult parasitoids of ash bark beetle (*Lesperisinus varius*) met in the field there was an aggressive display of wing-beating by both individuals. They spent 0.4 - 1.5% of their time fighting, and were estimated to have 12 - 39 fights per day, depending on species (Mills 1991). Parasitoid fights, in general, can include biting to the neck, legs, wings and ovipositors (Lawrence 1981). Pteromalid and eurytomid searching strategies enable them frequently to encounter other parasitoids probing through the bark for hosts. Some of the species involved appear to be facultative cleptoparasitoids, actively displacing adults of other species from sites where they have located hosts. Such cleptoparasitoids are more efficient at locating competitors than hosts (Mills 1991, 1994b). Other encounters arise as a result of patch defence behaviour, a poorly studied aspect of parasitoid biology (Van Alphen & Jervis 1996). For example, some parasitoids of Tephritidae fruit flies defend fruits from other parasitoids, some Scelionidae egg parasitoids defend egg batches and many Bethylidae remain with their brood after oviposition and defend it from other parasitoids (Waage 1992, Godfray 1994).

Interference with foraging can occur without physical contact, where there is chemical mediation. Some parasitoids avoid searching an area already visited by another species in response to a chemical mark left, or in response to an odour coming directly from a competitor species (Van Alphen & Jervis 1996). Amongst four coexisting species of Ichneumonidae attacking the same host, there was conspecific, congeneric and intergeneric recognition of odour trails left by searching parasitoids (Price 1970). The odour acts as an irritant and causes increased movement of the parasitoid. Thus the maximum level of abundance of such parasitoids might be determined not by host density but by concentration of trail odour. So, although the efficiency of individual parasitoids is increased, mortality of the host population is decreased (Price 1970).

Pseudo-interference can occur, without behavioural interactions, at higher parasitoid density, where search for prey is aggregated and parasitoids tend to encounter already-parasitized prey (Free, Beddington & Lawton 1977).

Hyperparasitism

Hyperparasitism occurs when larvae feed, not directly on the host, but on other parasitoid larvae already present on, or in, the host (Godfray 1994). Ectoparasitoids are often facultative hyperparasitoids, whilst larval endoparasitoids are usually not (Stiling & Rossi 1994). Hyperparasitism has been recorded from seventeen families of Hymenoptera and a few species of Diptera and Coleoptera; the Alloxystinae are exclusively obligate hyperparasitoids of aphids (Godfray 1994).

Many species of hyperparasitoid are associated with a single host species and a large percentage of primary parasitoids can be attacked. The tortricid caterpillar *Zeiraphera diniana* on conifers has 13 obligate and 6 facultative hyperparasitoid species (Delucchi 1982), walnut aphids have 5 species (Frazer & Van den Bosch 1973) and there can be 5 hyperparasitoid species of cereal aphids in the same field at the same time (Powell 1983, Wratten & Powell 1991). Hyperparasitoids are considered to be significant inhibitors of primary parasitoids of aphids on cabbage, apple, potato and rose (Hagen & Van den Bosch 1968). Hyperparasitism caused 68% of mummy mortality of the aphid *Macrosiphum euphorbiae* on potato (Brodeur & McNeil 1992) and usually more than 10% mortality of walnut aphid mummies (Frazer & Van den Bosch 1973). From 60 to 82 % of parasitized apple aphids were hyperparasitized (Bouchard et al. 1991). Hyperparasitism of parasitoids of Lepidoptera eggs is common but is mainly confined to large host eggs (Hirose 1994).

In the context of classical biological control, hyperparasitoids can add stability to a system by changing it from one with periodic pest outbreaks to one with a continuous sub-economic pest level (Lasalle 1993). On the other hand, there are recorded cases of hyperparasitoids and facultative autoparasitoids inhibiting classical biological control (Altieri et al. 1993, Rosenheim et al. 1995).

Competition

Competition is expected to be more intense between parasitoid species than between other natural enemies because individual hosts are usually sufficient for the development of only one parasitoid species (Mills 1994b). Interspecific competition is considered to be a potent force structuring parasitoid communities (Askew & Shaw 1986), and such competition between generalist idiobiont ectoparasitoids is thought to explain the similar mean size of parasitoid communities on endophytic and soil-inhabiting hosts throughout the world (Hawkins 1990). Not all parasitoid communities, however, are structured by competition. The abundances of parasitoid species attacking grass-feeding chalcid wasps in the UK fluctuated independently of each other in 90% of the cases studied, suggesting little competition (Dawah et al. 1995).

When two species of parasitoid compete, the main arena of competition is often within a host. The term multiple parasitism (also called multiparasitism or heterospecific superparasitism) was first coined by Smith (1929) to denote that "the same individual host is inhabited simultaneously by the young of two or more species of primary parasitoids". Many aphid species are attacked by several aphidiid and aphelinid parasitoids with overlapping host ranges (Mackauer 1990) and multiparasitism may be expected in such situations. In extreme cases multiparasitism can lead to competitive displacement of one species by another (Mackauer 1990), but in other cases mechanisms have evolved that enable parasitoids to avoid multiparasitism. Under laboratory conditions, *Aleochara bilineata* (a generalist staphylinid predator and a parasitoid) and *Trybliographa rapae* (a specialist cynipid parasitoid) compete in a complex manner during parasitism of cabbage root fly (*Delia radicum*) (Finch 1996), but multiparasitism is very rare in the field and it appears that *A. bilineata* has some mechanism for avoiding hosts already parasitized by *T. rapae* (Jones et al. 1993). Some parasitoids can recognise chemical marks on the host made by sibling species and thereby avoid multiparasitism (Godfray 1994).

Cleptoparasitism is a special case of multiparasitism where the wasp deliberately searches for and oviposits in hosts parasitised by other species (Mackauer 1990). Cleptoparasitoids may sometimes function as primary parasitoids; they are usually poor at finding the host but good at finding the competitor parasitoid against which they are always superior (Mills 1994b). Some cleptoparasitoids of insects in dead wood lack a boring ovipositor and have to wait for another parasitoid to drill the hole (Godfray 1994).

Efficient parasitoids in terms of host search and attack may nevertheless be poor competitors with other parasitoids (Ehler & Miller 1978, Waage & Mills 1992). The four primary parasitoids of a cecid midge *Rhopalomyia californica* have reproductive capacities negatively related to their competitive abilities. When the habitat is disturbed the poor competitors initially do well then gradually give way to the better competitors (Force 1974). A similar situation was reported by Price (1973) in relation to the parasitoids of the pine sawfly *Neodiprion swainei*. However, if the competing parasitoid species have different and complementary advantages in finding hosts and competing in hosts, "counterbalanced competition" can occur (Waage & Mills 1992) and this may enable the coexistence of many parasitoid species attacking the same host (Askew & Shaw 1986). An example is provided by the Arrowhead Scale on citrus in Japan which is controlled by two competing parasitoids. Under uniform conditions in the laboratory one species displaces the other, but in field environments that are changing spatially and temporally, counterbalanced competition is observed, which controls the pest and permits coexistence of the parasitoids (Takagi & Hirose 1994).

Host-feeding by adult parasitoids could provide a mechanism to reduce competition, but it is not yet known whether the decision to destructively host-feed, rather than oviposit, depends on whether the host is already parasitized (Ehler 1994).

In some hymenopteran species, such as *Aphidius smithi*, the foraging adult can detect the presence of another species of parasitoid within the host (heterospecific discrimination) (Mackauer 1990), but Tachinidae do not discriminate against hosts already attacked by heterospecific parasitoids (Belshaw 1994). The relative timing of attack on the host is one of the main factors affecting the outcome of competition and it is common for endoparasitoids of early host larvae to be curtailed by ectoparasitoids of late host larvae (Mills 1994b). Mixed multiparasitism involving hymenopteran and dipteran parasitoids can occur, and although a tachinid and hymenopteran parasitoid occasionally emerge from the same host individual, there is usually competition, with the outcome being determined by the relative timing of attack (Belshaw 1994). Although timing of attack by each species is usually crucial to the outcome of the interaction, younger larvae can sometimes successfully attack older larvae of another species (Mackauer 1990). Some species of parasitoid appear to be superior competitors to others regardless of the order of arrival in the host (Jones et al. 1994, Mills 1994b). For example, under most conditions, *Aphidius ervi* is a superior larval competitor to *Aphidius smithi*, in *Acyrthosiphon pisum* (Bueno *et al.* 1993).

Interference competition is not confined to interactions between larvae within the host. Some adult ectoparasitoids destroy eggs of other species present on the body of the host (Godfray 1994). It is thought that *Ephedrus* species inject a venom during oviposition that kills newly-laid eggs of *Aphidius* species (Hagvar 1988, Godfray 1994) and stinging by *Ephedrus*, without oviposition, caused the death of *Aphidius* eggs present in pea aphids (Pijls et al. 1995). Larvae of some parasitoid species secrete cytolytic chemicals that destroy eggs of competing species within the host (Godfray 1994). For example, larval *Apanteles fumiteranae* inhibited development of eggs of *Colypta fumiteranae* in spruce budworm larvae (Lewis 1960). The outcome of inter-larval competition may be determined by their relative speed of development. In multiparasitised cassava mealybugs, the parasitoid *Apoanagyrus lopezi* was four times more likely to survive than its congener *A. diversicornis* because *A. lopezi* develops faster (Pijls et al. 1995). Larvae of one species of endoparasitoid may asphyxiate younger larvae of another species (Godfray 1994) or suppress their competitors by toxic secretions, starvation or physical attack (Schmid-Hempel & Schmid-Hempel 1996). Conopid larvae (Diptera) seem capable of physical attacks in the first instar, which demonstrates a convergent pattern between the Diptera and Hymenoptera (Schmid-Hempel & Schmid-Hempel 1996).

Superparasitism, including multiparasitism, may seem wasteful in biological control projects because of loss of reproductive potential, searching time or both (Mackauer 1990). On the other hand, the habitat disturbance, that occurs commonly in agriculture, can promote coexistence between competing parasitoids (Miller & Ehler 1990), and the resulting greater diversity of parasitoid species might be beneficial for biological control. There was competition between *Opius humilis* and *Diachasma tryoni* (both Braconidae) in attacking the Mediterranean Fruit Fly *Ceratitis capitata* on coffee in Hawaii, but combined parasitism (44%) was greater than parasitism achieved by *Opius* alone (32%) (Smith 1929).

In classical biological control, if an inferior control agent is already established, it may prevent the subsequent establishment of better species (Waage & Mills 1992). This type of phenomenon, and community assembly rules (Drake 1990) in general, may also apply to endemic communities of annual crops, because of the short timescale involved. Thus, Cornell & Hawkins (1994) report that the disturbance of agricultural habitats does not reduce the rate of parasitoid community assembly and that endemic parasitoids may attack introduced herbivores within a year. The whitefly *Parabemisia myricae*, for example, that invaded California from Japan, was controlled by the USA endemic *Eretmocerus tabaci*, a parasitoid that switched from its native host to the novel host (Lasalle 1993). Because of the severe effect of some farming operations, such as ploughing, natural enemy communities on annual crops typically build up over time from an initial depauperate-population. The possibility that the *order* in which species arrive could affect the outcome of biological control on an annual timescale, does not seem to have been studied. The order in which species arrive is likely to have a large stochastic component, but might be amenable to manipulation, to some extent, by careful choice of which crops are planted adjacent to each other (Hirose 1994, Burel & Baudry 1995).

Predator-parasitoid interactions

In addition to the predation of adult parasitoids (Rosenheim et al. 1995), predators may interfere with the foraging of parasitoids or remove them indirectly by feeding on parasitised pests. In some cases, control of pests by parasitoids may be disrupted by the addition of a predator, but successful control of the Winter Moth (*Operophtera brumata*) in Canada seems to have resulted from the joint action of endemic predators and an introduced parasitoid (Rosenheim et al. 1995).

Interference with foraging

Little information is available on whether the efficiency of parasitoids foraging for herbivore hosts is reduced by encounters with predators, but it can be when the predators themselves are the target host. Some spiders within the Linyphiidae (Van Baarlen et al. 1994), Oxyopidae (Guillebeau & All 1989), Theridiidae (Evans 1969), Clubionidae, Salticidae and Thomisidae (Austin 1985) guard their eggsacs against parasitoid attack.

Predation of moribund parasitised pests

While the pest is moribund, in the period between being parasitized and dying, it may be more or less vulnerable to predators than are healthy pests. Its distribution, phenology, development rate, behaviour (e.g. activity level, defence reactions and abiotic preferences), size, and semiochemical signals etc, may be altered, and these changes are likely to affect attack rates by predators differentially, depending on the foraging strategy of the predators concerned.

Parasitised pests often move to vertical strata in the crop that are different from those occupied by healthy pests, and this may alter their encounter rate with predators, which are also vertically stratified. To the degree that this behaviour is driven by selection pressures acting on the parasitoid, a balance has to be struck between moving out of hyperparasitoid danger zones and into predator-rich areas; this will vary with the system under study. However, it has also been hypothesised that non-reproductive moribund pests (especially aphids and caterpillars) "commit suicide" by deliberately exposing themselves to predation or other mortality factors, to rid their immediate kin of parasitoids (Shapiro 1976, Smith Trail 1980, McAllister et al. 1990), but this hypothesis remains controversial (Godfray 1994).

Laboratory observations showed that many individuals of the cereal aphid *Metopolophium dirhodum* parasitized by *Toxares deltiger* left the host plant prior to mummification (Powell 1980). Kuo-Sell et al. (1988) reported that, depending upon aphid species, 35-89 % of parasitised cereal aphids left their feeding sites before mummification. Movement from feeding sites is also affected by the species of parasitoid that the cereal aphids contain. Parasitoids with relatively long development times are more likely to induce movement of moribund aphids (Höller 1991). In general, hosts of koinobiont parasitoids (with longer development times than idiobionts) are often not killed until they have prepared cryptic pupation retreats (Askew & Shaw 1986). Development time is also affected by diapause, and on potato in the laboratory, *Macrosiphum euphorbiae* containing non-diapausing *Aphidius nigriceps* left the colony and mummified on the upper surface of leaves, but aphids containing diapausing parasitoids left the host plant and mummified in concealed sites (Brodeur & McNeil 1989). In potato fields, aphids are found mostly on the undersides of leaves, but moribund aphids move to the upper surfaces to mummify where parasitoid survival is greater, due to reduced hyperparasitism and predation (Brodeur & McNeil 1992). Similarly, parasitised pea aphids move to the upper surfaces of alfalfa leaves where predation may be reduced (Frazer & Gilbert 1976). Some leafhoppers (Cicadellidae) parasitised by dryinid wasps embed their rostrum in the plant prior to death; it may be in the interest of the parasitoid that the host is not dislodged (Godfray 1994). *Acyrthosiphon pisum* parasitized by *Aphidius ervi* may drop off the plant when approached by a coccinellid (McAllister & Roitberg 1987), and the instar at which the host is parasitized (and thus its reproductive potential) determines the likelihood of this response (McAllister, Roitberg & Weldon 1990).

Dying parasitised caterpillars often climb plants and die at the top of the vegetation (Stamp 1981). The Common Armyworm *Leucania separata* parasitised by *Apanteles kariyai* crawl up the plant until the parasitoid emerges, whereas healthy larvae pupate on the ground (Sato et al. 1983). In this system, the parasitoid also prolongs the period of the last instar of the host (Sato et al. 1983), thus increasing its exposure to predation. Other cases have been reported of parasitized larvae remaining on plants for longer periods than unparasitized larvae (Hopper & King 1984, Ryan 1985). Parasitized green cloverworm larvae (*Plathypena scabra*) leave the plant and wander on the ground before the parasitoid emerges (Yeargan & Braman 1986). Spruce budworm larvae parasitised by *Apanteles fumiteranae* or *Colypta fumiteranae* become photonegative (healthy caterpillars are photopositive). Their pattern of movement on trees is different to that of healthy larvae and they emerge from hibernacula ten days later than healthy larvae (Lewis 1960, Nealis & Regniere 1987). Similarly, parasitized casebearer larvae (*Coleophora laricella*) are more likely than unparasitized larvae to descend from larch foliage, and they also seek different pupation sites (Ryan 1985).

Birds and mammals tend to feed less on moribund than on healthy caterpillars, probably because of reduced activity of the former (Fritz 1982). In both the laboratory and field, shrews and mice preferred healthy cocoons of the European Pine Sawfly, *Neodiprion sertifer*, to parasitised ones; this resulted from a mixture of innate and learnt behaviour (Holling 1955). Mammals do not, however, always prefer healthy prey. Pupae of the Gypsy Moth *Lymantria dispar* parasitised by the tachinid *Blepharipa pratensis* were eaten by small mammals three times as often as were healthy pupae (Weseloh 1990).

Invertebrate predators often feed more on moribund than on healthy pests. *Coccinella 7-punctata*, consumed parasitized *Schizaphis graminum* faster than non-parasitised individuals (T.J. Kring personal communication to Obrycki 1992). All large unmummified walnut aphids carried by Argentine Ants were found to be parasitised by *Trioxys pallidus*; the ant selectively removed them and was considered to be a threat to biological control of walnut aphid in California (Frazer & Van den Bosch 1973). On pine, 69% of *Neodiprion sertifer* sawfly larvae attacked by the pentatomid bug *Podisus modestus* were already parasitised (Tostowaryk 1971). The stage of development of the parasitoid may, however, affect the preference shown by the predator. The coccinellid *Delphastus pusillus* ate whiteflies (*Bemisia argentifolii*) parasitised by *Encarsia formosa*, but was less likely to attack hosts containing parasitoids in an advanced stage of development (Heinz & Nelson 1996).

The success of biological control can be affected by whether predators prefer healthy or moribund pests, as indicated in the following two examples. In laboratory studies, *Plodia interpunctella* was reduced by 74% over four weeks by the braconid *Bracon hebetor*, but by only 53% by a combination of *B. hebetor* and the anthocorid *Xylocoris flavipes*. This was because *X. flavipes* killed not just *Plodia* but also *Bracon* larvae (Press et al. 1974). In contrast, staphylinid beetles tended to avoid Winter Moth pupae (*Operophtera brumata*) parasitised by the tachinid *Cyzenis albicans,* and total mortality from parasitism plus predation was greater than if the two factors had acted independently (Roland 1990).

Predation of mummies

Predators that kill aphid mummies could have a detrimental effect on biological control by reducing the size of the next generation of parasitoids. In contrast, if mummies that contain developing hyperparasitoids are attacked (e.g. hyperparasitised walnut aphid mummies preyed upon by chrysopids and coccinellids (Frazer & Van den Bosch 1973)), a positive effect on biological control can be expected. It is quite common for predators to attack moribund parasitised pests, yet refuse pests in the later stages of parasitism (Rosenheim et al. 1995). Nevertheless, mummies are attacked by predators belonging to a wide range of taxonomic groups, as indicated in the examples below.

Mummified pea aphids, *Acyrthosiphon pisum,* on alfalfa were eaten by *Chrysopa* sp., two coccinellid species, a nabid bug and five species of mirid bug (Wheeler et al. 1968, Frazer & Gilbert 1976, Wheeler 1977). Wheeler (1974) considered that the mirid predators of pea aphid mummies function as the ecological equivalent of hyperparasitoids and play potentially as large a role in limiting primary parasitoid populations in alfalfa as do some species of hyperparasitoid. In the laboratory, aphid mummies were eaten readily by a range of beetle predators, including large Carabidae (Powell et al. 1987), and by the European earwig, *Forficula auricularia* (Ravensberg 1981). *Macrosiphum euphorbiae* mummies on potato were found to have characteristic holes made by coccinellids, chrysopids and nabids. This enabled Brodeur & McNeil (1992) to estimate that predators accounted for 15% of mummy mortality. Mummies of the walnut aphid, *Chromaphis juglandicola*, parasitised by *Trioxys pallidus*, were frequently observed being sucked out by chrysopid larvae. They were also attacked by coccinellids and ants, and overall *c.* 10% of mummies were killed by predators (Frazer & Van den Bosch 1973). Effects on biological control will be greater if predators remove mummies selectively. For example, field observations on sycamore in summer showed that, although less than 10% of *Drepanosiphum platanoides* were mummified, mummies formed 10-70% of the aphids killed and consumed by two species of *Anthocoris* bugs (Dixon & Russel 1972). Similarly, mummies of *C. juglandicola* on walnut were selectively removed by Argentine Ants (Frazer & Van den Bosch 1973).

Predator-pathogen interactions
Predation of moribund diseased pests
It has already been noted that developing parasitoid larvae can cause changes in the physiology, behaviour and distribution of their hosts and in the palatability of those hosts to predators (see above). Similar phenomena have been observed amongst moribund pests attacked by pathogens. Dying caterpillars attacked by viruses and entomogenous fungi often climb plants and die at the top (Shapiro 1976, Ohbayashi & Iwabuchi 1991), or lose their gregarious habit and wander randomly before falling to the ground (Smirnoff 1960). Aphids infected with some species of fungi are more likely to fall off the plant (Hagen & Van den Bosch 1968). Moribund carrot flies infected with *Entomophthora* move 4 - 6 metres above ground into hedges (Eilenberg 1987) which brings them potentially into contact with a different set of natural enemies.

In laboratory preference tests, three species of carabid did not discriminate between healthy larvae of the cabbage moth, *Mamestra brassicae*, and larvae infected with an NPV (Vasconcelos et al. 1996). In contrast, the predatory bug *Nabis roseipennis* preferred NPV-infected to healthy *Anticarsia gemmatalis* larvae, probably because of a reduced defensive response by diseased larvae (Young & Kring 1991). Shrews and mice, however, preferred healthy cocoons of the European pine sawfly, *Neodiprion sertifer*, compared to cocoons infected with fungi (Holling 1955).

Dissemination of pathogens by predators

Some predators will consume, not only moribund pests, but also dead diseased pests, and may then carry the pathogens on the surface of their bodies or spread them via their faeces.

Coccinellids and mirids have been observed, in alfalfa, to eat *Hylemya* flies killed by *Entomophthora muscae* (Wheeler 1971, 1974, 1977), but it is not known if the fungus is spread by these predators. Larvae of *C. 7-punctata*, however, will eat *A. pisum* in which *Erynia neoaphidis* are sporulating, and the larvae then act as passive vectors of *E. neoaphidis* to healthy *A. pisum* populations (Pell & Pluke 1994, Roy & Pell 1995).

The carabid, *Calosoma* sp., is known to excrete viable Gypsy Moth NPV particles in its faeces, but the extent of dissemination has not been measured in the field (Weseloh 1990). Infectivity of an NPV to larvae of cabbage moth *Mamestra brassicae* was maintained after passage through the carabid gut and carabids continued to pass infective virus for at least 15 days after a meal. In this case, however, carabids transferred enough virus to the soil to cause low levels of cabbage moth mortality in the field (Vasconcelos et al. 1996). Heteropteran predators will also consume NPV's and spread them (Abbas & Boucias 1984), causing epizootics in prey populations (Biever et al. 1982). Baculoviruses are also spread by Orthoptera, Dermaptera, Anthicidae, Coccinellidae and *Polistes* wasps (Gröner 1990).

Viral and bacterial diseases of pests can be spread through the faeces of predators (Capinera & Barbosa 1975) and microsporidia are also known to be eaten (Ragsdale & Oien 1996) and disseminated (Young & Hamm 1985) by predators. Spores of *Nosema pyrausta* (Microsporidia), accumulated when *Chrysoperla carnea* larvae ate *Nosema*-infected caterpillars of *Ostrinia nubilalis*, were eliminated at eclosion and were still infective (Sajap & Lewis 1989).

Less than 20% of predator attacks on prey are successful (Malcolm 1992), but the wounds resulting from unsuccessful attacks expose the pest to infection by bacteria, protozoa and viruses. Predatory Muscidae (Diptera) larvae can wound leatherjackets (Tipulidae) (Griffiths et al. 1984), and Doane et al. (1985) recorded 6% of soldier fly larvae (Diptera) carrying wounds caused by predatory wireworm larvae (Elateridae).

Pathogens of predators

The performance of predators may be impaired by eating diseased prey, or the predator may be attacked directly by pathogens, including species that also attack their prey.

Chrysoperla carnea larvae fed on eggs of *Ostrinia nubilalis* infected with *Nosema pyrausta* (Microsporidia) developed normally and there were no deleterious effects on reproduction or longevity (Sajap & Lewis 1989). Although Microsporidia have been described from many species of predator, predators are often resistant to infection by protozoan pathogens of their prey (Young & Hamm 1985, Vinson 1990, Oien & Ragsdale 1993). However, certain strains of *Bacillus thuringiensis* can have sublethal effects on Chrysopidae, Coccinellidae (Salama et al. 1982, Melin & Cozzi 1990) and Anthocoridae (Salama et al. 1991).

Some predators can be killed by eating prey infected with the entomopathogenic nematode (IPN) *Steinernema feltiae* (Vinson 1990), but *Heterorhabditis heliothidis* and *S. feltiae* were claimed to cause little infection of some species of Carabidae, Cicindelidae, Staphylinidae or labidurid earwigs (Georgis & Wojcik 1987, Vinson 1990). Larvae and pupae of other species of Carabidae and Staphylinidae were, however, killed by IPN's in the laboratory (Pölking & Heimbach 1992), and the carabids *Agonum dorsale* and *Amara similata* were reduced by IPN's in field plots, but not consistently in different years (Bathon 1996). Spiders, harvestmen and pseudoscorpions were infected by very high doses of IPN's in the laboratory, but not at field rates (Akhurst 1990). IPN's are, in turn, eaten by a wide range of predators, including mites and insects. They also exhibit a range of interactions (dissemination, avoidance of competition, competition etc) with fungi, viruses, bacteria and protozoa (Kaya & Koppenhöfer 1996), similar to those recorded in this review as occurring between predators, parasitoids and pathogens.

Carabid and rove beetles suffer mortality from both hyphomycete and entomophthorealean fungi (Thiele 1977, Steenberg et al. 1995) and from gregarine protozoa (Thiele 1977). Epizootics of the fungus *Beauveria bassiana* are common amongst hibernating coccinellids (Goettel et al. 1990) and *B. bassiana, Metarhizium anisopliae* and *Paecilomyces fumosoroseus* caused high levels of mortality of coccinellid larvae, *Hippodamia convergens*, under laboratory conditions (James & Lighthart 1994). Spiders are susceptible to non-entomophthoralean fungal pathogens (Evans & Samson 1987, Greenstone et al. 1987, Goettel et al. 1990) and carry bacterial and viral infections (Legendre & Morel 1980, Rollard 1984).

The predatory bug *Nabis roseipennis* feeding on soybean looper, *Pseudoplusia includens*, infected with an NPV, suffered the sublethal effects of reduced longevity and fecundity and increased preoviposition period (Ruberson et al. 1991).

Parasitoid-pathogen interactions
Predisposition for attack
Parasitism can predispose the host to attack by pathogens and, similarly, a diseased host may have a greater probability of being parasitised compared with a healthy host. Parasitism can also render hosts more resistant to attack by

viruses, fungi and bacteria (Brooks 1993). Various physiological and behavioural mechanisms operate to cause these effects.

Parasitism can render hosts more susceptible to attack by fungi (Vinson 1990). For example, the infection rate of the cereal aphid *Metopolophium dirhodum* by *Erynia neoaphidis* was greater for aphids parasitized by *Aphidius rhopalosiphi* than for healthy aphids (Powell et al. 1986). Similarly, larvae of *Helicoverpa zea* parasitised by the braconid *Microplitis croceipes* were predisposed to infection by *Nomuraea rileyi* (King & Bell 1978). One mechanism mediating predisposition is that parasitism can induce changes in the host cuticle (Goettel et al. 1990), which facilitates attacks by fungi such as *Beauveria bassiana* (Fuhrer et al. 1978). Parasitoid-induced movement of the host can also increase its exposure to fungal pathogens. Foraging by the parasitoid *Diadegma semiclausum* caused increased movement by larvae of the diamondback moth, *Plutella xylostella,* thereby increasing their contact with, and infection by, spores of the entomogenous fungus *Zoophthora radicans* (Furlong & Pell 1996). Some populations of lepidopteran caterpillars become more heavily parasitized after an application of *Bacillus thuringiensis* (*B.t.*) (Hamel 1977; Mellin & Cozzi 1990, Vinson 1990, Soares et al. 1994), and *B.t.* infection can increase the rate at which individual caterpillars become immobilised by parasitoids (Temerak 1980). Sublethal doses of *B.t.* extended the development time of the Gypsy Moth, *Lymantria dispar*, and thus increased its availabity to parasitoids, such as *Cotesia* (=*Apanteles*) *melanoscelus* (Weseloh et al 1983). Larvae of *Spodoptera exigua* and *Helicoverpa zea* that received sublethal doses of *B.t.* were available in a stage suitable for parasitism for longer periods than were healthy larvae, and they were more heavily parasitized by *Cotesia marginiventris* (Soares et al. 1994). But, conversely, parasitized hosts were more susceptible to *B.t.* than healthy hosts (Vinson 1990). Parasitism can increase (Hunter & Stoner 1975) or decrease (Hopper & King 1984, Coop & Berry 1986) host feeding rates (depending on host species) making them more or less likely to ingest pathogens, such as virus particles (Rahman 1970, Vinson 1990). Moribund diseased pests (e.g. virus-infected caterpillars) are usually less active than healthy pests, and this can reduce (Versoi & Yendol 1982) or increase (Soares et al. 1994) oviposition rates by parasitoids, depending on the species.

Avoidance of direct competition
Some parasitoids selectively refrain from ovipositing in hosts infected with fungi, *Bacillus thuringiensis*, viruses, microsporidia and nematodes; this may require ovipositor insertion to determine the state of the host (Versoi & Yendol 1982, Hotchkin & Kaya 1983, Harper 1986, Mellin & Cozzi 1990, Brooks 1993). *Cotesia* (=*Apanteles*) *glomerata* appeared to be able to select *Pieris brassicae* larvae that were free of the microsporidian *Nosema bombycis* (Vinson 1990).

Avoidance usually occurs when the hosts are in an advanced stage of infection, and seems to be largely independent of the type of pathogen involved (Brooks 1993). Such behavioural mechanisms for avoiding direct competition are positive attributes for biological control as they tend to maximise pest mortality. Direct competition may also be avoided by asynchrony of the attacks by parasitoids and pathogens. Bacterial and fungal diseases usually reduced the population of the cutworm *Euxoa ochrogaster* after parasitoids had completed development, so there was little negative interaction between these natural enemies. In fact, a high incidence of disease increased the percentage parasitism in the following year (King & Atkinson 1928). Negative relationships between rates of attack of pests by parasitoids and entomogenous fungi or microsporidia in the field (Andreadis 1982, Los & Allen 1983, Powell et al. 1986, Rosenheim et al. 1995) might be achieved by mechanisms that avoid or result from direct competition (see below). When the negative relationship is temporally (rather than spatially) determined, it could result from the parasitoids and pathogens having different optimal abiotic conditions (Goettel et al. 1990).

Competition
Although inter-kingdom competition, which occurs between parasitoids and pathogens, may be one of the commonest forms of interaction in nature (Hochberg & Lawton 1990), it has been studied relatively little to date (Lawton 1994). Competition between organisms that are widely separated, taxonomically, is more likely to be assymmetrical, than competition between organisms within the same genus (Diamond 1987). When parasitoids and pathogens compete for a host, such assymetry is common, and pathogens usually impair the performance of parasitoids, rather than *vice versa* (amensalism). Death of parasitoids in pathogen-infected hosts is often due to the pathogen killing the host rather than the parasitoid (Los & Allen 1983, Powell et al. 1986, Akhurst 1990, Vinson 1990, Brooks 1993). Direct mortality of the parasitoid can, however, be caused by protozoa, bacteria and fungi (Brooks 1993) and also, under laboratory conditions, by nematodes (Brooks 1993). Some *Nosema* species can attack both the host and the parasitoid (Andreadis 1980, Oien & Ragsdale 1993, Rosenheim et al. 1995, Ragsdale & Oien 1996), and even the hyperparasitoid (Brooks 1993). Parasitoids may die within hosts due to poisoning by viral toxins (Kaya 1970, Harper 1986, Gröner 1990, Brooks 1993). The scale of this effect can be considerable. For example, 100% of NPV-killed caterpillars of *Trichoplusia ni* in collections from some fields in California contained dead parasitoids (Harper 1986). Sarcophagid larvae may leave virus-infected hosts and pupate, but the adult parasitoid fails to develop (Gröner 1990). Parasitoid (e.g. Tachinidae) emergence is sometimes reduced when ovipositing into nematode-infected larvae (Mracek & Spitzer 1983), but a proportion of some hymenopterous parasitoids

complete development normally (Kaya & Hotchkin 1981). Adult *Compsilura concinnata* (Tachinidae) are susceptible to nematode infection on emerging from their pupal cases in the soil (Akhurst 1990). Parasitoids seem to have little effect on nematode development (Kaya 1978).

Parasitoids may out-compete fungi within hosts, providing the parasitoid is well-established before fungal attack occurs. If the aphid, *Metopolophium dirhodum*, is attacked by *Erynia neoaphidis* less than four days after being parasitized by *Aphidius rhopalosiphi* the parasitoid does not complete development, but if more than four days elapses, development of the fungus is impaired (Powell et al. 1986). When larvae of *Helicoverpa zea* parasitised by the braconid *Microplitis croceipes* were exposed to infection by *Nomuraea rileyi*, the fungus halted development of the parasitoid only if it attacked up to one day after parasitism (King & Bell 1978). For the alfalfa aphid *Therioaphis trifolii f. maculata* parasitised by the braconid *Trioxys complanatus* and exposed to the fungus *Zoophthora radicans*, the fungus out-competed early instar parasitoids but more mature parasitoid larvae were able to complete their development (Milner et al. 1984). Thus, just as for competition between different species of parasitoid within a host (see above), the relative timing of attack is an important factor in determining the outcome of competition between fungi and parasitoids. Some species of parasitoid larvae (e.g. some Braconidae and Ichneumonidae), if well established in the host, can produce anal secretions containing antibiotic and fungistatic substances which inhibit the development of entomogenous fungi (Fuhrer et al. 1978, Goettel et al. 1990, Vinson 1990).

If competition does not result in mortality of one of the competitors, there can still be sub-lethal effects on either or both of the competitors. Viruses, protozoa, and bacteria can have a serious impact on parasitoids in hosts, by affecting longevity, development rate, fecundity (Andreadis 1980, Temerak 1980, Mellin & Cozzi 1990, Brooks 1993) and nutrient uptake (Ragsdale & Oien 1996). Sub-lethal effects can also occur if the parasitoids have to compete with viruses for host nutrients (Hamm et al. 1983, 1985). However, Hochberg & Lawton (1990) noted that the parasitoid *Apanteles glomeratus* in *Pieris brassicae* reduced virus production by 28%.

Under field conditions, the incidence of pathogen attack is sometimes so high that parasitoid populations are severely affected. The fungus *Pandora neoaphidis* attacks *Acyrthosiphon kondoi* and *A. pisum* and directly reduces the efficiency of *Aphidius smithi*. Bueno *et al.* (1993) considered this fungus to be a key species determining the composition of *Acyrthosiphon* species and their parasitoids (*A. smithi* and *A. ervi*) in fields of alfalfa in California. In years when epizootics of the fungus *Zoophthora phytonomi* occur, complete decimation of the host weevil (*Hyperica postica*) population can be nearly as detrimental to populations of the parasitoid *Bathyplectes anurus* (and many other natural enemies in alfalfa) as an insecticide application (Los & Allen 1983). *Nosema pyrausta* is suspected of causing the gradual disappearance of a parasitoid of the corn borer, *Ostrinia nubilalis*, in the USA (Brooks 1993).

Dissemination of pathogens by parasitoids
Parasitoids can be mechanical vectors of insect viruses (Hamm et al. 1985) (as distinct from mutualistic polydnaviruses that are essential to some parasitoids for successful parasitism (Whitfield 1994)) and of *Bacillus thuringiensis* (Salama et al. 1982). Vectors of viruses can be either parasitoids developing in infected hosts or healthy parasitoids that have picked up virus particles by ovipositing in infected hosts (Brooks 1993). Although there are some instances of protozoa, bacteria, viruses and fungi being transmitted via the ovipositor (Brooks 1993), there are insufficient data to know whether this is a common form of transmission (Hochberg & Lawton 1990, Gröner 1990). The main mechanism for virus transmission seems to be by parasitoids contaminating host food as they walk, defaecate or void meconia (Harper 1986). The braconid *Microplitis croceipes* transmits *Heliothis* NPV to *Heliothis virescens* on soybean (Young & Yearian 1990). The virus is picked up initially when *Microplitis* oviposits in infected larvae. When the parasitoid moves about on the leaf near the host it is likely to contaminate the host's food. Young & Yearian (1990) considered that release of artificially contaminated parasitoids might be a useful tactic in IPM programmes. Parasitoids and hyperparasitoids can transmit protozoan (e.g. *Nosema*) diseases of pests mechanically (Hamm et al. 1983, Vinson 1990, Ragsdale & Oien 1996) and transovarially (Brooks & Cranford 1972, Vinson 1990, Brooks 1993, Ragsdale & Oien 1996). There are few recorded cases of entomogenous fungi being transmitted via the ovipositors of parasitoids (Vinson 1990), an exception being the fungus *Aschersonia aleyrodis* transmitted to glasshouse whitefly, *Trialeurodes vaporariorum,* by *Encarsia formosa* (Brooks 1993). Parasitic Diptera usually penetrate the host as larvae, causing wounds that provide routes of infection for bacteria, protozoa and viruses (Vinson 1990). Many hymenopterous parasitoids also cause wounding of pests, especially hemipterous pests (see references in Sunderland 1996).

Indirect interactions and the role of alternative prey
One natural enemy may have an effect on another natural enemy through sharing a prey or host. The action of one natural enemy species may alter the size structure of the prey population, making the survivors more or less available to other natural enemy species that have distinct prey size requirements (Strauss 1991). Similarly, parasitism by an early attacking guild of parasitoids may alter the size structure of the host population and disrupt synchronisation of host attack by later parasitoid guilds (Mills 1994b).

The availability of alternative prey and alternative hosts is an important aspect of the system in relation to the functioning of polyphagous predators and polyphagous parasitoids, and it has implications for the biological control of pests. There is some field evidence that polyphagous parasitoids can exert more influence on a target host species (even driving it to local extinction) when alternative host species are present in the habitat (Lawton 1986, Settle & Wilson 1990). Alternative hosts can also improve parasitoid distribution, reduce interspecific competition and improve the synchrony between parasitoid and host, providing the individual parasitoid is capable of host-switching (Powell 1986). Similar considerations apply to pathogens. For example, *Pandora neoaphidis* (Entomophthorales) was the main cause of a high mortality of *Acyrthosiphon pisum* in Californian alfalfa during wet periods, even at *A. pisum* densities unfavourable for an epizootic. This was thought to be made possible due to a reservoir of inoculum in *Acyrthosiphon kondoi* (Pickering and Gutierrez 1991). Similarly, bacterial diseases of the caterpillar, *Euxoa ochrogaster*, were more likely to cause epizootics when *Euxoa tessellata* was also present (King & Atkinson 1928).

Although the food range of aphidophagous ladybirds is much narrower than that of polyphagous predators, alternative food is nevertheless of significance in the ecology of these beetles too. Aphid honeydew is an arrestant for coccinellids (Carter & Dixon 1984) and sprays of artificial honeydew or hydrolysed proteins can be used to arrest coccinellids in the sprayed areas (Smith 1971, Evans & Swallow 1993). Fungal spores (e.g. *Alternaria* spp.) can be very common in the guts of coccinellids, even when aphid densities are high. Triltsch (1997) found that 86.4% of *Coccinella 7-punctata* adults in a winter wheat field had fungal spores present in the gut, even at an aphid density of 4.6 aphids per tiller.

Polyphagous predator populations should benefit from a wide range of alternative prey, but the value of a diverse array of alternative prey species depends on how readily predators switch their choice of victim species (Russell 1989). Knowledge is currently scarce, both on qualitative and quantitative aspects of alternative food types (Petersen & Holst 1997), and on the frequency of prey-switching by polyphagous predators under field conditions (Russell 1989). Alternative prey can contribute significantly to the survival and reproduction of polyphagous predators. In California vineyards tydeid mites, which are harmless to viticulure, sustain large populations of predatory phytoseiid mites, which are, in turn, important biological control agents of tetranychid mites (Lindquist 1983). Similarly, *Phytoseius fotheringhamiae* is sustained by eriophyid, tydeid and *Bryobia* mites in Australian apple orchards and this enables it to become numerous enough to control developing populations of *Tetranychus urticae* (Overmeer 1985). In general, a mixed diet results in greater fecundity of polyphagous predators than a single-species diet (Toft 1995, Sunderland et al., 1996). The reproductive success of generalist predators attacking the Gypsy Moth, *Lymantria dispar*, was largely dependent on the availability of alternative prey, and this enabled them to cause significant mortality of *L. dispar* even when moth population densities were very low (Weseloh 1990). Alternative prey may sometimes have a role in improving the synchronisation of predators and pests and maximising attack rates on pest species. Collembola, which are thought to be an important group of alternative prey for polyphagous predators in temperate crops (Sunderland 1975, Sunderland et al. 1986b, De Keer & Maelfait 1987, Alderweireldt 1994), tend to be most numerous in the spring and autumn (Roske 1989). This may be because reproduction of major groups of Collembola, such as the Isotomidae, can be inhibited in dry microclimates (Basedow 1994). Collembola may stimulate numerical responses (Potts & Vickerman 1974, Van Wingerden 1977, Gravesen & Toft 1987) of predators in the spring, then as Collembola populations collapse in summer the predators may switch to feeding predominantly on pests. Axelsen et al. (1997) modelled a system similar to this, incorporating linyphiid spiders, Collembola and cereal aphids, and the output suggested that spiders were ineffective against aphids at low levels of Collembola but their impact was greater when spider numbers built up on abundant Collembola in the spring. An analogous system has been found to operate in irrigated tropical rice, with detritivores as the alternative prey (Settle et al. 1996). In extreme cases, target prey species may be eliminated entirely from a community by populations of polyphagous enemies sustained by alternative prey (Jeffries & Lawton 1984).

Models of pest control by natural enemy communities

As indicated above, considerable system complexity arises from the numerous interactions between species involved in pest control in agroecosystems. Qualitative and semi-quantitative investigations of the functioning of isolated components of such complex systems can be carried out experimentally, but modelling is probably the only practicable means of exploring the dynamics of the system as a whole. A brief overview of the evolution of such models is given below.

Models involving two or three species
Hassell (1986) showed that three-species systems (host, generalist parasitoid, specialist parasitoid) can have a wider range of dynamic properties than two-species systems, and that counter-intuitive dynamics can also result (Hassell &

Godfray 1992). A similarly wide range of dynamics is exhibited by three-species systems involving host, specialist parasitoid and pathogen (Hassell & Godfray 1992). Jones et al. (1993) modelled a host-generalist-specialist interaction (cabbage root fly, *Aleochara bilineata*, *Trybliographa rapae*) and found that there could be three-species stable states. In general, modelled three-species systems tend to be more stable than two-species systems (Jones et al. 1994). However, in the context of pest control, and especially on annual crops, stability of pest populations is less important than their reduction below the economic threshold (Dempster & Coaker 1974), and local extinction (including by unstable dynamics) is acceptable.

Crop growth and pest population dynamics simulation models (e.g. Carter 1994) can be the starting point at which predator sub-models are added, as has been done, for example, for cereal aphids and coccinellids (Freier et al. 1996ab, Skirvin et al. 1996, Triltsch & Rossberg 1997).

Simulation models involving a larger number of species

The metabolic pool approach, that utilises supply-demand ratios to model trophic links, was used by Gilbert et al. (1976) to investigate plant-herbivore interactions and by Gutierrez et al. (1981) for prey-predator interactions. The currency of such models can be numbers, biomass or energy and the priority order of assimilate use is set as respiration followed by reproduction, then growth. Demand rates (for respiration, growth etc) are usually incorporated on an age-specific basis (Gutierrez & Baumgaertner 1984ab). The functional response for searching success at the recipient trophic level (e.g. a predator acquiring prey) can incorporate donor and recipient densities or biomasses, recipient demand rate, recipient search rate and an increment of time in day degrees (physiological time scale) (Gutierrez et al. 1981). The metabolic pool approach has been used to evaluate the impact of natural enemies on pest populations on a range of crops including alfalfa, apple, cassava, bean, cotton, grape, rice and tomato (Gutierrez et al. 1990). It is currently being used to model the natural control of cereal aphids (Holst & Ruggle, in press) incorporating the impact of selected parasitoids (Ruggle & Holst, in press), carabid beetles (Petersen & Holst 1997), linyphiid spiders and coccinellids (Axelsen et al. 1997) and Entomophthoralean fungi (Dromph et al. 1997). Such models cannot usually be used predictively because of uncontrolled variables such as weather and migration rates, so they are aimed at gaining a better understanding of ecosystem interactions (Gutierrez 1992). Fortunately, the great flexibility of the metabolic pool approach allows for the relatively easy incorporation of new knowledge into existing models (Ruggle & Holst 1997).

Conclusions and future research

This brief survey of pest control by a community of natural enemies has drawn together disparate information about interactions between natural enemies, which, *in toto*, suggests that such interactions may have a very significant influence on the success of biological pest control. The extent to which this is true, however, cannot be assessed currently, because information is sparse and, in many cases, is qualitative rather than quantitative. Good quantitative field data are needed, especially on pest dislodgement, limitation of parasitoid populations by predators, pathogen dissemination by predators and parasitoids, and alternative food sources for generalist predators.

It was noted, in the first section of this review, that natural enemy communities are often large and complex. It would be intractable to carry out detailed studies of the hundreds of species in a natural enemy community. Guild analysis (Ehler 1992), however, would provide a logical means of reducing the size of the task and would produce an extremely valuable grouping of species into a smaller number of functional units. Pest control may be enhanced by complementary natural enemies (i.e. species attacking different stages of pest, or at different times), so that the pest does not have an unattacked age class or a temporal refuge (Murdoch 1990). Complementary natural enemies are likely to belong to different guilds, and so guild analysis could provide an efficient structure for research aimed at increasing natural enemy effectiveness, for example by the development of integrated farming systems (Holland et al. 1994, Ogilvy et al. 1994, Wratten & Van Emden 1995). Other aspects of complementary action (such as pest dislodgement, avoidance of replaceable mortality, and dissemination of pathogens) could also be taken into account in the development of IPM systems.

Recent developments in simulation modelling (Holst & Ruggle 1997) open up the prospect of investigating the dynamics of pest control by entire natural enemy communities. Although the metabolic pool approach can readily accommodate the addition of new components (Ruggle & Holst 1997), it would be unwieldy (and inefficient) to include in the model the hundreds of natural enemy, pest and alternative prey species found in the community under study. It might, however, prove possible to include a component for each guild, and so guild analysis would have value in this context too. Other modelling approaches (Freier et al. 1997) strive to define beneficial thresholds (i.e. the density of a group of predators necessary to keep a pest under control), for a wide range of predators in field crops, by the accumulation of "predator units (PU's)". PU's are based on feeding rates of predators on pests at constant temperature in the laboratory. These models would also benefit from guild information to guide the consolidation of PU's from a wide range of predator species. Current models do not incorporate the full complexity of natural enemy interactions that this review has shown can occur; the onus is first on experimenters to establish which of these interactions are of sufficient quantitative significance in pest control to be worth modelling. This review has documented many categories of replaceable mortality (also sometimes called dispensible mortality), i.e. where a pest already under imminent control by one natural enemy is attacked by another natural enemy. Gutierrez (1992) suggested that dispensible and indispensible mortality could be modelled in the same way as for pre- and post-compensation point damage to plants, but some information on the ratio of dispensible to indispensible mortality in real systems will be needed to guide modellers.

In metabolic pool models the ratio of resource acquisition to demand is crucial because it regulates all birth, death, net immigration, ageing, and net growth processes (Gutierrez 1992). Resource acquisition is, in turn, heavily affected by the functional response equations, which incorporate search rate and predator and prey density. A limitation of the current models is that they are not spatially explicit. Natural enemy search success (number of prey attacked or consumed), however, is likely to be determined by the fine structure of the spatio-temporal distribution of predator and prey (summarised as encounter rate) rather than by gross densities. Encounter rates will be affected by many of the categories of natural enemy interactions (e.g. pest dislodgement, redistribution of moribund pests) explored in this review. Free et al. (1977) considered that spatial complexity and aggregation will often be dominant factors affecting the outcome of interactions between natural enemies and their hosts or prey. Mols (1993), in models of the searching behaviour of the carabid beetle *Pterostichus coerulescens* (= *Poecilus versicolor*), showed that spatial distribution had a marked effect on the form of the functional response. Kareiva (1990) stated that most pest – natural enemy models (including metabolic pool models) do not adequately address the consequences of patchiness and movement. They rely heavily on functional response data that are often derived from spatially simplistic universes. The development of spatially explicit models capable of dealing with the complexity of real agroecosystems (e.g. a spatially explicit metabolic pool model) would be a great step forward. Another advantage of having spatially defined community models would be their potential for integration with other models operating at different spatial scales. Topping et al. (1997) have pointed out that hierarchical methods might eventually be used to integrate spatially explicit single-species, community and landscape models. Thus the resources utilised to develop a community model concerned with pest control would then also be contributing, cost-effectively, to investigations of landscape manipulations.

Acknowledgements

This publication was made possible through an EU Concerted Action "Enhancement, dispersal and population dynamics of beneficial insects in integrated agroecosystems". The review was initiated by a Final Workshop Discussion among the authors listed. We are grateful for the help of the HRI Library staff and for informal comments on the MS by Dr David Chandler (HRI), Dr Stan Finch (HRI) and Dr Mark Tatchell (HRI). KDS and WP are supported financially by the UK Ministry of Agriculture, Fisheries and Food. JAA, KD, NH, MKP and PR were supported by the Centre for Agricultural Biodiversity of the Danish Environmental Research Programme.

References

Abbas, M.S.T. & Boucias, D.G. 1984. Interaction between nuclar polyhedrosis virus-infected *Anticarsia gemmatalis* (Lepidoptera: Noctuidae) larvae and a predator *Podisus maculiventris* (Say) (Hemiptera: Pentatomidae). *Environ. Entomol.* 13: 599-602.

Adams, J. 1985. The definition and interpretation of guild structure in ecological communities. *J. Anim. Ecol.* 54: 43-59.

Agnew, C.W. & Smith, J.W. 1989. Ecology of spiders (Araneae) in a peanut agroecosystem. *Environ. Entomol.* 18: 30-42.

Akhurst, R.J. 1990. Safety to nontarget invertebrates of nematodes of economically important pests. In: Laird, M., Lacey, L.A. & Davidson, E.W. (eds.) *Safety of Microbial Insecticides,* pp 233-240. CRC Press, Boca Raton, USA.

Alderweireldt, M. 1994. Prey selection and prey capture strategies of linyphiid spiders in high-input agricultural fields. *Bull. Br. arachnol. Soc.* 9: 300-8.

Altieri, M.A., Cure, J.R. & Garcia, M.A. 1993. The role and enhancement of parasitic Hymenoptera biodiversity in agroecosystems. In: Lasalle, J. & Gauld, I.D. (eds) *Hymenoptera and Biodiversity,* pp 257-75. CAB International, Wallingford, UK.

Anderson, D.J. & Kikkawa, J. 1986. Development of concepts. In: Kikkawa, J. & Anderson, D.J. (eds.) *Community Ecology: Pattern and Process,* pp 3-16. Blackwell Scientific Publications, London.

Andreadis, T.G. 1980. *Nosema pyrausta* infection in *Macrocentrus grandii,* a braconid parasite of the European Corn Borer, *Ostrinia nubilalis. J. Invert. Pathol.* 35: 229-33.

Andreadis, T.G. 1982. Impact of *Nosema pyrausta* on field populations of *Macrocentrus grandii*, an introduced parasite of the European Corn Borer, *Ostrinia nubilalis. J. Invert. Pathol.* 39: 298-302.

Askew, R.R. & Shaw, M.R. 1986. Parasitoid communities: their size, structure and development. In: Waage, J.K. & Greathead, D.J. (eds.) *Insect Parasitoids* pp 225-64. Academic Press, London.

Austin, A.D. 1985. The function of spider eggsacs in relation to parasitoids and predators, with special reference to the Australian fauna. *J. Nat. Hist.* 19: 359-76.

Axelsen, J.A., Ruggle, P., Holst, N. & Toft, S. 1997. Modelling natural control of cereal aphids III. Linyphiid spiders and coccinellids. In: Powell, W. (ed.) *Arthropod natural enemies in arable land. III. Acta Jutlandica* 72 (2): 221-31 (this volume)

Basedow, T. 1994. Phenology and egg production in *Agonum dorsale* and *Pterostichus melanarius* (Col., Carabidae) in winter wheat fields of different growing intensity in Northern Germany. In: Desender, K., Dufrene, M., Loreau, M., Luff, M.L. & Maelfait, J.P. (eds.) *Carabid Beetles: Ecology and Evolution*, pp 101-7. Kluwer Academic Publishers, Dordrecht.

Bathon, H. 1996. Impact of entomopathogenic nematodes on non-target hosts. *Biocontrol Sci. Technol.* 6: 421-34.

Begon, M., Harper, J.L. & Townsend, C.R. 1990. *Ecology: Individuals, Populations and Communities.* Blackwell Scientific Publications, London.

Belshaw, R. 1994. Life history characteristics of Tachinidae (Diptera) and their effect on polyphagy. In: Hawkins, B.A. & Sheehan, W. (eds.) *Parasitoid Community Ecology*, pp 145-62. Oxford University Press, Oxford.

Berdegue, M., Trumble, J.T., Hare, J.D. & Redak, R.A. 1996. Is it enemy-free space ? The evidence for terrestrial insects and freshwater arthropods. *Ecol. Entomol.* 21: 203-17.

Bhuiuyan, M.S.I. & Wratten, S.D. 1994. Grain aphid populations and their fall-off rate on different cultivars of wheat. *Bull. IOBC/WPRS* 17: 27-35.

Biever, K.D., Andrews, P.L. & Andrews, P.A. 1982. Use of a predator, *Podisus maculiventris*, to distribute virus and initiate epizootics. *J. Econ. Entomol.* 75: 150-2.

Blaustein, L. & Margalit, J. 1994. Mosquito larvae (*Culiseat longiareolata*) prey upon and compete with toad tadpoles (*Bufo viridis*). *J. Anim. Ecol.* 63: 841-50.

Bogya, S. & Mols, P.J.M. 1996. The role of spiders as predators of insect pests with particular reference to orchards: a review. *Acta Phytopath. Entomol. Hung.* 31: 83-159.

Bouchard, D., Pilon, J.G. & Tourneur, J.C. 1991. Importance of parasitism of aphids in Quebec apple orchards and the impact of hyperparasitism on parasite effectiveness. In: Polgar, L., Chambers, R.J., Dixon, A.F.G. & Hodek, I. (eds.) *Behaviour and Impact of Aphidophaga*, pp 29-33. S.P.B. Academic Publishers, The Hague, The Netherlands.

Brodeur, J. & McNeil, J.N. 1989. Seasonal microhabitat selection by an endoparasitoid through adaptive modification of host behaviour. *Science,* 244: 226-28.

Brodeur, J. & McNeil, J.N. 1992. Host behaviour modification by the endoparasitoid *Aphidius nigriceps*: a strategy to reduce hyperparasitism. *Ecol. Entomol.* 17: 97-104.

Brodsky, L.M. & Barlow, C.A. 1986. Escape responses of the pea aphid, *Acyrthosiphon pisum* (Harris)(Homoptera: Aphididae): influence of predator type and temperature. *Can. J. Zool.* 64: 937-39.

Brooks, W.M. 1993. Host-parasitoid-pathogen interactions. In: Beckage, N.E., Thompson, S.N. & Federici, B.A. (eds.) *Parasites and Pathogens of Insects*, Vol. 2: pp.231-72. Academic Press, San Diego, USA.

Brooks, W.M. & Cranford, J.D. 1972. Microsporidioses of the hymenopterous parasites, *Campoletis sonorensis* and *Cardiochiles nigriceps,* larval parasites of *Heliothis* species. *J. Invert. Pathol.* 20: 77-94.

Bueno, B.H.P., Gutierrez, A.P. & Ruggle, P. 1993. Parasitism by *Aphidius ervi* (Hym.: Aphidiidae): preference for pea aphid and blue alfalfa aphid (Hom.: Aphididae) and competition with *A. smithi. Entomophaga* 38: 273-84.

Burel, F. & Baudry, J. 1995. Farming landscapes and insects. In: Glen, D.M., Greaves, M.P. & Anderson, H.M. (eds.) *Ecology and Integrated Farming Systems*, pp 203-20. John Wiley & Sons, Chichester.

Capinera, J.L. & Barbosa,P. 1975. Transmission of a nuclear polyhedrosis virus to gypsy moth larvae by *Calosoma sycophanta*. *Ann. Ent. Soc. Amer.* 68: 593-94.

Carter, N. 1994. Cereal aphid modelling through the ages. In: Leather, S., Walters, K., Mills, N. & Watt, A. (eds.) *Individuals, Populations and Patterns in Ecology*, pp 129-138. Intercept, Andover, UK.

Carter, M.C. & Dixon, A.F.G. 1984. Honeydew: an arrestant stimulus for coccinellids. *Ecol. Entomol.* 9: 383-87.

Chandler, D., Hay, D. & Reid, A.P. (in press). Sampling and ocurrence of entomopathogenic fungi and nematodes in UK soils. *Appl. Soil Ecol.*

Charnley, A.K. 1989. Mycoinsecticides; present use and future prospects. In: *Progress and Prospects for Insect Control*, BCPC Monog. 43: 165-81.

Clegg, J.M. & Barlow, C.A. 1982. Escape behaviour of the pea aphid *Acythosiphon pisum* (Harris) in response to alarm pheromone and vibration. *Can. J. Zool.* 60: 2245-52.

Cloudsley-Thompson, J.L. 1995. A review of the anti-predator devices of spiders. *Bull. Br. arachnol. Soc.* 10: 81-96.

Cloutier, C. & Johnson, S.G. 1993. Predation by *Orius tristicolor* (Hemiptera: Anthocoridae) on *Phytoseiulus persimilis* (Acarina: Phytoseiidae): testing for compatibility between biocontrol agents. *Environ. Entomol.* 22: 477-82.

Coop, L.B. & Berry, R.E. 1986. Reduction in variegated cutworm (Lepidoptera: Noctuidae) injury to peppermint by larval parasitoids. *J. Econ. Entomol.* 79: 1244-48.

Cornell, H.V. & Hawkins, B.A. 1994. Patterns of parasitoid accumulation on introduced herbivores. In: Hawkins, B.A. & Sheehan, W. (eds.) *Parasitoid Community Ecology*, pp 77-89. Oxford University Press, Oxford.

Craig, T.P. 1994. Effects of intraspecific plant variation on parasitoid communities. In: Hawkins, B.A. & Sheehan, W. (eds.) *Parasitoid Community Ecology*, pp 205-27. Oxford University Press, Oxford.

Crawley, M.J. 1992. Population dynamics of natural enemies and their prey. In: Crawley, M.J. (ed.) *Natural Enemies*, pp 40- 89. Blackwell Scientific Publications, London.

Currie, C.R., Spence, J.R. & Niemela, J. 1996. Competition, cannibalism and intraguild predation among ground beetles (Coleoptera: Carabidae): a laboratory study. *Coleopt. Bull.* 50: 135-48.

Dawah, H.A., Hawkins, B.A. & Claridge, M.F. 1995. Structure of the parasitoid communities of grass-feeding chalcid wasps. *J. Anim. Ecol.* 64: 708-20.

Dean, D.A., Nyffeler, M. & Sterling, W.L. 1988. Natural enemies of spiders: mud dauber wasps (Hymenoptera: Sphecidae) in East Texas. *Southw. Ent.* 13: 283-90.

Dean, G.J. 1973. Aphid colonisation of spring cereals. *Ann. Appl. Biol.* 75: 183-93.

Dean, G.J.W. & Wilding, N. 1971. *Entomophthora* infecting the cereal aphids *Metapolophium dirhodum* and *Sitobion avenae. J. Invert. Pathol.* 18: 169-76.

De Keer, R. & Maelfait, J.P. 1987. Life history of *Oedothorax fuscus* (Blackwall, 1834) (Araneae, Linyphiidae) in a heavily grazed pasture. *Rev. Ecol. Biol. Sol.* 24: 171-85.

Delucchi, V. 1982. Parasitoids and hyperparasitoids of *Zeiraphera diniana* [Lep., Tortricidae] and their role in population control in outbreak areas. *Entomophaga* 27: 77-92.

Dempster, J.P. 1971. The population ecology of the cinnabar moth *Tyria jacobaeae* L.

(Lepidoptera, Arctiidae). *Oecologia (Berl.)* 7: 26-67.

Dempster, J.P. & Coaker, T.H. 1974. Diversification of crop ecosystems as a means of controlling pests. In: Price Jones, D. & Solomon, M.E. (eds.) *Biology in Pest and Disease Control*, pp 106-14. John Wiley & Sons. New York.

Diamond, J.M. 1987. Competition among different taxa. *Nature* 326: 241.

Dill, M.L., Fraser, A.H.G. & Roitberg, B.D. 1990. The economics of escape behaviour in the pea aphid, *Acythosiphon pisum*. *Oecologia (Berl.)* 83: 473-78.

Dixon, A.F.G. 1987. Cereal aphids as an applied problem. *Agric. Zool. Rev.*, Intercept, Dorset, 2, 1-57.

Dixon, A.F.G. & Russel, R.J. 1972. The effectiveness of *Anthocoris nemorum* and *A. confusus* (Hemiptera: Anthocoridae) as predators of the sycamore aphid, *Drepanosiphum platanoides*. II. Searching behaviour and the incidence of predation in the field. *Entomol. Exp. Appl.*, 15: 35-50.

Dixon, P.L. & McKinlay, R.G. 1992. Pitfall trap catches of and aphid predation by *Pterostichus melanarius* and *Pterostichus madidus* in insecticide treated and untreated potatoes. *Entomol. Exp. Appl.* 64: 63-72.

Doane, J.F., Scotti, P.D., Sutherland, O.R.W. & Pottinger, R.P. 1985. Serological identification of wireworm and staphylinid predators of the Australian soldier fly (*Inopus rubriceps*) and wireworm feeding on plant and animal food. *Entomol. Exp. Appl.*, 38: 65-72.

Drake, J.A. (1990). The mechanics of community assembly and succession. *J. Theor. Biol.* 147: 213-33.

Dromph, K., Holst, N. & Eilenberg, J. 1997. Modelling natural control of cereal aphids. V. Entomophthoralean fungi. In: Powell, W. (ed.) *Arthropod natural enemies in arable land. III. Acta Jutlandica* 72 (2): 247-58 (this volume).

Edgar, W.D. 1969. Prey and predators of the Wolf spider *Lycosa lugubris*. *J. Zool. Lond.* 159: 405-11.

Ehler, L.E. 1992. Guild analysis in biological control. *Environ. Entomol.* 21: 26-40.

Ehler, L.E. 1994. Parasitoid communities, parasitoid guilds, and biological control. In: Hawkins, B.A. & Sheehan, W. (eds.) *Parasitoid Community Ecology*, pp 418-36. Oxford University Press, Oxford.

Ehler, L.E. & Miller, J.C. 1978. Biological control in temporary agroecosystems. *Entomophaga* 23: 207-12.

Eilenberg, J. 1987. Abnormal egg-laying behaviour of female carrot flies (*Psila rosae*) induced by the fungus *Entomophthora muscae*. *Entomol. Exp. Appl.* 43: 61-65.

Evans, E.W. 1991. Intra versus interspecific interactions of ladybeetles (Coleoptera: Coccinellidae) attacking aphids. *Oecologia (Berl.)* 87: 401-8.

Evans, E.W. & Swallow, J.G. 1993. Numerical responses of natural enemies to artificial honeydew in Utah alfalfa. *Environ. Entomol.* 22: 1392-1401.

Evans, H.F. 1976. The role of predator-prey size ratio in determining the efficiency of capture by *Anthocoris nemorum* and the escape reactions of its prey, *Acyrthosiphon pisum*. *Ecol. Entomol.* 1: 85-90.

Evans, R.E. 1969. Parasites on spiders and their eggs. *Proc. Birmingham Nat. Hist. and Phil. Soc.* 21: 156-68.

Evans, H.C. 1982. Entomogenous fungi in tropical forest ecosystems; an appraisal. *Ecol. Entomol.* 7: 47-60.

Evans, H.C. & Samson, R.A. 1987. Fungal pathogens of spiders. *Bull. Br. Mycol. Soc.* 21: 152-59.

Faeth, S.H. 1994. Induced plant responses: effects on parasitoids and other natural enemies of phytophagous insects. In: Hawkins, B.A. & Sheehan, W. (eds.) *Parasitoid Community Ecology*, pp 245-60. Oxford University Press, Oxford.

Feng, M.G., Nowierski, R.M., Scharen, A.L. & Sands, D.C. 1991. Entomopathogenic fungi (Zygomycotina: Entomophthorales) infecting cereal aphids (Homoptera: Aphididae) in Montana. *Pan-Pacific Entomol.* 67: 55-64.

Ferran, A. & Dixon, A.F.G. 1993. Foraging behaviour of ladybird larvae (Coleoptera, Coccinellidae). *Eur. J. Entomol.* 90: 383-402.

Ferran, A., Iperti, G., Lapchin, L. & Rabasse, J.M. 1991. La localisation, le comportement et les relations "proie-predateur" chez *Coccinella septempunctata* L. dans un champ de ble. *Entomophaga* 36: 213-25.

Finch, S.K. 1996. A review of the progress made to control the cabbage root fly (*Delia radicum*) using parasitoids. In: Booij, C.J.H. & den Nijs, L.J.M.F. (eds.) *Arthropod natural enemies in arable land. II. Survival, reproduction and enhancement,* Acta Jutlandica 71 (2): 227-39.

Force, D.C. 1974. Ecology of insect host-parasitoid communities. *Science* 184: 624-32.

Frazer, B.D. & Gilbert, N. 1976. Coccinellids and aphids (A quantitative study of the impact of adult ladybirds (Coleoptera: Coccinellidae) preying on field populations of pea aphids (Homoptera: Aphididae)). *J. Ent. Soc. Br. Columbia* 73: 33-55.

Frazer, B.D. & Van Den Bosch, R. 1973. Biological control of the walnut aphid in California: the interrelationship of the aphid and its parasite. *Environ. Entomol.* 2: 561-68.

Free, C.A., Beddington, J.R. & Lawton, J.H. 1977. On the inadequacy of simple models of mutual interference for parasitism and predation. *J. Anim. Ecol.* 46: 543-54.

Freier, B., Triltsch, H. & Rossberg, D. 1996a. GTLAUS – A model of wheat-cereal aphid-predator interaction and its use in complex agroecological studies. *J. Plant Dis. and Prot.* 103: 543-54.

Freier, B., Möwes, M., Triltsch, H. & Rappaport, V. 1996b. Investigations on the predatory effect of coccinellids in winter wheat fields and problems of situation-related evaluation. *Bull. IOBC/WPRS.* 19(3): 41-52.

Freier, B., Möwes, M. & Triltsch, H. 1997. Beneficial thresholds for *Coccinella 7-punctata* L. as a predator of cereal aphids in winter wheat – results of population investigations and computer simulations. *J. Appl. Entomol.* 121, (in press).

Fritz, R.S. 1982. Selection for host modification by insect parasitoids. *Evolution* 36: 283-88.

Fuhrer, E., Elsufty, R. & Willers, D. 1978. Antibiotic effects of entomophagous endoparasites against micro-organisms within the host body. *4th Int. Cong. Parasitol. Warszawa 1978,* Sect. F, p 100.

Furlong, M.J. & Pell, J.K. 1996. Interactions between the fungal entomopathogen *Zoophthora radicans* Brefeld (Entomophthorales) and two hymenopteran parasitoids attacking the diamondback moth, *Plutella xylostella* L. *J. Invert. Pathol.* 68: 15-21.

Georgis, R. & Wojcik, W. 1987. The effect of entomogenous nematodes *Heterorhabditis heliothidis* and *Steinernema feltiae* on selected predatory soil insects. *J. Nematol.* 19: 523.

Gilbert, N., Gutierrez, A.P., Frazer, B.D. & Jones, R.E. 1976. *Ecological Relationships.* W.H. Freeman, San Francisco, USA.

Godfray, H.C. J. 1994. *Parasitoids: Behavioral and Evolutionary Ecology.* Intercept, Andover.

Goettel, M.S., Poprawski, T.J., Vandenberg, J.D., Li, Z. & Roberts, W. 1990. Safety to nontarget invertebrates of fungal biocontrol agents. In: Laird, M., Lacey, L.A. & Davidson, E.W. (eds.) *Safety of Microbial Insecticides*, pp 209-31. CRC Press, Boca Raton, USA.

Goff, A.M. & Nault, L.R. 1974. Aphid cornicle secretions ineffective against attack by parasitoid wasps. *Environ. Entomol.* 3: 565-66.

Gowling, G.R. & Van Emden, H.F. 1994. Falling aphids enhance impact of biological control by parasitoids on partially aphid-resistant plant varieties. *Ann. Appl.Biol.* 125: 233-242.

Gravesen, E. & Toft, S. 1987. Grass fields as reservoirs for polyphagous predators (Arthropoda) of aphids (Homopt., Aphididae). *Z. angew. Entomol.* 104: 461-73

Greenstone, M.H., Ignoffo, C.M. & Samson, R.A. 1987. Susceptibility of spider species to the fungus *Nomuraea atypicola. J. Arachnol.* 15: 266-68.

Griffiths, C., Carter, J.B. & Overend, J. 1984. *Phaonia signata* (Meigen) (Diptera: Muscidae) larvae predatory upon leatherjackets, *Tipula paludosa* (Meigen) (Diptera: Tipulidae) larvae. *Entomol. Gaz.* 35: 53-55.

Griffiths, E., Wratten, S.D. & Vickerman, G.P. 1985. Foraging by the carabid *Agonum dorsale* in the field. *Ecol. Entomol.* 10: 181-89.

Groner, A. 1990. Safety to nontarget invertebrates of baculoviruses.In: Laird, M., Lacey, L.A. & Davidson, E.W. (eds.) *Safety of Microbial Insecticides*, pp 135-67. CRC Press, Boca Raton, USA.

Gross, H.R. 1987. Conservation and enhancement of entomophagous insects – a perspective. *J. Entomol. Sci.* 22: 97-105.

Guillebeau, L.P. & All, J.N. 1989. *Geocoris* spp. (Hemiptera: Lygaeidae) and the striped lynx spider (Araneae: Oxyopidae): cross predation and prey preferences. *J. Econ. Entomol.* 82: 1106-10.

Gutierrez, A.P. 1992. The ecological basis for crop protection: theory and practice. *Proc. Brighton Conf. – Pests & Diseases 1992*, BCPC, Farnham, UK, pp 955-64.

Gutierrez, A.P. & Baumgaertner, J.U. 1984a. Age-specific energetics models – pea aphid *Acyrthosiphon pisum* (Homoptera: Aphididae) as an example. *Can. Ent.* 116: 924-32.

Gutierrez, A.P. & Baumgaertner, J.U. 1984b. A realistic model of plant – herbivore – parasitoid – predator interactions. *Can. Ent.* 116: 933-49.

Gutierrez, A.P., Baumgaertner, J.U. & Hagen, K.S. 1981. A conceptual model for growth, development and reproduction in the ladybird beetle, *Hippodamia convergens* (Coleoptera: Coccinellidae). *Can. Ent.* 113: 21-33.

Gutierrez, A.P., Hagen, K.S. & Ellis. C.K. 1990. Evaluating the impact of natural enemies: a multitrophic perpective. In: Mackauer, M., Ehler. L.E. & Roland, J. (eds.) *Critical Issues in Biological Control*, pp 81-109. Intercept, Andover, UK.

Hagen, K.S. & Van Den Bosch, R. 1968. Impact of pathogens, parasites and predators on aphids. *Ann. Rev. Entomol.* 13: 325-84.

Hagvar, E.B. 1988. Multiparasitism of the green peach aphid , *Myzus persicae*: Competition in the egg stage between *Aphidius matricariae* and *Ephedrus cerasicola. Entomol. Exp. Appl.* 47: 275-82.

Hamel, D.R. 1977. The effects of Bacillus thuringiensis on parasitoids of the western spruce budworm, *Choristoneura occidentalis* (Lepidoptera: Tortricidae), and the spruce coneworm, *Dioryctria reniculloides* (Lepidoptera: Pyralidae), in Montana. *Can. Ent.* 109: 1409-15.

Hamm, J.J., Nordlund, D.A. & Mullinix, B.G. 1983. Interaction of the microsporidian *Vairimorpha* sp. with *Microplitis croceipes* (Cresson) and *Cotesia marginiventris* (Cresson) (Hymenoptera: Braconidae), two parasitoids of *Heliothis zea* (Boddie) (Lepidoptera: Noctuidae). *Environ. Entomol.* 12: 1547-50.

Hamm, J.J., Nordlund, D.A. & Marti, O.C. 1985. Effects of a nonoccluded virus of *Spodoptera frugiperda* (Lepidoptera: Noctuidae) on the development of a parasitoid, *Cotesia marginiventris* (Hymenoptera: Braconidae). *Environ. Entomol.* 14: 258-61.

Hance, T. (1987). Predation impact of carabids at different population densities on *Aphis fabae* development in sugar beet. *Pedobiologia* 30: 251-62.

Hare, J.D. 1992. Effects of plant variation on herbivore-natural enemy interactions. In: Fritz, R.S. & Simms, E.L. (eds.) *Plant Resistance to Herbivores and Pathogens: Ecology, Evolution and Genetics*, pp 278-98. University of Chicago Press, Chicago.

Harper, J.D. 1986. Interactions between baculoviruses and other entomopathogens, chemical insecticides and parasitoids. In: Granados, R.R. & Federici, B.A. (eds.) *The Biology of Baculoviruses*, Vol. 2: pp 133-55. CRC Press, Boca Raton, USA.

Hassell, M.P. 1986. Parasitoids and population regulation. In: Waage, J. & Greathead, D. (eds.) *Insect Parasitoids*, pp 201-24. Academic Press, London.

Hassell, M.P. & Godfray, H.C.J. 1992. The population biology of insect parasitoids. In: Crawley, M.J. (ed.) *Natural Enemies*, pp 265-92. Blackwell Scientific Publications, London.

Hawkins, B.A. 1990. Global patterns of parasitoid assemblage size. *J. Anim. Ecol.* 59: 57-72.

Hawkins, B.A. 1993. Refuges, host population dynamics and the genesis of parasitoid diversity. In: Lasalle, J. & Gauld, I.D. (eds.) *Hymenoptera and Biodiversity*, pp 235-56. CAB International, Wallingford, UK.

Hawkins, B., & Lawton, J.H. 1987. The determinants of species richness for the parasitoids of British phytophagous insects. *Nature* 326: 788-90.

Hawkins, C.P. & MacMahon, J.A. 1989. Guilds: the multiple meanings of a concept. *Ann. Rev. Entomol.* 34: 423-51.

Hawkins, B.A. & Mills, N.J. 1996. Variability in parasitoid community structure. *J. Anim. Ecol.* 65: 501-16.

Heinz, K.M. & Nelson, J.M. 1996. Interspecific interactions among natural enemies of *Bemisia* in an inundative biological control program. *Biological Control*, 6: 384-93.

Hemptinne, J.-L., Dixon, A.F.G. & Coffin, J. 1992. Attack strategy of ladybird beetles (Coccinellidae): factors shaping their numerical response. *Oecologia (Berl.)* 90: 238-45.

Hemptinne, J.L., Doucet, J.L. & Gaspar, C. 1994. How do ladybirds and syrphids respond to aphids in the field ? *Bull. IOBC/WPRS..* 17(4): 101-11.

Hight, S.C., Eikenbary, R.D., Miller, R.J. & Starks, K.J. 1972. The greenbug and *Lysiphlebus testaceipes*. *Environ. Entomol.* 1: 205-9.

Hirose, Y. 1994. Determinants of species richness and composition in egg parasitoid assemblages of Lepidoptera. In: Hawkins, B.A. & Sheehan, W. (eds.) *Parasitoid Community Ecology*, pp 19-29. Oxford University Press, Oxford.

Hochberg, M.E. & Hawkins, B.A. 1994. The implications of population dynamics theory to parasitoid diversity and biological control. In: Hawkins, B.A. & Sheehan, W. (eds.) *Parasitoid Community Ecology*, pp 451-71. Oxford University Press, Oxford.

Hochberg, M.E. & Lawton, J.H. 1990. Competition between kingdoms. *TREE* 5: 367-371.

Hockland, S.H., Dawson, G.W., Griffiths, D.C., Marples, B., Pickett, J.A. & Woodcock, C.M. 1986. The use of aphid alarm pheromone (E)-beta-farnesene to increase effectiveness of the entomophilic fungus *Verticillium lecanii* in controlling aphids on chrysanthemums under glass. In: Samson, R.A., Vlak, J.M. & Peters, R. (eds.) *Fundamental and Applied Aspects of Invertebrate Pathology*, p. 252. The Foundation of the Fourth International Colloquium of Invertebrate Pathology, Wageningen, The Netherlands,

Hoffmeister, T.S. & Vidal, S. 1994. The diversity of fruit fly (Diptera: Tephritidae) parasitoids. In: Hawkins, B.A. & Sheehan, W. "*Parasitoid Community Ecology*, pp 47-76. Oxford University Press, Oxford.

Holland, J.M., Frampton, G.K., Cilgi, T. & Wratten, S.D. 1994. Arable acronyms analysed – a review of integrated arable farming systems research in Western Europe. *Ann. Appl. Biol.* 125: 399-438.

Höller, C. 1991. Movement away from the feeding site in parasitized aphids: host suicide or an attempt by the parasitoid to escape hyperparasitism? In: Polgar, L., Chambers, R.J., Dixon, A.F.G. & Hodek, I. (eds.) *Behaviour and Impact of Aphidophaga*, pp 45-49. S.P.B. Academic Publishers, The Hague, The Netherlands.

Holling, C.S. 1955. The selection by certain small mammals of dead, parasitized and healthy prepupae of the European pine sawfly, *Neodiprion sertifer* (Geoff.). *Can. J. Zool.* 33: 404-19.

Holst, N. & Ruggle, P. 1997. Modelling natural control of cereal aphids I. The metabolic pool model, winter wheat and cereal aphids. In: Powell, W. (ed.) *Arthropod natural enemies in arable land. III. Acta Jutlandica* 72 (2): 195-206 (this volume).

Hominick, W.M., Reid, A.P., Bohan, D.A. & Briscoe, B.R. 1996. Entomopathogenic nematodes: biodiversity, geographical distribution and Convention on Biological Diversity. *Biocont. Sci. Tech.* 6: 317-31.

Hopper, K.R. & King, E.G. 1984. Feeding and movement on cotton of *Heliothis* species (Lepidoptera: Noctuidae) parasitized by *Microplitis croceipes* (Hymenoptera: Braconidae). *Environ. Entomol.* 13, 1654-60.

Hotchkin, P.G. & Kaya, H.K. 1983. Interactions between two baculoviruses and several insect parasites. *Can. Ent.* 115: 841-46.

Howell, J.O. & Pienkowski, R.L. 1971. Spider populations in alfalfa, with notes on spider prey and effect of harvest. *J. Econ. Entomol.* 64: 163-68.

Hunter, K.W. & Stoner, A. 1975. *Copidosoma tuncatellum*: effect of parasitization on food consumption of larval *Trichoplusia ni*. *Environ. Entomol.* 4: 381-82.

Ito, Y. & Miyashita, K. 1968. Biology of *Hypantria cunea* Drury (Lepidoptera; Arctiidae) in Japan. V. Preliminary life tables and mortality data in urban areas. *Res. Popul. Ecol.* 10: 177-209.

Jackson, R.R. & Pollard, S.D. 1996. Predatory behaviour of jumping spiders. *Ann. Rev. Entomol.* 41: 287-308.

James, R.R. & Lighthart, B. 1994. Susceptibility of the Convergent Lady Beetle (Coleoptera: Coccinellidae) to four entomogenous fungi. *Environ. Entomol.* 23: 190-92.

Jeffries, M.J. & Lawton, J.H. 1984. Enemy free space and the structure of ecological communities. *Biol. J. Linn. Soc.* 23: 269-86.

Jmhasly, P. & Nentwig, W. 1995. Habitat management in winter wheat and evaluation of subsequent spider predation on insect pests. *Acta Oecol.* 16: 389-403.

Jones, D. 1982. Predators and parasites of temporary row crop pests: Agents of irreplaceable mortality or scavengers acting prior to other mortality factors? *Entomophaga* 27: 245-66.

Jones, T.H., Hassell, M.P. & Pacala, S.W. 1993. Spatial heterogeneity and the population dynamics of a host-parasitoid system. *J. Anim. Ecol.* 62: 251-62.

Jones, T.H., Hassell, M.P. & May, R.M. 1994. Population dynamics of host-parasitoid interactions. In: Hawkins, B.A. & Shechan, W. (eds.) *Parasitoid Community Ecology*, pp 371-94. Oxford University Press, Oxford.

Kareiva, P. 1990. The spatial dimension in pest-enemy interactions. In: Mackauer, M., Ehler, L.E. & Roland, J. (eds.) *Critical Issues in Biological Control*, pp 213-27. Intercept, Andover, UK.

Kaya, J.K. 1970. Toxic factor produced by a granulosis virus in armyworm larva: effect on *Apanteles militaris. Science* 168: 251-53.

Kaya, H.K. 1978. Interactions between *Neoaplectana carpocapsae* (Nematoda: Steinernematidae) and *Apanteles militaris* (Hymenoptera: Braconidae), a parasitoid of the armyworm *Pseudaletia unipuncta. J. Invert. Pathol.* 31: 358-64.

Kaya, H.K. & Hotchkin, P.G. 1981. The nematode *Neoaplectana carpocapsae* Weiser and its effect on selected ichneumonoid and braconid parasites. *Environ. Entomol.* 10: 474-78.

Kaya, H.K. & Koppenhofer, A.M. 1996. Effects of microbial and other antagonistic organisms and competition on entomopathogenic nematodes. *Biocont. Sci. Tech.* 6: 357-71.

Kennedy, T.F. 1990. *A study of the spider fauna of Irish cereal fields with particular reference to the role of Linyphiidae as aphid predators*. PhD thesis, National University of Ireland.

Kidd, N.A.C. & Jervis, M.A. 1996. Population dynamics. In: Jervis, M.A. & Kidd, N.A.C. (eds.) *Insect Natural Enemies*, pp 293-374. Chapman and Hall, London, UK.

King, K.M. & Atkinson, N.J. 1928. The biological control factors of the immature stages of *Euxoa ochrogaster* Gn. [Lepidoptera, Phalaenidae] in Saskatchewan. *Ann. Entomol. Soc. Amer.* 21: 167-88.

King, E.G. & Bell, J.V. 1978. Interactions between a braconid, *Microplitis croceipes*, and a fungus, *Nomuraea rileyi*, in laboratory-reared bollworm larvae. *J. Invert. Pathol.* 31: 337-40.

Kitching, R.L. 1986. Prey-predator interactions. In: Kikkawa, J. & Anderson, D.J. (eds.) *Community Ecology: Pattern and Process*, pp. 214-39. Blackwell Scientific Publications, London.

Klingauf, F. 1967. Abwehr- und Meidereaktionen von Blattlausen (Aphididae) bei Bedrohung durch Raubern und Parasiten. *Z. angew. Entomol.* 60: 269-317.

Krieg, A. 1971. Interactions between pathogens. In: Burges, H.D. & Hussey, N.W. (eds.) *Microbial Control of Insects and Mites*, pp. 459-68. Academic Press, London.

Kuo-Sell, H.L., Wilhelms, A. & Holthusen, C. 1988. Wirt-parasitoid-beziehungen zwischen getreideblattlausen und *Ephedrus plagiator* (Nees) (Hymenoptera: Aphidiidae). *Med.*

Fac. Landouwwet. Rijksuniv. Gent, 53: 1045-1053.

Lasalle, J. & Gauld, I.D. 1993. *Hymenoptera and Biodiversity.* CAB International, Wallingford, UK.

Lasalle, J. & Gauld, I.D. 1993. Hymenoptera: their diversity, and their impact on the diversity of other organisms. In: Lasalle, J. & Gauld, I.D. (eds.) *Hymenoptera and Biodiversity,* pp 1-26. CAB International, Wallingford, UK.

Lasalle, J. 1993. Parasitic Hymenoptera, biological control and biodiversity. In: Lasalle, J. & Gauld, I.D. (eds.) *Hymenoptera and Biodiversity,* pp 197-215. CAB International, Wallingford, UK.

Lawrence, P.O. 1981. Interference competition and optimal host selection in the parasitic wasp *Biosteres longicaudatus. Ann. Entomol. Soc. Amer.* 74: 540-44.

Lawton, J.H. 1986. The effect of parasitoids on phytophagous insect communities. In: Waage, J.K. & Greathead, D.J. (eds.) *Insect Parasitoids,* pp 265-87. Academic Press, London.

Lawton, J.H. 1994. Parasitoids as model communities in ecological theory. In: Hawkins, B.A. & Sheehan, W. (eds.) *Parasitoid Community Ecology,* pp 492-506. Oxford University Press, Oxford.

Legendre, R. & Morel, G. 1980. Data on the role of rickettsial and viral diseases in the regulation of arachnid populations. *Proc. 8th Int. Arachnol. Cong.,* pp 183-85.

Lewis, F.B. 1960. Factors affecting assessment of parasitization by *Apanteles fumiteranae* (Vier.) and *Colypta fumiteranae* (Vier.) on spruce budworm larvae. *Can. Ent.* 92: 881-91.

Lindquist, E.E. (1983). Some thoughts on the potential use of mites in biological control, including a modified concept of "Parasitoids". In: Hoy, M.A., Cunningham, G.L. & Knutson, L. (eds.) *Biological Control of Pests by Mites,* pp 12-20. University of California Special Publication 3304, USA.

Longley, M. & Jepson, P.C. 1996. Effects of honeydew and insecticide residues on the distribution of foraging aphid parasitoids under glasshouse and field conditions. *Entomol. Exp. Appl.* 81: 189-98.

Los, L.M. & Allen, W.A. 1983. Incidence of *Zoophthora phytonomi* [Zygomycetes: Entomophthorales] in *Hypera postica* [Coleoptera: Curculionidae] larvae in Virginia. *Environ. Entomol.* 12: 1318-21.

Loughridge, A.H. & Luff, M.L. 1983. Aphid predation by *Harpalus rufipes* (DeGeer) (Coleoptera: Carabidae) in the laboratory and field. *J. Appl. Ecol.* 20: 451-62.

Lövei, G.L. & Sunderland, K.D. 1996. Ecology and behaviour of ground beetles (Coleoptera: Carabidae). *Ann. Rev. Entomol.* 41: 231-56.

Lys, J.A. 1995. Observation of epigeic predators and predation on artificial prey in a cereal field. *Entomol. Exp. Appl.* 75: 265-72.

Mackauer, M. 1990. Host discrimination and larval competition in solitary endoparasitoids. In: Mackauer, M., Ehler, L.E. & Roland, J. (eds.) *Critical Issues in Biological Control,* pp 41-62. Intercept, Andover, UK.

MacRae, I.V. & Croft, B.A. 1993. Influence of temperature on interspecific predation and cannibalism by *Metaseiulus occidentalis* (Nesbitt) and *Typhlodromus pyri* Scheuten (Acarina: Phytoseiidae). *Environ. Entomol.* 22: 770-75.

Malcolm, S.B. 1992. Prey defence and predator foraging. In: Crawley, M.J. (eds.) *Natural Enemies,* pp 458-75. Blackwell Scientific Publications, London.

Mann, J.A., Tatchell, G.M., Dupuch, M.J., Harrington, R., Clark, S.J. & McCartney, H.A. 1995. Movement of apterous *Sitobion avenae* (Homoptera: Aphididae) in response to leaf disturbances caused by wind and rain. *Ann. Appl. Biol.* 126: 417-27.

Mansour, F., Rosen, D. & Shulov, A. 1981. Disturbing effect of a spider on larval aggregations of *Spodoptera littoralis. Entomol. Exp. Appl.* 29: 234-37.

McAllister, M.K. & Roitberg, B.D. 1987. Adaptive suicidal behaviour in pea aphids. *Nature,* 328: 797-99.

McAllister, M.K., Roitberg, B.D. & Weldon, L. 1990. Adaptive suicide in pea aphids: decisions are cost sensitive. *Anim. Behav.* 40: 167-75.

McClure, M.S. 1995. *Diapterobates humeralis* (Oribatida: Ceratozetidae): an effective control agent of hemlock woolly adelgid (Homoptera: Adelgidae) in Japan. *Environ. Entomol.* 24: 1207-15.

McConnell, J.A. & Kring, T.J. 1990 Predation and dislodgement of *Schizaphis graminum* (Homoptera: Aphididae) by adult *Coccinella septempunctata* (Coleoptera: Coccinellidae). *Environ. Entomol.* 19: 1798-1802.

Melin, B.E. & Cozzi, E.M. 1990. Safety to nontarget invertebrates of Lepidopteran strains of *Bacillus thuringiensis* and their beta-exotoxins. In: Laird, M., Lacey, L.A. & Davidson, E.W. (eds.) *Safety of Microbial Insecticides*, pp 149-67. CRC Press, Boca Raton, USA.

Memmott, J. & Godfray, H.C.J. 1993. Parasitoid webs. In: Lasalle, J. & Gauld, I.D. (eds.) *Hymenoptera and Biodiversity*, pp. 217-34. CAB International, Wallingford, UK.

Miller, J.C. 1980. Niche relationships among parasitic insects occurring in a temporary habitat. *Ecology* 61: 270-75.

Miller, J.C. & Ehler. L.E. 1990. The concept of parasitoid guild and its relevance to biological control. In: Mackauer, M., Ehler, L.E. & Roland, J. (eds.) *Critical Issues in Biological Control*, pp 159-69. Intercept, Andover, UK.

Mills, N.J. 1991. Searching strategies of the parasitoids of the ash bark beetle (*Leperisinus varius* F.) and its relevance to biological control. *Ecol. Entomol.* 16: 461-70.

Mills, N.J. 1993. Species richness and structure in the parasitoid complexes of tortricoid hosts. *J. Anim. Ecol.* 62: 45-58.

Mills, N.J. 1994a. Parasitoid guilds: a comparative analysis of the parasitoid communities of tortricids and weevils. In: Hawkins, B.A. & Sheehan, W. (eds.) *Parasitoid Community Ecology* pp 30-46. Oxford University Press, Oxford.

Mills, N.J. 1994b. The structure and complexity of parasitoid communities in relation to biological control. In: Hawkins, B.A. & Sheehan, W. (eds.) *Parasitoid Community Ecology*, pp 397-417. Oxford University Press, Oxford.

Milner, R.J., Lutton, G.G. & Bourne, J. 1984. A laboratory study of the interaction between aphids, fungal pathogens and parasites. *Proc. 4th Aust. Appl. Entomol. Res. Conf.,* Adelaide 1984, pp 375-81.

Mols, P.J.M. 1993. Foraging behaviour of the carabid beetle *Pterostichus coerulescens* L. (= *Poecilus versicolor* Sturm) at different densities and distributions of the prey. *Wageningen Agric. Univ. Papers* 93 - 95: 105-201.

Mols, P.J.M. 1996. Do natural enemies control woolly apple aphid? *Bull. IOBC/WPRS* 19: 203-7.

Mols, P.J.M. & Diederik, D. 1996. INSIM – a simulation environment for pest forecasting and simulation of pest-natural enemy interaction. *Acta Hort.* 416: 255-62.

Montgomery, M.E. & Nault, L.R. 1977 Comparative response of aphids to the alarm pheromone (E)-beta-farnesene. *Entomol. Exp. Appl.* 22: 236-42.

Moran, V.C. & Southwood, T.R.E. 1982. The guild composition of arthropod communities in trees. *J. Anim. Ecol.* 51: 289-306.

Mracek, Z. & Spitzer, K. 1983. Interactions of the predators and parasitoids of the sawfly, *Cephalcia abietis* (Pamphilidae: Hymenoptera) with its nematode *Steinernema kraussei. J. Invert. Pathol.* 42: 397-99.

Muller, F.P. 1983. Differential alarm response between different strains of the aphid *Acyrthosiphon pisum. Entomol. Exp. Appl.* 34: 347-48.

Murdoch, W.W. 1990. The relevance of pest-enemy models to biological control. In: Mackauer, M., Ehler, L.E. & Roland, J. (eds.) *Critical Issues in Biological Control*, pp 1-24. Intercept, Andover, UK.

Nakasuji, F., Yamanaka, K. & Kiritani, K. 1973. The disturbing effect of micryphantid spiders on larval aggregation of the tobacco cutworm. *Kontyu* 41: 220-27.

Nault, L.R., Edwards, L.J. & Styer, W.E. 1973. Aphid alarm pheromones: secretion and reception. *Environ. Entomol.* 2: 101-5.

Nealis, V. & Regniere, J. 1987. The influence of parasitism by *Apanteles fumiferanae* (Hymenoptera: Braconidae) on spring dispersal and changes in the distribution of larvae of the spruce budworm (Lepidoptera: Tortricidae). *Can. Ent.* 119: 141-46.

Nentwig, W. 1995. Ackerkrautstreifen als Systemansatz fur eine umweltfreundliche Landwirtschaft. *Mitt. Dtsch. Ges. Allg. Angew. Ent.* 9: 679-83.

Nyffeler, M. & Benz, G. 1979. Overlap of niches concerning space and prey of crab spiders (Araneae: Thomisidae) and wolf spiders (Araneae: Lycosidae) in cultivated meadows. *Rev. Suisse de Zool.* 86: 855-65.

Nyffeler, M., Breene, R.G., Dean, D.A. & Sterling, W.L. 1990. Spiders as predators of arthropod eggs. *J. Appl. Entomol.* 109: 490-501.

Nyffeler, M., Sterling, W.L. & Dean, D.A. 1994. Insectivorous activities of spiders in United States field crops. *J. Appl. Entomol.* 118: 113-28.

Obrycki, J.J. 1992. Techniques for evaluation of predators of Homoptera. *Florida Entomol.* 75: 472-76.

Ogilvy, S.E., Turley, D.B., Cook, S.K., Fisher, N.M., Holland, J., Prew, R.D. & Spink, J. 1994. Integrated farming – putting together systems for farm use. *Asp. Appl. Biol.* 40: 53-60.

Ohbayashi, T. & Iwabuchi, K. 1991. Abnormal behaviour of the common armyworm *Pseudaletia separata* (Walker)(Lepidoptera: Noctuidae) larvae infected with an entomogenous fungus, *Entomophaga aulicae*, and a nuclear polyhedrosis virus. *Appl. Ent. Zool.* 26: 579-85.

Oien, C.T. & Ragsdale, D.W. 1993. Susceptibility of nontarget hosts to *Nosema furnacalis* (Microsporidia: Nosematidae), a potential biological control agent of the European Corn Borer, *Ostrinia nubilalis* (Lepidoptera: Pyralidae). *Biol. Cont.* 3: 323-28.

Overmeer, W.P.J. 1985. Alternative prey and other food resources. In: Helle, W. & Sabelis, M.W. (eds.) *Spider Mites their Biology, Natural Enemies and Control*, Vol 1B, pp 131-39. Elsevier, Amsterdam, The Netherlands.

Pell, J.K. & Pluke, R. 1994. Interactions between two aphid natural enemies, the entomophthoralean fungus *Erynia neoaphidis* and the predatory beetle *Coccinella septempunctata. Abst. VIth Int. Coll. Inv. Pathol. & Microb. Cont.* p 114.

Petersen, M.K. & Holst, N. 1997. Modelling natural control of cereal aphids. II. The carabid *Bembidion lampros.*In: Powell, W. (ed.) *Arthropod natural enemies in arable land. III. Acta Jutlandica* 72 (2): 207-19 (this volume).

Pickering, J. & Gutierrez, A.P. 1991. Differential impact of the pathogen *Pandora neoaphidis* (R. & H.) Humber (Zygomycetes: Entomophthorales) on the species composition of *Acyrthosiphon* aphids in alfalfa. *Can. Ent.* 123: 315-20.

Pijls, J.W.A.M., Hofker, K.D., Van Staalduinen, M.J. & Van Alphen, J.J.M. 1995. Interspecific host discrimination and competition in *Apoanagyrus* (*Epidinocarsis*) *lopezi* and *A.* (*E.*) *diversicornis*, parasitoids of the cassava mealybug *Phenacoccus manihoti. Ecol. Entomol.* 20: 326-32.

Pimentel, D. & Wheeler, A.G. 1973. Species and diversity of arthropods in the alfalfa community. *Environ. Entomol.* 2: 659-68.

Pimm, S.L. & Lawton, J.H. 1978. On feeding in more than one trophic level. *Nature* 275: 542-43.

Pimm, S.L. & Kitching, R.L. 1987. The determinants of food-chain length. *Oikos* 50: 302-307.

Pimm, S.L., Lawton, J.H. & Cohen, J.E. 1991. Food web patterns and their consequences. *Nature* 350: 669-74.

Polis, G.A., Myers, C.A. & Holt, R.D. 1989. The ecology and evolution of intraguild predation: potential competitors that eat each other. *Ann. Rev. Ecol. Syst.* 20: 297-330.

Polking, A. & Heimbach, U. 1992. Wirkung einiger biologischer Pflanzenschutzmittel auf zwei rauberische Kaferarten in Laborversuchen. *Mitt. Biol. Bundes. Land- Forst.*283: 378.

Potts, G.R. & Vickerman, G.P. 1974. Studies on the cereal ecosystem. *Adv. Ecol. Res.* 8: 107-97.

Powell, W. 1980. *Toxares deltiger* (Haliday)(Hymenoptera: Aphidiidae) parasitising the cereal aphid, *Metopolophium dirhodum* (Walker)(Hemiptera, Aphididae), in Southern England: a new host-parasitoid record. *Bull. Ent. Res.* 70: 407-9.

Powell, W. 1983. The role of parasitoids in limiting cereal aphid populations. In: Cavalloro, R. (ed.) *Aphid Antagonists*, pp 50-56. A.A. Balkema, Rotterdam.

Powell, W. 1986. Enhancing parasitoid activity in crops. In: Waage, J.K. & Greathead, D.J. (eds.) *Insect Parasitoids*, pp 319-40. Academic Press, London.

Powell, W., Ashby, J. & Wright, A.F. 1987 Encouraging polyphagous predators. *Rothamsted Exp. Stn, Rep. for 1986*: 90.

Powell, W., Wilding, N., Brobyn, P.J. & Clark, S.J. 1986. Interference between parasitoids [Hym.: Aphidiidae] and fungi [Entomophthorales] attacking cereal aphids. *Entomophaga,* 31: 293-302.

Press, J., Flaherty, R. & Arbogast, R. 1974. Interactions among *Plodia interpunctella, Bracon hebetor,* and *Xylocoris flavipes. Environ. Entomol.* 3: 183-84.

Price, P.W. 1970. Trail odours: recognition by insects parasitic in cocoons. *Science* 170: 546-47.

Price, P.W. 1973. Parasitoid strategies and community organisation. *Environ. Entomol.* 2: 623-26.

Price, P.W. 1994. Evolution of parasitod communities. In: Hawkins, B.A. & Sheehan, W. (eds.) *Parasitoid Community Ecology*, pp 472-91. Oxford University Press, Oxford.

Ragsdale, D.W. & Oien, C.T. 1996. An environmental risk assessment for release of an exotic microsporidium for European corn borer control. In: Symondson, W.O.C. & Liddell, J.E. (eds.) *The Ecology of Agricultural Pests*, pp 401-17. Chapman & Hall, London.

Rahman, M. 1970. Effect of parasitism on food consumption of *Pieris rapae* larvae. *J. Econ. Entomol.* 63: 820-21.

Randall, J.B. 1982. Prey records of the green lynx spider, *Peucetia viridans* (Hentz) (Araneae, Oxyopidae). *J. Arachnol.* 10: 19-22.

Ravensberg, W.J. 1981. The natural enemies of the woolly apple aphid, *Eriosoma lanigerum* (Hausm.) (Homoptera: Aphididae), and their susceptibility to diflubenzuron. *Med. Fac. Landbouww. Rijksuniv. Gent*, 46: 437-41.

Raw, A. 1988. Social wasps (Hymenoptera: Vespidae) and insect pests of crops of the Surui and Cinta Larga indians in Rondonia, Brazil. *The Entomologist* 107: 104-9.

Redfearn, A. & Pimm, S.L. 1987. Insect outbreaks and community structure. In: Barbosa, P. & Schulz, J.C. (eds.) *Insect Outbreaks*, pp 99-133. Academic Press, San Diego.

Redfearn, A. & Pimm, S.L. 1992. Natural enemies and community dynamics. In: Crawley, M.J. (ed.) *Natural Enemies*, pp 395-411. Blackwell Scientific Publications, London.

Riechert, S.E. 1992. Spiders as representative 'sit-and-wait' predators. In: Crawley, M.J. (ed.) *Natural Enemies*, pp 313-28. Blackwell Scientific Publications, London.

Riechert, S.E. & Bishop, L. 1990. Prey control by an assemblage of generalist predators: spiders in garden test systems. *Ecology* 71: 1441-50.

Roitberg, B.D. & Myers, J.H. 1978a. Adaptation of alarm pheromone responses of the pea aphid *Acyrthosiphon pisum* (Harris). *Can. J. Zool.* 56: 103-8.

Roitberg, B.D. & Myers, J.H. 1978b. Effect of adult Coccinellidae on the spread of a plant virus by an aphid. *J. Appl. Ecol.* 15: 775-79.

Roitberg, B.D. & Myers, J.H. 1979. Behavioural and physiological adaptations of pea aphids (Homoptera: Aphididae) to high ground temperatures and predator disturbance. *Can. Ent.* 111: 515-19.

Roitberg, B.D., Myers, J.H. & Frazer, B.D. 1979 The influence of predators on the movement of apterous pea aphids between plants. *J. Anim. Ecol.* 48: 111-22.

Roland, J. 1990. Interaction of parasitism and predation in the decline of winter moth in Canada. In: Watt, A.D., Leather, S.R. Hunter, M.D. & Kidd, N.A.C. (eds.) *Population Dynamics of Forest Insects*, pp 289-302. Intercept, Andover, UK.

Rollard, C. 1984. Composition et structure de la biocenose consommatrice des Araneides. *Rev. Arachnol.* 5: 211-37.

Root, R.B. 1967. The niche exploitation pattern of the blue-gray gnatcatcher. *Ecol. Monogr.* 37: 317-50.

Rosenheim, J.A., Wilhoit, L.R. & Armer, A. 1993. Influence of intraguild predation among generalist insect predators on the suppression of an herbivore population. *Oecologia (Berl.)* 96: 439-49.

Rosenheim, J.A., Kaya, H.K., Ehler, L.E., Marois, J.J. & Jaffee, B.A. 1995. Intraguild predation among biological control agents: Theory and evidence. *Biol. Cont.* 5: 303-35

Roske, H. 1989. Collembola fauna on different types of agriculturally used soil. *3rd Int. Seminar on Apterygota, Ed. by R. Dallai, University of Sienna, Italy*, pp 283-90.

Roy, H.E. & Pell, J.K. 1995. Feeding behaviour of fourth instar *Coccinella septempunctata* larvae on *Acyrthosiphon pisum* aphids infected with *Erynia neoaphidis. Abst. SIP 28th Ann. Meet., Cornell Univ. N.Y.*, p 53.

Ruberson, J.R., Young, S.Y. & Kring, T.J. 1991. Suitability of prey infected by nuclear polyhedrosis virus for development, survival and reproduction of the predator *Nabis roseipennis* (Heteroptera: Nabidae). *Environ. Entomol.* 20: 1475-79.

Ruggle, P. & Holst, N. 1997. Modelling natural control of cereal aphids: IV. Aphidiid and aphelinid parasitoids. In: Powell, W. (ed.) *Arthropod natural enemies in arable land. III. Acta Jutlandica* 72 (2): 233-45 (this volume).

Russell, E.P. 1989. Enemies hypothesis: a review of the effect of vegetational diversity on predatory insects and parasitoids. *Environ. Entomol.* 18: 590-99.

Ruth, W.E., McNew, R.W., Caves, D.W. & Eikenbary, R.D. 1975. Greenbugs (Hom.: Aphididae) forced from host plants by *Lysiphlebus testaceipes* (Hym.: Braconidae). *Entomophaga*, 20: 65-71.

Ryan, R.B. 1985. A hypothesis for decreasing parasitization by larch casebearer (Lepidoptera: Coleophoridae) on larch foliage by *Agathis pumila*. *Can. Ent.* 117: 1573-74.

Salama, H.S., Zaki, F.N. & Sharaby, A.F. 1982. Effect of *Bacillus thuringiensis* Berl. on parasites and predators of the cotton leafworm *Spodoptera littoralis* (Boisd.). *Z. angew. Entomol.* 94: 498-504.

Salama, H.S., El-Moursy, A., Zaki, F.N., Aboul-Ela, R. & Razek, Abdel-A. 1991. Parasites and predators of the meal moth *Plodia interpunctella* Hbn. as affected by *Bacillus thuringiensis* Berl. *J. Appl. Entomol.* 112: 244-53.

Samu, F., Sunderland, K.D., Topping, C.J. & Fenlon, J.S. 1996. A spider population in flux: selection and abandonment of artificial web-sites and the importance of intraspecific interactions in *Lepthyphantes tenuis* (Araneae: Linyphiidae) in wheat. *Oecologia (Berl.)* 106: 228-39.

Sajap, A.S. & Lewis, L.C. 1989. Impact of *Nosema pyrausta* (Microsporidia: Nosematidae) on a predator, *Chrysoperla carnea* (Neuroptera: Chrysopidae). *Environ. Entomol.* 18: 172-76.

Sato, Y., Tanaka, T., Imafuku, M. & Hidaka, T. 1983. How does diurnal *Apanteles kariyai* parasitise and egress from a nocturnal host larva? *Kontyu* 51: 128-39.

Schmid-Hempel, R. & Schmid-Hempel, P. 1996. Larval development of two parasitic flies (Conopidae) in the common host *Bombus pascuorum*. *Ecol. Entomol.* 21: 63-70.

Schoener, T.W. 1974. Resource partitioning in ecological communities. *Science* 185: 27-38.

Schoener, T.W. 1983. Field measurements on interspecific competition. *Amer. Nat.* 122: 240-85.

Schoener, T.W. 1986. Resource partitioning. In: Kikkawa, J. & Anderson, D.J. (eds.) *Community Ecology: Pattern and Process*, pp 91-126. Blackwell Scientific Publications, London.

Settle, W.H. & Wilson, L.T. 1990. Behavioural factors affecting differential parasitism by *Anagrus epos* (Hymenoptera: Mymaridae), of two species of Erythroneuran leafhoppers (Homoptera: Cicadellidae). *J. Anim. Ecol.* 59: 877-91.

Settle, W.H., Ariawan, H., Astuti, E.T., Cahyana, W., Hakima, A.L., Hindayana, D., Lestari, A.S. & Pajarningsih, S. 1996. Managing tropical rice pests through conservation of generalist natural enemies and alternative prey. *Ecology*, 77: 1975-88.

Severinghaus, W.D. 1981. Guild theory development as a mechanism for assessing environmental impact. *Environ. Manag.* 5: 187-90.

Shapiro, A. 1976. Beau geste? *Amer. Nat.* 110: 900-902.

Sheehan, W. 1994. Parasitoid community structure: effect of host abundance, phylogeny and ecology. In: Hawkins, B.A. and Sheehan, W. (eds.) *Parasitoid Community Ecology*, pp 90-107. Oxford University Press, Oxford.

Skirvin, D.J., Perry, J.N. & Harrington, R. 1996. A model to describe the effect of climate change on aphid and coccinellid population dynamics. *Bull. IOBC/WPRS,* 19(3): 30-40.

Smirnoff, W.A. 1960. Observations on the migration of larvae of *Neodiprion swainei* Midd. (Hymenoptera: Tenthredinidae). *Can. Ent.* 92: 957-58.

Smith, B.C. 1971. Effects of various factors on the local distribution and density of coccinellid adults on corn (Coleoptera: Coccinellidae). *Can. Ent.* 103: 1115-20.

Smith, H.S. 1929. Multiple parasitism: its relation to the biological control of insect pests. *Bull. Ent. Res.* 20: 141-49.

Smith Trail, D.R. 1980. Behavioral interactions between parasites and host: Host suicide and the evolution of complex life cycles. *Amer. Nat.* 116: 77-91.

Soares, G.G., Lewis, W.J., Strong-Gunderson, J.M., Waters, D.J. & Hamm, J.J. 1994. Integrating the use of MVP[R] Bioinsecticide, a unique *B.t*-based product, with natural enemies of noctuid pests : a novel concept in cotton IPM. In: Akhurst, R.J. (ed.) *Proceedings of the 2nd Canberra Meeting on Bacillus thuringiensis*, pp. 133-45. CSIRO, Australia.

Soper, R.S. & MacLeod, D.M. 1981. Descriptive epizootiology of an aphid mycosis. *USDA Science & Education Administration. Tech. Bull.* 1632, 17 pp.

Sopp, P.I., Sunderland, K.D. & Coombes, D.S. 1987. Observations on the number of cereal aphids on the soil in relation to aphid density in winter wheat. *Ann. Appl. Biol.* 111: 53-57.

Stadler, B., Weisser, W.W. & Houston, A.I. 1994. Defense reactions in aphids: the influence of state and future reproductive success. *J. Anim. Ecol.* 63: 419-30.

Stamp, N.E. 1981. Behaviour of parasitized aposematic caterpillars: Advantages to the parasitoid or the host? *Amer. Nat.* 118: 715-25.

Steenberg, T. & Eilenberg, J. 1995. Natural occurrence of entomopathogenic fungi on aphids at an agricultural field site. *Czech Mycol.* 48: 89-96.

Steenberg, T., Langer, V. & Esbjerg, P. 1995. Entomopathogenic fungi in predatory beetles (Col.: Carabidae and Staphylinidae) from agricultural fields. *Entomophaga* 40: 77-85.

Stiling, P. & Rossi, A.M. 1994. The window of parasitoid vulnerability to hyperparasitism: template for parasitoid complex structure. In: Hawkins, B.A. and Sheehan, W. (eds.) *Parasitoid Community Ecology*, pp 228-44. Oxford University Press, Oxford.

Strauss, S.Y. 1991. Indirect effects in community ecology: their definition, study and importance. *TREE* 6: 206-210.

Sunderland, K.D. 1975. The diet of some predatory arthropods in cereal crops. *J. Appl. Ecol.* 12: 507-15.

Sunderland, K.D. 1996. Progress in quantifying predation using antibody techniques. In: Symondson, W.O.C. & Liddell, J.E. (eds.) *The Ecology of Agricultural Pests*, pp 419-55. Chapman & Hall, London.

Sunderland, K.D. & Chambers, R.J. 1983. Invertebrate polyphagous predators as pest control agents: Some criteria and methods. In: Cavalloro, R. (ed.) *Aphid Antagonists*, pp 100-108. A.A. Balkema, Rotterdam.

Sunderland, K.D., Chambers, R.J. & Carter, O.C.R. 1988. Potential interactions between varietal resistance and natural enemies in the control of cereal aphids. In: Cavalloro, R.

& Sunderland, K.D. (eds.) *Integrated Crop Protection in Cereals*, pp 41-56. A.A. Balkema, Rotterdam.

Sunderland, K.D., Chambers, R.J., Helyer, N.L. & Sopp, P.I. 1992. Integrated pest management of greenhouse crops in Northern Europe. *Hort. Rev.* 13: 1-66.

Sunderland, K.D., Ellis, S.J., Weiss, A., Topping, C.J. & Long, S.J. 1994. The effects of polyphagous predators on spiders and mites in cereal fields. *Proc. Bright. Crop Prot. Conf. – Pests and Diseases*, BCPC, Farnham, Surrey, pp 1151-56.

Sunderland, K.D., Fraser, A.M. & Dixon, A.F.G. 1986a. Field and laboratory studies on money spiders (Linyphiidae) as predators of cereal aphids. *J. Appl. Ecol.* 23: 433-47.

Sunderland, K.D., Fraser, A.M. & Dixon, A.F.G. 1986b. Distribution of linyphiid spiders in relation to capture of prey in cereal fields. *Pedobiologia* 29: 367-75.

Sunderland, K.D., Bilde, T., Den Nijs, L.J.F.M., Dinter, A., Heimbach, U., Lys, J.A., Powell, W. & Toft, S. 1996. Reproduction of beneficial predators and parasitoids in agroecosystems in relation to habitat quality and food availability. In: Booij, C.J.H. & den Nijs, L.J.M.F. (eds.) *Arthropod natural enemies in arable land. II. Survival, reproduction and enhancement, Acta Jutlandica* 71 (2): 117-53.

Takagi, M. & Hirose, Y. 1994. Building parasitoid communities: the complementary role of two introduced parasitoid species in a case of successful biological control. In: Hawkins, B.A. & Sheehan, W. (eds.) *Parasitoid Community Ecology*, pp 437-48. Oxford University Press, Oxford.

Tamaki, G., Halfhill, J.E. & Hathaway, D.O. 1970. Dispersal and reduction of colonies of pea aphids by *Aphidius smithi* (Hymenoptera: Aphidiidae). *Ann. Ent. Soc. Amer.* 63: 973-80.

Temerak, S.A. 1980. Detrimental effects of rearing a braconid parasitoid on the pink borer larvae inoculated by different concentrations of the bacterium, *Bacillus thuringiensis* Berliner. *Z. angew. Entomol.* 89: 315-19.

Thiele, H.U. 1977. *Carabid Beetles in their Environmemts*. Springer Verlag, Berlin.

Thompson, J.N. 1984. Insect diversity and the trophic structure of communities. In: Huffaker, C.B. & Rabb, R.L. (eds.) *Ecological Entomology*, pp 591-606. John Wiley, New York.

Thompson, W.R. 1955. Mortality factors acting in a sequence. *Can. Ent.* 87: 264-75.

Toft, S. 1988. Interference by web take-over in sheet-web spiders. In: Haupt, J. (ed.) *XI Europaisches Arachnologisches Colloquium, Berlin*, pp 48-59. Technische Universitat Berlin Dokumentation Kongresse und Tagungen, Berlin.

Toft, S. 1989. Aspects of the ground-living spider fauna of two barley fields in Denmark: species richness and phenological synchronization. *Ent. Meddr.* 57: 157-68.

Toft, S. 1990. Interactions among two coexisting *Linyphia* spiders. *Act. Zool. Fenn.* 190: 367-72.

Toft, S. 1995. Value of the aphid *Rhopalosiphum padi* as food for cereal spiders. *J. Appl. Ecol.* 32: 552-60.

Topping, C.J., Booij, C.J.H., Daamen, R.A., Heimbach, U., Kennedy, P.J., Langer, V., Perry, J.N., Powell, W., Skirvin, D., Stilmant, D., Thomas, G. & Winstone, L. 1997. The use of single-species spatially explicit models. In: Powell, W. (ed.) *Arthropod natural enemies in arable land. III. Acta Jutlandica* 72 (2): 171-91 (this volume).

Tostowaryk, W. 1971. Relationship between parasitism and predation of diprionid sawflies. *Ann. Ent. Soc. Amer.* 64: 1424-27.

Triltsch, H. 1997. Gut contents in field sampled adults of *Coccinella septempunctata* L. (Col., Coccinellidae). *Entomophaga* (in press).

Triltsch, H. & Rossberg, D. 1997. Cereal aphid feeding by the ladybird *Coccinella septempunctata* L. (Coleoptera: Coccinellidae) – Inclusive simulation in the model GTLAUS. In: Powell, W. (ed.) *Arthropod natural enemies in arable land. III. Acta Jutlandica* 72 (2): 259-70 (this volume).

Tscharntke, T. 1992. Coexistence, tritrophic interactions and density dependence in a species-rich parasitoid community. *J. Anim. Ecol.* 61: 59-67.

Turnbull, A.L. 1973. Ecology of the true spiders (Araneomorphae). *Ann. Rev. Entomol.* 18: 305-48.

Turner, M. & Polis, G.A. 1979. Patterns of co-existence in a guild of raptorial spiders. *J. Anim. Ecol.* 48: 509-20.

Uetz, G.W. 1977. Coexistence in a guild of wandering spiders. *J. Anim. Ecol.* 46: 531-41.

Van Alphen, J.J.M. & M.A. Jervis. 1996. Foraging behaviour. In: Jervis, M. & Kidd, N. (eds.) *Insect Natural Enemies,* pp 1-62. Chapman and Hall, London, UK.

Van Baarlen, P., Sunderland, K.D. & Topping, C.J. 1994. Eggsac parasitism of money spiders (Araneae, Linyphiidae) in cereals, with a simple method for estimating percentage parasitism of *Erigone* spp. eggsacs by Hymenoptera. *J. Appl. Entomol.* 118: 217-23.

Van Driesche, R.G.1983. Meaning of "percent parasitism" in studies of insect parasitoids. *Environ. Entomol.* 12: 1611-22.

Vanninen, I., Husberg, G.B. & Hokkanen, H.M.T. 1989. Occurrence of entomopathogenic fungi and entomoparasitic nematodes in cultivated soils in Finland. *Acta Entomol. Fennica* 53:: 65-71.

Van Wingerden, W.K.R.E. 1977. *Population dynamics of* Erigone arctica *(White) (Araneae, Linyphiidae).* Thesis, Free University of Amsterdam, The Netherlands.

Vasconcelos, S.D.,   Williams, T., Hails, R.S. & Cory, J.S. 1996. Prey selection and baculovirus dissemination by carabid predators of Lepidoptera. *Ecol. Entomol.* 21: 98-104.

Versoi, P.L. & Yendol, W.G. 1982. Discrimination by the parasite, *Apanteles melanoscelus* between healthy and virus-infected gypsy moth larvae. *Environ. Entomol.* 11: 42-45.

Vinson, S.B. 1990. Potential impact of microbial insecticides on beneficial arthropods in the terrestrial environment. In:Laird, M., Lacey, L.A. & Davidson, E.W. (eds.) *Safety of Microbial Insecticides*, pp 43-64. CRC Press, Boca Raton, USA.

Waage, J.K. 1992. Sib-mating and sex ratio strategies in scelionid wasps. *Ecol. Entomol.* 7: 102-12.

Waage, J.K. & Mills, N.J. 1992. Biological control. In: Crawley, M.J. (ed.) *Natural Enemies*, pp 412-30. Blackwell Scientific Publications, London.

Watt, K.E.F. 1965. Community stability and the strategy of biological control. *Can. Ent.* 97: 887-95.

Weseloh, R.M. 1990. Gypsy moth predators: an example of generalist and specialist natural enemies. In: Watt, A.D., Leather, S.R., Hunter, M.D. & Kidd, N.A.C. (eds.) *Population Dynamics of Forest Insects*, pp 233-43. Intercept, Andover, UK.

Weseloh, R.M., Andreadis, T.G., Moore, R.E.B., Anderson, J.F., Dubois, N.R. & Lewis, R.B. 1983. Field confirmation of a mechanism causing synergism between *Bacillus thuringiensis* and the gypsy moth parasitoid *Apanteles melanoscelus. J. Invert. Pathol.* 41: 99-103.

Wheeler, A.G. 1971. Studies on the arthropod fauna of alfalfa. Insect feeding on *Hylemya* flies (Diptera: Anthomyiidae) killed by a phycomycosis. *J. N.Y. Entomol. Soc.* 79: 225-27.

Wheeler, A.G. 1974. Studies on the arthropod fauna of alfalfa. VI. Plant bugs (Miridae). *Can. Ent.* 106: 1267-75.

Wheeler, A.G. 1977. Studies on the arthropod fauna of alfalfa. VII. Predaceous insects. *Can. Ent.* 109: 423-27.

Wheeler, A.G., Hayes,J.T. & Stephens, J.L. 1968. Insect predators of mummified pea aphids. *Can. Ent.* 100: 221-22.

Whitcomb, W.H. & Bell, K. 1964. Predaceous insects, spiders and mites of Arkansas cotton fields. *Bull. Arkansas Agric. Exp. Stn.* 690, 84 pp.

Whitfield, J.B. 1994. Mutualistic viruses and the evolution of host ranges in endoparasitoid Hymenoptera. In: Hawkins, B.A. & Sheehan, W. (eds.) *Parasitoid Community Ecology*, pp 163-76. Oxford University Press, Oxford.

Winder, L. 1990. Predation of the cereal aphid Sitobion avenae by polyphagous predators on the ground. *Ecol. Entomol.* 15: 105-10.

Winder, L., Hirst, D.J., Carter, N., Wratten, S.D. & Sopp, P.I. 1994. Estimating predation of the grain aphid *Sitobion avenae* by polyphagous predators. *J. Appl. Ecol.* 31: 1-12.

Winder, L., Wratten, S.D. & Carter, N. 1997. Spatial heterogeneity and predator searching behaviour – can carabids detect patches of their prey? In: Powell, W. (ed.) *Arthropod natural enemies in arable land. III. Acta Jutlandica* 72 (2): 47-62 (this volume).

Wratten, S.D. 1976. Searching by *Adalia bipunctata* L. (Coleoptera: Coccinellidae) and escape behaviour of its aphid and cicadellid prey on lime (*Tilia x vulgaris* Hayne). *Ecol. Enomol.* 1: 139-42.

Wratten, S.D. & W. Powell. 1991. Cereal aphids and their natural enemies. In: Firbank, L.G., Carter, N., Darbyshire, J.F. & Potts, G.R. (eds.) *The Ecology of Temperate Cereal Fields*, pp 233-58. Blackwell Scientific Publications, Oxford.

Wratten, S.D. & Van Emden, H.F. 1995. Habitat management for enhanced activity of natural enemies of insect pests. In: Glen, D.M., Greaves, M.P. & Anderson, H.M. (eds.) *Ecology and Integrated Farming Systems*, pp 117-45. John Wiley & Sons, Chichester.

Yao, D.S. & Chant, D.A. 1989. Population growth and predation interference between two species of predatory phytoseiid mites (Acarina: Phytoseiidae) in interactive systems. *Oecologia (Berl.)* 80: 443-55.

Yeargan, K.V. & Braman, S.K. 1986. Life-history of the parasite *Diolcogaster facetosa* (Weed) (Hymenoptera: Braconidae) and its behavioural adaptation to the defensive response of a lepidopteran host. *Ann. Ent. Soc. Amer.* 79: 1029-33.

Yeargan, K.V. & Braman, S.K. 1989. Life-history of the hyperparasitoid *Mesochorus discitergus* (Hymenoptera: Ichneumonidae) and tactics used to overcome the defensive behaviour of the green cloverworm (Lepidoptera, Noctuidae). *Ann. Ent. Soc. Amer.* 82: 393-98

Young, O.P. & Edwards, J.P. 1990. Spiders in United States field crops and their potential effect on crop pests. *J. Arachnol.* 18: 1-27.

Young, O.P. & Hamm, J.J. 1985. The compatibility of two fall armyworm pathogens with a predaceous beetle, *Calosoma sayi* (Coleoptera: Carabidae). *J. Entomol. Sci.* 20: 212-18.

Young, S.Y. & Kring, T.J. 1991. Selection of healthy and nuclear polyhedrosis virus infected

Anticarsia gemmatalis (Lep.: Noctuidae) as prey by nymphal *Nabis roseipennis* (Hemiptera: Nabidae) in laboratory and on soybean. *Entomophaga* 36: 265-73.

Young, S.Y. & Yearian, W.C. 1990. Transmission of nuclear polyhedrosis virus by the parasitoid *Microplitis croceipes* (Hymenoptera: Braconidae) to *Heliothis virescens* (Lepidoptera: Noctuidae) on soybean. *Envir. Entomol.* 19: 251-56.

Zhang, Z.Q. & Croft, B.A. 1995. Interspecific competition and predation between immature *Amblyseius fallacis, Amblyseius andersoni, Typhlodromus occidentalis* and *Typhlodromus pyri* (Acari: Phytoseiidae). *Exp. Appl. Acarol.* 19: 247-57.